ADVANCED NEUROMUSCULAR EXERCISE PHYSIOLOGY

ADVANCED EXERCISE PHYSIOLOGY SERIES

Phillip F. Gardiner, PhD
University of Manitoba

Human Kinetics

Library of Congress Cataloging-in-Publication Data

Gardiner, Phillip F., 1949-
    Advanced neuromuscular exercise physiology / Phillip F. Gardiner.
        p. ; cm. -- (Advanced exercise physiology series)
    Includes bibliographical references and index.
    ISBN-13: 978-0-7360-7467-4 (hard cover)
    ISBN-10: 0-7360-7467-8 (hard cover)
    1. Exercise--Physiological aspects. 2. Neurophysiology. 3. Motor neurons.
    4. Muscles. I. Title. II. Series: Advanced exercise physiology series.
    [DNLM: 1. Muscle Contraction--physiology. 2. Exercise--physiology. 3. Muscle
    Fatigue--physiology. 4. Resistance Training. WE 500]
    QP301.G365 2011
    612'.044--dc22
                                                          2010047297
ISBN-10: 0-7360-7467-8 (print)
ISBN-13: 978-0-7360-7467-4 (print)

The Web addresses cited in this text were current as of October 2010, unless otherwise noted.

Acquisitions Editor: Michael S. Bahrke, PhD, and Amy N. Tocco; **Developmental Editor:** Christine M. Drews; **Assistant Editors:** Antoinette Pomata and Brendan Shea; **Graphic Artist:** Dawn Sills; **Copyeditor:** Jocelyn Engman; **Indexer:** Betty Frizzell; **Permission Manager:** Dalene Reeder; **Graphic Designer:** Joe Buck; **Cover Designer:** Bob Reuther; **Art Manager:** Kelly Hendren; **Associate Art Manager:** Alan L. Wilborn; **Illustrations:** ©Human Kinetics; **Printer:** Sheridan Books

Printed in the United States of America   10 9 8 7 6 5 4 3 2

The paper in this book is certified under a sustainable forestry program.

Human Kinetics
Web site: www.HumanKinetics.com

United States: Human Kinetics
P.O. Box 5076
Champaign, IL 61825-5076
800-747-4457
e-mail: humank@hkusa.com

Canada: Human Kinetics
475 Devonshire Road Unit 100
Windsor, ON N8Y 2L5
800-465-7301 (in Canada only)
e-mail: info@hkcanada.com

Europe: Human Kinetics
107 Bradford Road
Stanningley
Leeds LS28 6AT, United Kingdom
+44 (0) 113 255 5665
e-mail: hk@hkeurope.com

Australia: Human Kinetics
57A Price Avenue
Lower Mitcham, South Australia 5062
08 8372 0999
e-mail: info@hkaustralia.com

New Zealand: Human Kinetics
P.O. Box 80
Torrens Park, South Australia 5062
0800 222 062
e-mail: info@hknewzealand.com

E4477

I dedicate this book to the most important people in my life—
my wife and best friend, Kalan; my sons, Patrick and
Matthew; and my mother and father, Verlie and Charles.

Contents

Series Preface

Having a detailed knowledge of the effects of exercise on specific physiological systems and under various conditions is essential for advanced-level exercise physiology students. For example, students should be able to answer questions such as these: What are the chronic effects of a systematic program of resistance training on cardiac structure and function, vascular structure and function, and hemostatic variables? How do different environments influence the ability to exercise, and what can pushing the body to its environmental limits tell us about how the body functions during exercise? When muscles are inactive, what happens to their sensitivity to insulin, and what role do inactive muscles play in the development of hyperinsulinemia and type 2 diabetes? These questions and many others are answered in the books in Human Kinetics' Advanced Exercise Physiology Series.

Beginning where most introductory exercise physiology textbooks end their discussions, each book in this series describes in detail the effects of exercise on a specific physiological system or the effects of external conditions on exercise. Armed with this information, students will be better prepared both to conduct the high-quality research required for advancing scientific knowledge and to make decisions in real-life scenarios such as the assement of health and fitness or the formulation of effective exercise guidelines and prescriptions.

Although many graduate programs and some undergraduate programs in exercise science and kinesiology offer specific courses on advanced topics in exercise physiology, there are few good options for textbooks to support those classes. Some instructors adopt general advanced physiology textbooks, but such books focus almost entirely on physiology without emphasizing *exercise* physiology.

Each book in the Advanced Exercise Physiology Series addresses the effects of exercise on a certain physiological system (e.g., cardiovascular or neuromuscular) or in certain contexts (e.g., in various types of environments). These textbooks are intended primarily for students, but researchers and practitioners will also benefit from the detailed presentation of the most recent research regarding topics in exercise physiology.

Preface

The field of neuromuscular exercise physiology is changing rapidly. A major part of this is due to the relatively young age of the field of neuroscience compared with the study of other physiological systems such as the cardiovascular and respiratory systems and even the study of muscle physiology. For example, we did not know the mechanisms involved in the generation of the action potential until the 1940s, and recordings from mammalian neurons did not occur until late into that decade. The Society for Neuroscience, whose annual meeting attracts more than 30,000 delegates, held its first meeting in 1971, with less than 1,400 delegates attending. Another important dimension of science that has changed our way of thinking about neuromuscular exercise physiology is molecular biology. We learned of the structure of DNA in the 1950s and have been systematically measuring mRNAs as an index of altered gene expression only since the 1980s. As a consequence, our ideas of how the nerves and muscles collaborate during acute and chronic exercise have evolved as our knowledge of the basic function of nervous and muscular systems has allowed us to understand exercise more thoroughly.

The basis for this text was formed some 30 years ago when, as a professor at the University of Montreal, I decided to create a graduate course titled Neuromuscular Aspects of Physical Activity, or, more precisely, Aspects neuromusculaires de l'activité physique (the course was in French, since Université de Montréal is a francophone institution). There was no available text on this subject at the time. The course material included excerpts from the neuroscience texts that were just becoming available as well as research articles relating to the effects of training and disuse and the causes of fatigue in the neuromuscular system. My postdoctoral years at UCLA, spent with Dr. Reggie Edgerton, had already sensitized me to the idea that physical activity is not merely an issue of turning entire muscles and muscle groups on and off and that the nervous system plays an extremely important and complicated role in recruiting subvolumes of muscle, with various patterns and intensities of recruitment of motor units of different types and sizes, to correspond to the needs of the task. This was a major shift in my thinking—my doctoral research at the University of Alberta involved stimulating entire dog hind-limb muscles to fatigue and determining the biochemical changes that might explain why force was decreasing. I still look on the 2 years in Edgerton's laboratory as a major epiphany in my research life.

The text titled *Neuromuscular Aspects of Physical Activity,* authored by me and published by Human Kinetics in 2001, was prepared in order to allow students in my graduate course to have the material in text format as opposed to the 400+ pages of photocopied material that the course entailed by 2000. As it turned out, I taught the course only once using the text; I moved to the University of Manitoba in 2002. Since that time, my membership and research in the department of physiology at the Spinal Cord Research Center have extended my research horizons significantly. In particular, an appreciation for the complexities of

locomotion and other forms of voluntary movement has become a principal issue in my research, thanks to regular interaction with my colleagues in the Center, including the director David McCrea, Larry Jordan, Brian Schmidt, and Brent Fedirchuk.

This current text constitutes a significant extension of the 2001 text. An appreciable amount of material (in fact, several chapters) from research areas that have not evolved since 2001 has been removed. Perhaps more importantly, evidence emanating from important research advances since 2001 has been added and often highlighted in several text boxes in each chapter.

I start the text with a chapter on the relationships between motoneurons and the muscle fibers they innervate and the idea of motor unit types in order to demonstrate how the properties of each cell type seem to be matched appropriately. I also make reference to late adaptation as a phenomenon that might limit the excitability of motoneurons during prolonged exercise and bistability as a phenomenon by which the generation of self-sustaining currents might function to counteract this loss of excitability during exercise by supercharging motoneurons. This is followed by chapter 2, in which recruitment of different motor unit types during various types and intensities of movement is discussed. Here I describe the difficulties inherent in determining recruitment patterns during complex forms of exercise as well as the knowledge we have gained from well-controlled experiments. In this chapter I consider the possibilities that individuals might be able to rotate the recruitment of motor units to offset fatigue and that all individuals may or may not be actually capable of recruiting all of their motor units maximally during an all-out voluntary effort.

In chapter 3, muscle alone is the target. Here I present an overview of two issues that are currently of considerable interest in the research literature: the control of muscle blood flow and the metabolic pathways involved in control of metabolism, including AMP-activated protein kinase (AMPK), glucose sensitivity of muscle, and fatty acid transport into muscle. Special consideration is given to discussing what happens in type 2 diabetes, in which inactivity and obesity result in disruptions in glucose and fatty acid uptake and in fatty acid by-product accumulation in muscle. The effects of increased physical activity on these processes, including muscle blood flow, which is itself affected by type 2 diabetes, are discussed.

Fatigue is the topic of chapters 4 and 5. In chapter 4, I discuss the sites believed to be involved in the fatigue process that are more peripheral, such as the muscle fiber itself, the neuromuscular junction, and the motoneuron axon, with specific reference to the experimental evidence for their roles in fatigue. Chapter 5 is concerned with the currently available evidence that central fatigue, or physiological changes occurring at sites in the spinal cord level and above, contributes to decreased performance as exercise continues. Information in these two chapters suggests that there are many physiological processes that change during exercise and that no one site can be targeted as the single site of fatigue.

In chapters 6 and 7, the subject is aerobic endurance training. In chapter 6, I consider how endurance training changes the protein profiles of muscle fibers. Since whole-body endurance exercise is so complex, with different fibers in the same muscles being exposed to various degrees of relative overload, the use of the biopsy technique is problematic with respect to a mechanistic interpretation of results. For this reason, the literature on the effects of chronic electrical stimulation on muscle protein synthesis and degradation is referred to extensively. I consider the mechanisms and the metabolic signals involved in the changes in fiber phenotype that accompany aerobic endurance training. Chapter 7 is devoted to the changes that occur in the nervous system in response to aerobic endurance training. In this chapter, most of the information is gleaned from the literature on actual increases in voluntary activity as opposed to chronic electrical stimulation. Of particular interest in this chapter is the applicability of the research information to the exercise rehabilitation of individuals with compromised nervous system function, such as spinal cord injury, other trauma, and neuromuscular disease.

The last three chapters are concerned with resistance training. Chapter 8 summarizes the current knowledge regarding the molecular mechanisms that promote increased muscle mass and associated strength. While aerobic endurance training signals seem to be primarily metabolic, many resistance training signals have to do with mechanical perturbations associated with the acute exercise challenge. The signaling discussed here includes the production of insulin-like growth factor 1 (IGF-1); activation of the Akt and mTOR pathway; activation of phospholipases A and C (PLA and PLC); and activation of the kinase systems protein kinase C (PKC), focal adhesion kinase (FAK), and mitogen-activated protein kinase (MAPK). In chapter 9, I present the phenotypic responses of muscles to resistance training. This discussion includes the specific changes that occur with different forms of resistance training, including isometric, slow isotonic, lengthening, and plyometric types of training. Finally, chapter 10 constitutes an overview of the effects of resistance training on the nervous system, with discussion on the strength of each piece of evidence as it is presented. The clinical implications are obvious and are discussed within that chapter.

I cannot pretend that this book includes every piece of literature relating to the neuromuscular system and exercise that has ever been published. Indeed, I know full well that I have omitted many important and elegant pieces of research in each chapter, and I apologize if it is your work. Much of the information in this text is also based on opinion—in choosing to present research literature on the effects of chronic electrical stimulation of muscle as representative of aerobic endurance training, for example, I am expressing my opinion that the commonalities are more important than the differences between these two models of increased muscle activity. I hope that you appreciate my thought processes as you progress through the text. Please contact me if you have any thoughts about the material that is included or that is absent and that you think should be included or any thoughts about my interpretations, which might not always be correct.

Acknowledgments

Like everyone else, I am a product of my environment, and my life trajectory has been redirected by people whom I have interacted with along the way.

I would like to acknowledge the wisdom and patience of my mentors at the University of Windsor, where I began my university training and awakened to the possibility of research as a career path; at the University of Alberta, where I honed important research skills and attitudes; at the University of California, Los Angeles, where I learned from a master and was allowed and encouraged to develop research independence and confidence; and at the University of Amsterdam, where I attained a new level of appreciation for the absolute and indescribable beauty of the experimental process. I have been extremely fortunate to have had the opportunity to cross paths with these exceptional individuals.

I must acknowledge the trainees whom I have worked with during my career, ranging from undergraduate students to postdoctoral fellows. I remember each one of them (they number more than 100 now) for their energy, enthusiasm, and youthful optimism, and I cannot help but think that I have been, in the long run, the major beneficiary of my contact with each one of them, which now in many cases seems sadly evanescent.

I have enjoyed very much the collegial relationships with my colleagues at the Université de Montréal and the University of Manitoba during my 30-plus-year career. I cannot think of a better environment in which one can spend one's working life, interacting daily with individuals who are kindred spirits in their appreciation of the importance of the pursuit of knowledge and truth.

Finally, I would like to acknowledge the assistance of Maria Setterbom for the artwork in the book, and also for the many times that she has blessed us all at the Spinal Cord Research Center with her genius in graphics. Thanks, Maria!

Credits

Chapter 1

Figure 1.3 Reprinted from *Brain Research,* Vol. 204, D. Kernell and B. Zwaagstra, "Input conductance, axonal conduction velocity and cell size among hindlimb motoneurones of the cat," pgs. 311-326. Copyright 1981, with permission from Elsevier.

Figure 1.5 Reprinted from P.F. Gardiner, 1993, "Physiological properties of motoneurons innervating different muscle unit types in rat gastrocnemius," *Journal of Neurophysiology* 69: 1160-1170. Used with permission.

Figure 1.7 Reprinted from *Electroencephlalography and Clinical Neurophysiology/Evoked Potentials Section,* Vol. 85, "After potentials and control of repetitive firing in human motoneurones," L.P. Kudina and N.L. Alexeeva, pg. 9. Copyright 1992, with permission from Elsevier.

Figure 1.8 Reprinted from P.C. Schwindt and W.H. Calvin, 1972, "Membrane-potential trajectories between spikes underlying motoneuron firing rates," *Journal of Neurophysiology* 35: 311-325. Used with permission.

Figure 1.10 Reprinted from P.C. Schwindt and W.H. Calvin, 1972, "Membrane-potential trajectories between spikes underlying motoneuron firing rates," *Journal of Neurophysiology* 35:311-325. Used with permission.

Figure 1.11 Reprinted from A. Sawczuk, R.K. Powers, and M.D. Binder, 1995, "Spike frequency adaptation studied in hypoglossal motoneurons of the rat," *Journal of Neurophysiology* 73: 1799-1810. Used with permission.

Figure 1.12 Reprinted from D.C. Button et al., 2007, "Spike frequency adaptation of rat hindlimb motoneurons," *Journal of Applied Physiology* 102(3): 1041-1050. Used with permission.

Figure 1.13 Reprinted from R.H. Lee and C.J. Heckman, 1998b, "Bistability in spinal motoneurons in vivo: Systematic variations in rhythmic firing patterns," *Journal of Neurophysiology* 80: 572-582. Used with permission.

Figure 1.14 Reprinted from C.F. Hsiao et al., 1997, "Multiple effects of serotonin on membrane properties of trigeminal motoneurons in vitro," *Journal of Neurophysiology* 77: 2910-2924. Used with permission.

Figure 1.15 Reprinted from *Neuroscience Letters,* Vol. 247, M.A. Gorassini, D.J. Bennett, and J.F. Yang, "Self sustained firing of human motor units," pg. 4. Copyright 1998, with permission from Elsevier.

Chapter 2

Figure 2.1 Reprinted from A.W. Monster and H. Chan, 1977, "Isometric force production by motor units of extensor digitorum communis muscle in man," *Journal of. Neurophysiology* 40: 1432-1443. Used with permission.

Figure 2.2 Reprinted from *Brain Research,* Vol. 125, J.A. Stephens and T.P Usherwood, "The mechanical properties of human motor units with special reference to their fatiguability and recruitment threshold," pg. 7. Copyright 1977, with permission from Elsevier.

Figure 2.3 Reprinted from C.J. Heckman and M.D. Binder, 1993, "Computer simulations of motoneuron firing rate modulation," *Journal of Neurophysiology* 69: 1005-1008. Used with permission.

Figure 2.4 Reprinted from C.J. De Luca, P.J. Foley, and Z. Erim, 1996, "Motor unit control properties in constant force isometric contractions," *Journal of Neurophysiology* 76: 1503-1516. Used with permission.

Figure 2.5 Reprinted from P. Bawa et al., 2006, "Rotation of motoneurons during prolonged isometric contractions in humans," *Journal of Neurophysiology* 96: 1135-1140. Used with permission.

Figure 2.6 Reprinted from C.K. Thomas, B.H. Ross, and B. Calancie, 1987, "Human motor unit recruitment during isometric contractions and repeated dynamic movements," *Journal of Neurophysiology* 57: 311-324. Used with permission.

Figure 2.8 Reprinted from J.N. Howell et al., 1995, "Motor unit activity during isometric and concentric eccentric contractions of the human first dorsal interosseus muscle," *Journal of. Neurophysiology* 74: 901-904. Used with permission.

Chapter 3

Figure 3.3 Reprinted from M.E. Tschakovsky et al., 2004, "Immediate exercise hyperemia in humans is contraction intensity dependent: evidence for rapid vasodilation," *Journal of Applied Physiology* 96: 639-644. Used with permission.

Figure 3.5 Reproduced with permission, from J.F.P. Wojtaszewski and E.A. Richter, 2006, "Effects of acute exercise and training on insulin action and sensitivity: focus on molecular mechanisms in muscle," *Essays in Biochemistry* 42: 31-46. © The Biochemical Society. http://www.biochemi.org.

Figure 3.6 Reprinted from A. Bonen et al., 2007, "A null mutation in skeletal muscle FAT/CD36 reveals its essential role in insulin- and AICAR-stimulated fatty acid metabolism," *American Journal of Physiology* (Endocrinol.Metab.) 292: E1740-E1749. Used with permission.

Chapter 4

Figure 4.3 Reprinted from M.J. McKenna, J. Bangsbo, and J.M. Renaud, 2008, "Muscle K+, Na+, and Cl+ disturbances and Na+-K+ pump inactivation: implications for fatigue," *Journal of Applied Physiology* 104(1): 288-295. Used with permission.

Figure 4.4 Reprinted from C. Juel et al., 2000, "Interstitial K+ in human skeletal muscle during and after dynamic graded exercise determined by microanalysis," *American Journal of Physiology* (Regulatory Integrative Comp. Physiol.) 278: R400-R406. Used with permission.

Figure 4.5 Reprinted from D.G. Allen, G.D. Lamb, and H. Westerblad, 2008b, "Skeletal muscle fatigue: cellular mechanisms," *Physiological Reviews* 88: 287-332. Used with permission.

Figure 4.6 Reprinted from J.H. Kuei, R. Shadmehr, and G.C. Sieck, 1990, "Relative contribution of neurotransmission failure to diaphragm fatigue," *Journal of Applied Physiology* 68: 174-80. Used with permission.

Figure 4.9 Reprinted from J. Woods et al., 1987, "Evidence for a fatigue induced reflex inhibition of motoneuron firing rates," *Journal of Neurophysiology* 58: 125-137. Used with permission.

Chapter 5

Figure 5.6 Reprinted from J.A. Psek and E. Cafarelli, 1993, "Behavior of coactive muscles during fatigue," *Journal of Applied Physiology* 74: 170-175. Used with permission.

Chapter 6

Figure 6.3 Reprinted from F. Jaschinski et al, 1998, "Changes in myosin heavy chain mRNA and protein isoforms of rat muscle during forced contractile activity," *American Journal of Physiology - Cell Physiology* 274: C365-C370. Used with permission.

Figure 6.7 Reprinted from J. Henriksson et al., 1986, "Chronic stimulation of mammalian muscle: Changes in enzymes of six metabolic pathways," *American Journal of Physiology - Cell Physiology* 251: C614-C632. Used with permission.

Figure 6.11 Reprinted, by permission, from A.B. Jorgensen, T.E. Jensen, and E.A. Richter, 2007, "Role of AMPK in skeletal muscle gene adaptation in relation exercise," *Applied Physiology, Nutritional and Metabolism* 32(5): 904-911.

Chapter 7

Figure 7.2 Reprinted from R. Panenic and P.F. Gardiner, 1998. The case for adaptability of the neuromuscular junction to endurance exercise training. *Canadian Journal of Applied Physiology* 23:339-360. © 2008 NRC Canada. Or its licensors. Reproduced with permission.

Figure 7.3 Reprinted from *Neuroscience,* Vol. 89, R. Gharakhanlou, S. Chadan, and P.F. Gardiner, "Increased activity in the form of endurance training increases calcitonin gene related peptide content in lumbar motoneuron cell bodies and in sciatic nerve in the rat," pg. 11. Copyright 1999, with permission from Elsevier.

Table 7.1 Reprinted from A. Adam, C.J. De Luca, and Z. Erim,1998, "Hand dominance and motor unit firing behavior,' *Journal of Neurophysiology* 80: 1373-1382.

Figure 7.4 Reprinted from J.B. Munson et al., 1997, "Fast to slow conversion following chronic low frequency activation of medial gastronomies muscle in cats: 2. Motoneuron properties," *Journal of Neurophysiology* 77: 2605-2615. Used with permission.

Figure 7.5 Reprinted from *Trends in Neurosciences,* Vol. 13, J.R. Wolman and J.S. Carp, "Memory traces in spinal cord," pgs. 137-147. Copyright 1990, with permission from Elsevier.

Figure 7.6 Reprinted from R.D. De Leon et al., 1998, "Locomotors capacity attributable to step training versus spontaneous recovery after signalization in adult cats," *Journal of Neurophysiology* 79: 1329-1340. Used with permission.

Chapter 8

Figure 8.1 Reprinted from S.M. Phillips et al., 1997, "Mixed muscle protein synthesis and breakdown after resistance exercise in humans," *American Journal of Physiology-Endocrinology and Metabolism* 273: E99E107. Used with permission.

Figure 8.4 Reproduced with permission, from U. Seedorf, E. Leberer, B.J. Krischbaum, and D. Pette, 1986, "Neural control of gene expression in skeletal muscle. Effects of chronic stimulation on lactate dehydrogenase isoenzymes and citrate synthase," *Biochemistry Journal* 239: 115-120. © The Biochemical Society. http://www.biochemi.org.

Figure 8.5 Reproduced with permission, from N.J. Osbaldston, D.M. Lee, V.M. Cox, J.E. Hesketh, J.F.J. Morrison, G.E. Blair, and D.F. Goldspink, 1995, *Biochemical Journal,* Vol.308, pgs. 465-471. ©The Biochemical Society. http://www.biochemi.org.

Figure 8.6 Reprinted from K. Baar and K. Esser, 1999, "Phosphorylation of p70^{S6K} correlates with increased skeletal muscle mass following resistance exercise," *American Journal of Physiology: Cell Physiology* 276: C120-C1127. Used with permission.

Figure 8.7 Reprinted from M.B. Reid, 2005, "Response of the ubiquitin-proteasome pathway to changes in muscle activity," *American Journal of Physiology: Regulatory, Integrative and Comparative Physiology* 288: R1423-R1431. Used with permission.

Chapter 9

Figure 9.1 Reprinted from G.R. Adams et al., 1993, "Skeletal muscle myosin heavy chain composition and resistance training," *Journal of Applied Physiology* 74: 911-915. Used with permission.

Figure 9.2 Reprinted from J.D. MacDougall et al., 1984, "Muscle fiber number in biceps brachii in bodybuilders and control subjects," *Journal of Applied Physiology: Respiratory, Environmental and Exercise Physiology* 57: 1399-1403. Used with permission.

Figure 9.3 Reprinted from D.L. Allen et al., 1995, "Plasticity of myonuclear number in hypertrophied and atrophied mammalian skeletal muscle fibers," *Journal of Applied Physiology* 78: 1969-1976. Used with permission.

Figure 9.4 Reprinted from M.T. Woolstenhulme et al., 2006, "Temporal response to desmin and dystrophin proteins to progressive resistance exercise in human skeletal muscle," *Journal of Applied Physiology* 100: 1876-1882. Used with permission.

Figure 9.5 Reprinted from J. Duchateau and K. Hainaut, 1984, "Isometric or dynamic training: Differential effects on mechanical properties of a human muscle," *Journal of Applied Physiology* 56: 296-301. Used with permission.

Chapter 10

Figure 10.1 Reprinted from L.L. Ploutz et al., 1994, "Effect of resistance training on muscle use during exercise," *Journal of Applied Physiology* 76: 1675-1681. Used with permission.

Figure 10.2 Adapted from T. Hortobágyi et al., 1996, "Greater initial adaptations to submaximal muscle lengthening than maximal shortening," *Journal of Applied Physiology* 81: 1677-1682. Used with permission.

Figure 10.3 Reprinted from G. Yue and K. Cole, 1992, "Strength increases from the motor program: Comparison of training with maximal voluntary and imagined muscle contractions," *Journal of Neurophysiology* 67: 1114-1123. Used with permission.

Figure 10.4 Reprinted from B. Carolan and E. Cafarelli, 1992, "Adaptations in coactivation after isometric resistance training," *Journal of Applied Physiology* 73: 911-917. Used with permission.

Muscle Fibers, Motor Units, and Motoneurons

In this chapter, we discuss how muscle fibers, motor units, and motoneurons take part in voluntary movements and how the variability in their properties allows for variability in these movements. Muscle fibers demonstrate a coordinated set of biochemical and functional properties that are used to divide them into several types. The most frequently used nomenclature for distinguishing fiber types classifies the types according to their myosin heavy chain (MHC) composition, and we use that nomenclature in this chapter when referring to different muscle fiber types. The motor unit (muscle unit and its innervating motoneuron) exhibits a set of specific properties that provides a logical recruitment blueprint for the performance of simple isometric contractions. This is facilitated by the fact that all muscle fibers in the muscle unit have the same properties; therefore, heterogeneity at the muscle fiber level is transferred to the muscle unit, and thus motor unit, level. This recruitment blueprint is known as the *size principle,* which dictates that small, more excitable motoneurons are recruited before large, less excitable ones are recruited. For the motoneuron, properties that vary systematically among the motor units include rheobase current, input resistance, afterhyperpolarization duration, and propensity for late adaptation. Motoneurons also exhibit active properties such as self-sustained firing, which is modulated by several monoamines and peptides. These properties can acutely increase the excitability of the motoneuron—and thus the propensity of the motor unit to fire— and modulate its firing patterns once activated.

MUSCLE HETEROGENEITY

In a book about neuromuscular exercise physiology, an important first step is to discuss the ways in which contractile units (the motor units) combine to produce voluntary movements. This chapter assumes that in previous courses, you have learned about the basic mechanics of muscle function. You should already know the sliding fila-ment theory of muscle contraction; the basic structure of skeletal muscles and typical muscle cells, or fibers; the mechanisms by which action potentials are transduced into chemical, and then mechanical, events during excitation–contraction coupling; the twitch, unfused tetanic, and tetanic mechanical responses that take place when muscles are stimulated; and the events that take place at the neuromuscular junction. These topics have been covered very well in a variety of texts currently available, and you should consult these references if questions arise. Assuming that you already have this knowledge base allows us to delve more deeply into more timely issues sur-rounding our knowledge of muscle contribution to exercise in a variety of situations.

The gross anatomy of muscles is another factor that plays a major role in the pat-terns of movement the crossed joints are capable of producing. Factors such as muscle size and the point of origin and insertion of the muscle relative to the axis of rotation of the joint affect the strength and speed of movement. Muscles that have a pennate arrangement of short muscle fibers (meaning that the fibers are arranged at an angle not completely in line with the angle of force generation between the proximal and distal attachments of the muscle) are potentially stronger, albeit slightly slower in shortening speed, than muscles of the same weight in which fibers are fusiform (run the entire length of the muscle along its line of force generation). This is because fibers in a pennate muscle are shorter and more numerous. This arrangement allows for more fibers in parallel for force generation that are slower in shortening distance per unit time because of their shorter length. Human muscles vary widely in their degree of pennate arrangement, ranging from the fusiform (e.g., the sartorius) to the highly pen-nate (e.g., the quadriceps femoris, which is multipennate). This chapter assumes that you are familiar with the importance of these macrostructural anatomical issues, and they will be referred to in this chapter, and throughout the text, with this assumption.

Our understanding of how muscles respond during exercise would be greatly simpli-fied if muscles always contracted isometrically, via a twitch contraction or a maximal tetanic contraction (which are the types of contractions usually shown in textbooks), with all fibers responding simultaneously. Alas, they do not. In addition, often during exercise only some of the fibers in the muscle are used, with the proportion of maximal force generated among those recruited fibers ranging from low to high (relative to their maximum), and with contractile forces of individual fibers not synchronized, such that voluntary contractions are a combination of twitch, semifused tetanic, and tetanic contractions, all occurring out of synchrony. With respect to examining what is occurring in exercising muscles, the interpretation of data using techniques such as the muscle biopsy, in which samples are taken from the muscle belly to measure metabolic changes in the working muscle fibers, becomes problematic, especially when exercise is submaximal in intensity. The force of the recruited fibers also depends on the velocity at which they are shortening, or lengthening, according to the force–velocity relationship that you have learned previously. Finally, we know that all muscle fibers are not the same, which leads us to our currently used terminology of fiber types, and that there are systematic (as opposed to random or unorganized) ways in which these fibers are recruited during exercise.

One indication of the importance of muscle heterogeneity in the contractile perfor-mance of the whole muscle is the relationship between muscle fiber type composition and contractile performance during voluntary and stimulation-evoked contractions. Data from the literature describing this relationship are summarized in table 1.1.

Table 1.1 Correlations Between Percentage of Type II Fibers and Performance Measures

Performance measure	Correlation	Subjects	Reference
Absolute force			
Absolute isometric force of quadriceps	0.55	1	Jansson and Hedberg 1991
Absolute isometric force of quadriceps	0.59*	2	Kyrolainen et al. 2003
Maximum one-leg isometric strength	0.55 to 0.58*	1	Tesch and Karlsson 1978
Explosive speed and power			
Squat jump (centimeters) with 40-kg load	0.45	2	Häkkinen, Newton et al. 1998
Vertical jump, countermovement jump (centimeters)	0.49	2	Viitasalo, Häkkinen, and Komi 1981
Peak acceleration of unloaded knee extension	0.4	1	Houston, Norman, and Froese 1988
Peak acceleration of unloaded knee extension during maximum leg extension	0.48	2	Viitasalo, Häkkinen, and Komi 1981
Maximum rates of isometric force development	0.3 to 0.5	1	Viitasalo and Komi 1978
Maximum stair-climbing velocity	0.37	1, 2	Komi et al. 1977
40-m maximum running velocity from a flying start	0.73	2	Inbar, Kaiser, and Tesch 1981
Isokinetic torque or velocity, power			
% maximum isokinetic torque at 180°/s	0.50	1	Thorstensson, Grimby, and Karlsson 1976
Maximum isokinetic torque at 180°/s	0.52	2	Inbar, Kaiser, and Tesch 1981
Maximum isokinetic torque at 180°/s	0.91**	2	Johansson et al. 1987
Maximum isokinetic torque at 180°/s	0.48*	2	Thorstensson, Larsson, and Karlsson 1977
% maximum isokinetic torque at 115°/s to 400°/s of leg (hip and knee) extension	0.44 to 0.75	1	Coyle, Costill, and Lesmes 1979
Maximum isokinetic torque at 92°/s to 288°/s	0.55 to 0.7*	2	Gregor et al. 1979
Maximum isokinetic torque at 30°/s to 180°/s	0.87 to 0.90***	1	Ryushi and Fukunaga 1986
Knee extension peak power, optimal velocity estimated from isokinetic torque–velocity relationship	0.5 to 0.55	1	MacIntosh et al. 1993
Work, peak power, and rate of power production during isokinetic knee extensions at 6°/s to 300°/s	0.5 to 0.7	1	Ivy et al. 1981
Maximum isometric torque, maximum power during dynamic knee extensions	0.6 to 0.69	2	Tihanyi, Apor, and Fekete 1982

(continued)

Table 1.1 *(continued)*

Performance measure	Correlation	Subjects	Reference
Isokinetic torque or velocity, power			
Ratio of quadriceps torque 240°/s to 30°/s	0.74 to 0.81*	2	Gur et al. 2003
Shot put performance and triceps type II fiber area	0.70*	1	Terzis et al. 2003
Electromechanical delay during isokinetic knee extensions	−0.58	1	Viitasalo and Komi 1981
Anaerobic power and capacity			
Torque decline during repeated maximal knee extensions (50/min) at 189°/s	0.86	1	Thorstensson and Karlsson 1976
Torque decline during repeated maximal knee extensions (50/min) at 189°/s	0.57****	1	Tesch 1980
% increase in EMG/torque ratio during repeated maximal knee extensions (50/min) at 189°/s	0.84	1	Nilsson, Tesch, and Thorstensson 1977
Power decrease during Wingate test	0.68 to 0.76*	1	Froese and Houston 1987
Power decrease during Wingate test	0.67	2	Inbar, Kaiser, and Tesch 1981
Anaerobic power, Wingate test	0.59	1	Kaczkowski et al. 1982
Anaerobic capacity, Wingate test	0.81	1	Kaczkowski et al. 1982
Average power, Wingate test	0.59 to 0.72	2	Inbar, Kaiser, and Tesch 1981
Wingate peak 5-s power output, total work	0.76 to 0.89	1	Froese and Houston 1987
Wingate peak power	0.54	1	BarOr et al. 1980
Aerobic endurance			
Isometric endurance (leg extension) at 50% MVC	−0.51	1	Hulten and Karlsson 1974
Isometric endurance (leg extension) at 50% MVC	−0.51	1	Häkkinen and Komi 1983b
Workload corresponding to the onset of blood lactate accumulation (WOBLA, watts)	−0.57	1	Karlsson and Jacobs 1982
Velocity, 2,000-m run	−0.60	2	Inbar, Kaiser, and Tesch 1981
Running performance time, 1 to 6 mi (1.6-9.7 km; gastrocnemius)	0.52 to 0.55	2	Foster et al. 1978
9-min run distance	−0.58****	1	Jansson and Hedberg 1991
Gross efficiency (ergometer work/oxygen consumption) during an all-out 1-h ergometer ride	−0.75	2	Horowitz, Sidossis, and Coyle 1994
Gross efficiency during cycling at 52% to 71% of $\dot{V}O_2max$	−0.75	2	Coyle et al. 1992

Performance measure	Correlation	Subjects	Reference
Aerobic endurance			
Initial $\dot{V}O_2$ response to exercise ($\Delta\dot{V}O_2/\Delta$work rate)	−0.84	1	Barstow et al. 2000
Speed, amplitude of the $\dot{V}O_2$ response at onset of ergometer exercise	−0.57 to −0.74 1		Pringle et al. 2003
Energy cost in kJ · kg^{-1} · km^{-1} of a 45-min run at 3.33 m/s	0.6	2	Bosco et al. 1987
Time to fatigue at 88% $\dot{V}O_2$max	−0.62	2	Coyle et al. 1988
Whole-body metabolic economy	−0.68*	1	Hunter et al. 2001
Metabolic economy during plantar flexion	−0.53*	1	Hunter et al. 2001
Running economy at 7 m/s	−0.67*	2	Kyrolainen et al. 2003
Running economy at 7 m/s	−0.64****	2	Kyrolainen et al. 2003
Marathon running velocity	−0.64	2	Sjodin, Jacobs, and Karlsson 1981

1 = nonathletes; 2 = athletes.

*Percentage area of Type II fibers.

**Absolute total area of Type II fibers.

***Percentage area of Type IIa fibers.

****Percentage of Type IIb fibers.

In many cases, it is clear that fiber type, which varies considerably for the same muscle among individuals, can influence muscle performance capacity. Nonetheless, the extent to which fiber type is determined by genetics versus training is still the subject of controversy, although we know that some change in fiber type is possible with altered activity.

The concept of differing properties of muscle fibers and muscle units sets the stage for considering their systematic usage during voluntary movements. While we agree that some movements are performed more appropriately using slow, fatigue-resistant motor units, and others require fast, powerful movements, the logic alone does not make it happen in this logical way. It just so happens that the properties of the motoneurons that innervate the muscle units are organized with reference to the properties of their innervated fibers so that muscle fibers are used in appropriate combinations to accomplish the tasks that are requested by the nervous system. The motoneurons obey commands from the central nervous system and assist in deciding which motor units will be recruited, and to what extent they will be recruited. In this section, we consider this coordination of nerve and muscle properties, which is an important prerequisite to understanding how these properties are implicated in fatigue and in adaptation during exercise training. Knowledge of how motoneurons work and how their properties govern the types of movements that we may or may not be able to generate is a very important prequel to discussions of fatigue, recruitment, and training. A summary of the properties to be discussed and their differences among the motor unit types is presented in table 1.2.

Table 1.2 Properties of Cat Motoneurons Innervating Different Muscle Unit Types

Property	S	FR	FF	Source
Soma diameter (micrometers)	49	53	53	Burke et al. 1982
Total membrane area (micrometers)	249	323	369	Burke et al. 1982
Stem dendrite number	12	12.6	10	Burke et al. 1982
Rin (megohms)	1.6 to 2.6	0.9 to 1	0.6	Zengel et al. 1985, Burke et al. 1982
Irh (nanoamperes)	5.0	12.0	21.3	Zengel et al. 1985
Vth (millivolts)	14.4	18.5	20.1	Gustafsson and Pinter 1984a
AHP duration (milliseconds)	161	78	65	Zengel et al. 1985
Minimum firing frequency (impulses per second)*	10		22	Kernell 1979
Maximum firing frequency (impulses per second)*	20		70	Kernell 1979
Current–frequency slope (impulses per second per nanoampere)	1.4	1.4	1.4	Kernell 1965a, 1979, 1983, 1992
Late adaptation	+	+++	+++	Spielmann et al. 1993
Membrane bistability	++		+	Lee and Heckman 1998a, 1998b

*Primary range of firing.

ORDERLY MOTOR UNIT RECRUITMENT

Although the physical size of the motoneuron cell body appears to covary with motor unit type (see table 1.2), the relationship is weak, and the sizes of motoneurons of different types of motor units overlap considerably. Nonetheless, motor units in the triceps surae of decerebrate cats are recruited during stretch in the order of increasing axon conduction velocity and tetanic force and decreasing twitch contraction time and fatigue index in a very reproducible manner and with very few exceptions (Cope et al. 1997). In muscles where all muscle fiber types and motor unit types are the same (such as the slow soleus of the cat), there is evidence that recruitment occurs according to motoneuron size (Binder et al. 1983).

This orderly recruitment based on motoneuron size is known as the *size principle* (see corresponding text box) and appears to be the same during the stretch reflex in awake human subjects (Calancie and Bawa 1985). For example, the response of recruited motor units in human wrist extensor muscles to tendon taps (that excite primary spindle afferents) was shown to be consistent with the orderly recruitment of motor units based on muscle unit size, motoneuron size, and Ia afferent efficacy (Schmied et al. 1997; see their figure 2). In addition, both percutaneous electrical stimulation (Rothwell et al. 1991; Gandevia and Rothwell 1987) and transcranial magnetic stimulation (Bawa and Lemon 1993; Rothwell et al. 1991) of the cortex resulted in recruitment of motor units in hand and forearm muscles that is consistent with the size principle. Thus, the information showing that motoneuron size is strongly related to recruitment order, as measured in reduced preparations, seems to be pertinent to human voluntary movements.

HENNEMAN'S SIZE PRINCIPLE

Decerebrate cats were initially used to demonstrate the orderly recruitment of motor units according to the size of their motoneurons (measured by the amplitude of the action potentials in the ventral root). In these groundbreaking experiments (Henneman 1981; Henneman, Somjen, and Carpenter 1965), recruitment of hind-limb motoneurons was elicited by simply stretching the muscle. Since decerebration increases the tone of extensors, stretching the muscle, and thus stimulating the muscle stretch receptors, results in orderly recruitment—the greater the stretch, the greater the number of motoneurons recruited. Henneman and colleagues showed that the response of motoneurons to the stretching of their innervated muscle was a function of the amplitude of the action potential recorded in the ventral root and thus of motoneuron size—as the stretch amplitude increased, the motoneurons that joined the increasing recruitment had larger and larger amplitudes of action potentials recorded on the axon, signifying that larger and larger cells were being recruited. This phenomenon has become known as the *Henneman size principle* or *size principle* of motor unit recruitment. Furthermore, it appeared that motoneuron size, and not quality of synaptic input to motoneurons, was the determining factor in this stretch-induced recruitment order, since recruitment order was the same with a variety of excitatory and inhibitory stimuli (delivered both alone and in combination), including ipsilateral and contralateral sources, physiological and electrical stimulation, and mono- and polysynaptic input, alone and in combination. The size principle has stood the test of time. Although there have been purported violations of this principle, these have been few and have often involved relatively complicated muscles with anatomical or functional compartments or multiple functions (such as flexion and abduction).

But what do we mean exactly by the term *motoneuron size?* Motoneuron size has been expressed in a number of ways. In the original experiments of Henneman, motoneuron size was estimated from measurements taken at the axon level, since larger motoneurons possess larger axons with larger spike amplitudes and faster conduction velocities. Figure 1.1 shows that this estimation seems to be fairly accurate, and this estimation was the basis for Henneman's original proposal for the size principle, since he showed that motor axons were recruited in decerebrate cats in order of increasing amplitude of the axonal spike and thus inferred that larger axons, and thus larger motoneurons, were recruited after the smaller ones were recruited.

A second technique to estimate motoneuron size is to attempt to measure total surface membrane area histologically. This involves estimating the combined volumes of the cell body and the trunk dendrites near the soma and using these measurements to make some assumptions about the length of the dendrites. An example of this technique used in a study of cat hind-limb motoneurons is shown in figure 1.1. Unfortunately, we do not yet have data like these from human motoneurons.

Motoneuron size can be determined electrophysiologically by using electrophysiological techniques to measure cell capacitance. Motoneurons behave much like a resistor and capacitor in parallel, such that a sustained injection of subthreshold current produces a voltage change that is a combination of the response of these two elements (see figure 1.2). By measuring the time constant of the voltage change and

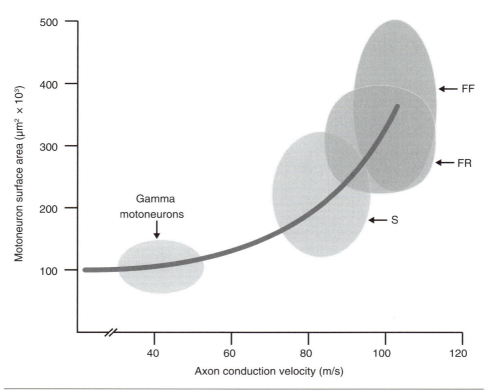

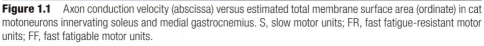

Figure 1.1 Axon conduction velocity (abscissa) versus estimated total membrane surface area (ordinate) in cat motoneurons innervating soleus and medial gastrocnemius. S, slow motor units; FR, fast fatigue-resistant motor units; FF, fast fatigable motor units.

Data from Burke et al. 1982.

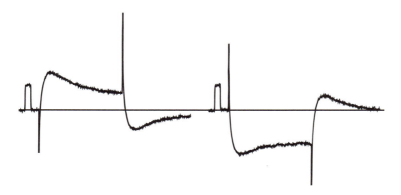

Figure 1.2 Intracellular membrane voltage response of the rat tibial motoneuron to (left) depolarizing and (right) hyperpolarizing current pulses of 1 nA lasting 150 ms. Notice the voltage overshoot followed by sag that occurs after initiation and termination of the current pulses (which are indicated by large on and off voltage artifacts). Calibration pulses at the beginning of the traces are 1 mV for 10 ms.

assuming that the motoneuron's soma and dendrites can be modeled as a cylinder, it is possible to determine the total cell capacitance. Since it is assumed that motoneuronal membranes possess a standard capacitance per unit of membrane area, it is possible to use total cell capacitance to estimate the total cell membrane area. Representing the motoneuron and its dendrites as a cylinder and making assumptions about the value and constancy of the specific membrane capacitance in different parts of the motoneuron limit the precision of this technique (Kernell and Zwaagstra 1989). For all intents and purposes, none of the techniques discussed is superior to the others in the determination of cell size: They are all estimates.

SMALLER MOTONEURONS ARE MORE EXCITABLE

Motoneurons are integrative transducers in that they integrate the voltage responses of the soma and dendrites to thousands of synaptic currents and transduce this integrated signal into a voltage change at the initial segment of the axon, which is the most excitable portion of the motoneuron. Here, if threshold conditions are met, a single action potential or a series is generated and propagated down the axon to the neuromuscular junctions of the muscle fibers. Thus, the intrinsic excitability of a motoneuron plays a major role in determining the probability of its being recruited during excitation of a motor pool. Motoneuron intrinsic excitability has been investigated using intracellular electrophysiological techniques in reduced animal preparations.

The most important factor dictating the intrinsic excitability of the motoneuron is input resistance *(Rin)*. *Rin* is simply Ohm's law applied to the motoneuron, which, because of its membrane composition, resists the current that flows through it:

$$Rin = V/I, \qquad (1.1)$$

where V is potential in volts and I is current in amperes.

Neuron *Rin* can be measured experimentally using intracellular recording techniques by injecting small, subthreshold depolarizing or hyperpolarizing currents into a motoneuron at or near its resting membrane potential and measuring the resultant voltage change. It is termed *input resistance* since it is a measure of the resistance to a current injected into the cell. In large cells, much of the injected current flows out of the cell through conductance channels, and the voltage change is minimal. In small cells, which have a small surface area and thus a smaller number of conductance channels through which the current can leave the cell, injection of the current results in a larger voltage change. Thus, a larger change in membrane potential in response to a small standard current injection indicates a more excitable cell. Small current amplitudes are used to avoid relatively large deviations from resting membrane potential that might bring into play voltage- and time-dependent nonlinearities in membrane response.

Cell *Rin* varies considerably within a motor pool and among motor unit types (see table 1.2). Zengel and colleagues (1985) suggest that in cat medial gastrocnemius, cell *Rin* may be one of the best passive membrane properties of motoneurons by which to predict the properties (speed, fatigue resistance) of the innervated muscle unit and thus that this property and the contractile characteristics (as well as the number) of fibers in the muscle unit are somehow linked.

MEMBRANE RESISTIVITY AND MOTONEURON SIZE

What are the prime determinants of cell *Rin?* The original thought was that cell surface area is the prime determinant and that specific membrane resistivity (expressed

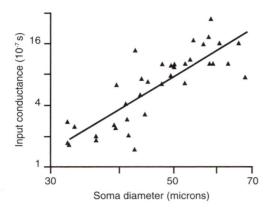

Figure 1.3   Double logarithmic plot of the relationship between input conductance and soma diameter in cat hind-limb motoneurons. The slope of the regression line is 3.3, which is twice as great as the value of 1.6 that would be expected if smaller and larger motoneurons had similar specific membrane resistivities.

Reprinted by permission from Kernell and Zwaagstra 1981.

in ohms times centimeters squared) is constant across all motoneurons. Thus, larger cells with more membrane and thus more conductance channels in parallel have lower resistances (because more current can flow out across the more numerous conductance channels in parallel). It appears, however, that specific membrane resistivity may vary systematically among motoneurons, with smaller cells having disproportionately higher resistance per unit of membrane area. This has been difficult to verify experimentally, since it involves obtaining detailed morphological and electrophysiological measurements on the same cells. These measurements have been made, however, and the results suggest that smaller cells have disproportionately higher membrane resistivity (Burke et al. 1982; Kernell and Zwaagstra 1981; see figure 1.3).

Perhaps the most convincing argument for systematic variation in specific membrane resistivity among motoneurons is simply the difference between the range of motoneuron size and the range of cell *Rin*. In fact, while we have seen that input resistance varies significantly and systematically among motoneurons of different muscle unit types, measurements of cell size using light microscopy indicate that the motoneurons innervating different muscle unit types are very similar in size (see figure 1.3 and compare the range of values for input conductance, which is the inverse of *Rin,* and motoneuron cell body diameter for cat motoneurons).

There are two major caveats for these conclusions about specific membrane resistivity and motoneuron size. First, when estimating total cell membrane area using either microscopic images or electrophysiology, several assumptions are necessary. Morphological analyses use the trunk diameter of each of the stem dendrites to estimate the total surface area of the membrane covering the dendrites; this is an important consideration, since together the dendrites constitute more than 90% of the neuron's surface membrane. Electrophysiological estimates rely on the use of a constant value for membrane capacitance, around 1 microfarad per square centimeter, in order to arrive at whole-cell capacitance and thus membrane area. A second caveat has to do with the assumption that membrane resistivity is constant throughout the cell. Current indications suggest that this assumption is false and that specific resistivity of dendritic membrane is higher than that of somal membrane (Rall et al. 1992). This issue raises the questions of whether measurements of whole-cell *Rin* are meaningful and whether the location of the recording electrode within the cell influences this measurement.

The main message is that whole-cell *Rin,* and therefore intrinsic motoneuron excitability, varies systematically across motor unit types and that this variability is a major factor in determining the recruitment order of motor units during exercise.

Why muscle units of different types are systematically controlled by motoneurons of differing excitability and why even the membrane resistivity varies among motor unit types are not yet known.

OTHER FACTORS DETERMINING ACTION POTENTIAL GENERATION

The motoneuron *Rin* greatly affects the voltage change that occurs in the motoneuron when current is injected into the cell (either through a microelectrode or through synaptic connections on the motoneuron). Factors that determine if a motoneuron will be excited enough to produce an action potential include the amount of current that is coming into the cell, the resting membrane potential (RMP), and the depolarization amplitude required to reach the threshold for spike generation (known as *threshold depolarization, Vth*). The amount of current that produces an action potential is known as *rheobase current (Irh)*.

As might be expected from our previous discussion, the amount of current needed to generate an action potential (assuming that RMP is the same for all motoneurons) is dictated to a large extent by *Rin:* The larger the *Rin*, the smaller the current required for generating an action potential:

$$Irh = Vth/Rin \qquad\qquad (1.2)$$

If variation in *Irh* among motoneurons could be explained completely by corresponding variations in *Rin*, the relationship between input conductance *(Cin*, the inverse of *Rin)* and *Irh* should be linear and proportional, with the slope of this relationship corresponding to the average value of *Vth* (Gustafsson and Pinter 1984a). In fact, these two independent measurements are very highly correlated, but the range of *Irh* exceeds that of *Cin* by a factor of around 2. In figure 1.4, it is apparent that this linear relationship is in fact not so linear; low-rheobase cells are mostly below the line of best fit, while high-rheobase cells are above the line. Since the slope of the line gives the average amount of depolarization from the RMP that is necessary for generating an action potential *(Vth*, which ranges from 10 to 25 millivolts in cat and rat motoneurons), the suggestion is that low-rheobase cells require less depolariza-

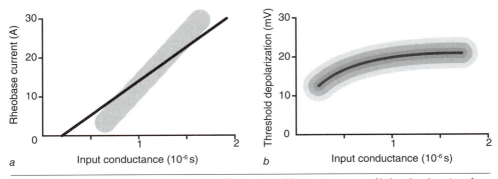

Figure 1.4 *(a)* Relationship between *Irh* and *Cin* in cat hind-limb motoneurons. Notice the departure from linearity (solid line) of data points (shaded area) at low and high rheobase values (the former are more to the right of the line, while the latter are to the left). *(b)* *Vth* (the depolarization from RMP, in millivolts, required to produce an action potential) versus *Cin* for the same study. *Vth* tends to be smaller for smaller cells.

Data from Gustafsson and Pinter 1984a.

tion to generate an action potential than high-rheobase cells require. This difference cannot be attributed to variation in RMP.

Another factor that contributes to motoneuronal excitability is nonlinearity in membrane response between RMP and the voltage at which the action potential is generated. For example, the product of *Irh* and *Rin* (known as *rheobase voltage*) should give *Vth* under linear conditions, but in fact the measured *Vth* is somewhat greater than the product of *Irh* and *Rin*. This is due to a persistent inward current (PIC), most likely calcium mediated, that occurs at subthreshold voltages between RMP and threshold (Gustafsson and Pinter 1984a; Binder, Heckman, and Powers 1996). This inward rectification, also known as *anomalous rectification,* appears to be slightly more significant in smaller cells than it is in larger cells, although this variability among cells is most likely secondary to variations in passive properties (Gustafsson and Pinter 1984a).

Thus, small cells may have an additional advantage over larger cells, and thus slow (type S) over fast fatigable (type FF) motor units, with regard to their susceptibility to be recruited. Their passive properties may accentuate the effect of inward currents that operate at depolarized voltages, and at the same time they may require less depolarization to generate an action potential (i.e., they may have a lower *Vth*). This latter consideration, assuming that there are no systematic differences in RMP among cells, suggests that the voltage at which action potentials occur (the firing level) is slightly (around 6 millivolts) more negative in small type S motoneurons. As with the differences in *Rin* and membrane resistivity, we do not know why these systematic differences exist among motor unit types.

MINIMAL FIRING RATES AND AFTERHYPERPOLARIZATION DURATIONS

Immediately following the somal-dendritic spike of the motoneuron, the repolarization continues at a level below RMP for a short time before returning to RMP. The temporary, prolonged hyperpolarization that follows the action potential is termed the *medium afterhyperpolarization (mAHP,* or simply *AHP),* and its duration varies systematically among motoneurons of different types (see figure 1.5). The AHP is attributed to the activation of a calcium-activated potassium conductance (GKCa) that gradually declines after the spike (Binder, Heckman, and Powers 1996; Kudina and Alexeeva 1992). Injection of a calcium chelator into the cell or exposure of the cell to the bee venom apamin (a specific GKCa channel blocker) or to cobalt or manganese significantly reduces the AHP, indicating that it is controlled by calcium influx. It appears to be the only electrophysiological variable that can unequivocally distinguish fast motor units from slow motor units in both the cat and the rat (Zengel et al. 1985; Bakels and Kernell 1993; Gardiner 1993). The reason for the difference in the AHP time course between slow and fast motoneurons is unknown but may involve differences in the density, location, or activation and deactivation kinetics of the GKCa channels or in the regulation of the intracellular calcium levels that activate these channels. The importance of the AHP for our discussion relates to how it plays a role in the rhythmic firing patterns of motoneurons in response to sustained excitation.

Figure 1.6, from Kernell's seminal work (1965c), shows the minimal firing frequency of cat hind-limb motoneurons in response to sustained currents injected through a microelectrode as a function of AHP duration (shown as the reciprocal of the time to

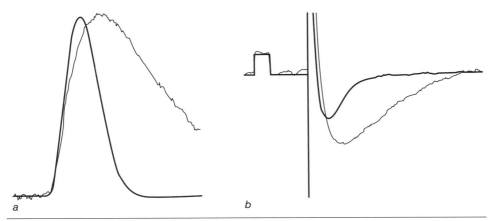

Figure 1.5 *(a)* Muscle unit isometric twitch and *(b)* corresponding AHP response of the innervating motoneuron from rat gastrocnemius motor units. The fast unit is designated by a thick line and the slow unit by a thin line. Calibration at the beginning of AHP is 10 ms (for both twitch and AHP) and 1 mV (for AHP only). Twitch forces have been normalized for ease of comparison.

Reprinted by permission from Gardiner 1993.

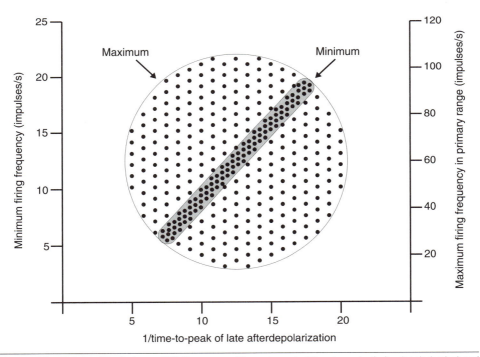

Figure 1.6 Relationships between AHP time course (the reciprocal of time to peak of afterdepolarization) and minimal and maximal firing frequencies. AHP time course relates more highly to minimal than to maximal firing rate.

Data from Kernell 1965c.

after depolarization). Kernell found significant correlations between AHP time course and maximal firing frequencies, although they were not as impressive $(r = 0.58\text{-}0.76)$. The author also demonstrated the relationships between the limits of motoneuron firing rates and the corresponding expected muscle fiber responses at those frequencies. The minimal firing frequency corresponded to a frequency that would result in slightly fused twitch responses of the innervated muscle fibers, while the maximal firing rates in the primary and secondary ranges corresponded to approximately 80% and 95%, respectively, of maximal tetanic force. This appears to be the same for fast and slow muscles; thus, the firing frequencies of slow and fast motoneurons, dictated to some extent by their AHP, appear to correspond quite conveniently to the contractile speed of the innervated muscle fibers. The fact that minimal firing frequencies of motoneurons produce not individual twitches but slightly fused contractions most likely has an implication for the smoothness of contractions at low force requirements, when few motor units are recruited at their lowest frequencies.

There are, however, factors other than AHP duration that control minimal firing rates of human motoneurons during voluntary movement. Kudina and Alexeeva (1992) estimated the characteristics of soleus and flexor carpi ulnaris motoneuron after-potentials in human subjects from the recovery curve of motoneuron excitability after a single discharge evoked by afferent stimulation or gentle voluntary muscle contraction. They found that the minimal rates of motoneuron firing were not correlated with that expected from the estimated AHP duration (figure 1.7). They suggested that other spinal or supraspinal mechanisms controlling motoneuron firing are masked in in situ experiments with animals in which anesthesia or spinalization is involved. They

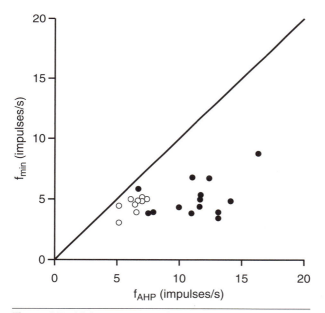

also found that some fast motoneurons demonstrate early recovery of excitability, which they attributed to the delayed depolarization that follows the spike, before the downstroke of the AHP. These motoneurons were capable of firing double discharges (doublets) with a relatively rapid interspike interval (5-15 milliseconds).

Additional evidence that motoneuron firing rates may not be controlled only by AHP time course comes from decerebrate cats in which fictive locomotion was induced via stimulation of the mesencephalic locomotor region. Under these conditions, AHP amplitudes in motoneurons during trains of impulses were significantly decreased compared with amplitudes

Figure 1.7   Minimum motor unit firing rates (ordinate) versus estimated minimum firing based on AHP duration (abscissa) for human soleus (open symbols) and flexor carpi ulnaris (filled symbols) motor units. Firing rates are lower than predicted from the AHP duration (line of identity).

Reprinted by permission from Kudina and Alexeeva 1992.

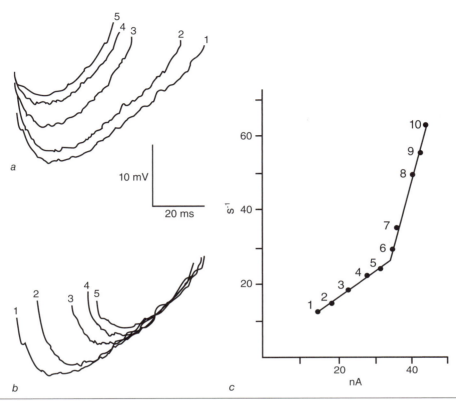

Figure 1.8 Membrane potential trajectories in a cat motoneuron firing at different rates in response to current injection. The frequency–current relationship is shown in *(c)*, in which the numbers correspond to the recordings in *(a)* and *(b)*. The alignment of AHPs in *(a)* is according to original direct current levels. In *(b)* the same recordings are aligned to a common firing level (i.e., the voltage at which the following spike occurs). Note how AHP gets shorter and smaller in amplitude as firing rate increases.

Reprinted by permission from Schwindt and Calvin 1972.

seen during comparable frequencies of firing induced by intracellular current injections. In addition, motoneurons showed a decreased gain (change in firing frequency per unit current injected) during locomotion compared with the resting condition (Brownstone et al. 1992). Thus, the AHP–firing frequency relationship may constitute a blueprint, evident in anesthetized animals, upon which influences from afferent sources can induce modifications.

The AHP is not an absolute refractory period in that a second action potential can be generated during the AHP if currents are strong enough. Figure 1.8 shows the evolution of the AHP in a motoneuron firing at low and high frequencies. What happens to the AHP when changing from a single spike to a series of spikes during rhythmic firing is dealt with in the next section.

MOTONEURON CURRENT–FREQUENCY RELATIONSHIP AND EXCITABILITY

Up to this point, we have considered recruitment during a relatively short excitation of the motoneuron pool. Let us consider now the additional factors that determine the behaviors of the system when more prolonged, sustained excitations are instituted.

In moving from generation of single action potentials to trains of impulses, we begin to see the influence of time-dependent and voltage-dependent membrane nonlinearities in which motoneuron excitability can change as a function of the time and intensity of sustained current injection. For example, when *Irh* is sustained, trains of impulses do not occur until the current intensity is increased approximately 50% (this is called the *current threshold for rhythmic firing;* Kernell and Monster 1981). A principal contributing factor is the sag property of the membrane in response to prolonged current injection (see figure 1.2). This is a time-dependent, mixed-cation conductance active around RMP that gradually alters membrane voltage in the hyperpolarizing direction during depolarizing pulses and in the depolarizing direction during hyperpolarizing pulses. Interestingly, this sag property appears to be more significant in motoneurons that innervate fast muscle units. Contribution of GKCa to this sag response becomes more significant with larger depolarizations. It has been proposed that this membrane sag may explain part of the difference between fast and slow motoneurons in the time course of their afterpotentials (Gustafsson and Pinter 1985).

The current–frequency relationship for a typical motoneuron is shown in figure 1.9. As current intensity increases above the current threshold for rhythmic firing, frequency of firing increases, as might be expected. What is less expected is that the current–frequency relationship has two linear portions, which are termed the *primary* and *secondary ranges of firing* (Granit, Kernell, and Shortess 1963a; Kernell 1965b). This phenomenon has been observed in anesthetized cat motoneurons and in preparations of several species of motoneurons in vitro. Whether or not this is a property of all motoneurons is not known; it may be that this property is lost in some motoneurons as a result of electrode impalement. The mechanism by which motoneurons switch from primary to secondary range of firing with increasing current injection is unknown. One hypothesis is that the increase in the threshold voltage for spike generation results in a membrane trajectory between spikes that is more depolarized

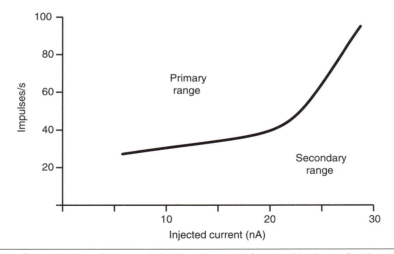

Figure 1.9 Steady discharge frequency (1.0-1.5 s after onset of current injection, ordinate) versus current strength (in nanoamperes, abscissa) for a cat motoneuron. Primary and secondary range slopes (in impulses per second per nanoampere) are shown.

Data from Kernell 1965b.

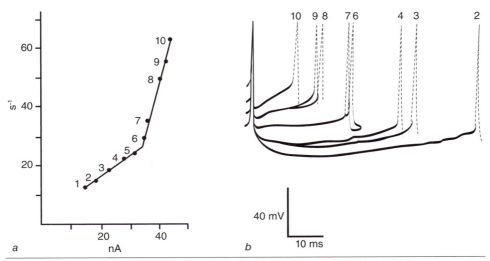

Figure 1.10   Change in firing level (voltage at which spike is generated) during repetitive firing in a cat motoneuron. Numbers in *(b)* correspond to their frequency in *(a)*. In *(b)*, recordings are aligned to the peak of the preceding spike. Note how the firing level, which is the voltage at which the steep upward voltage change toward the action potential peak occurs, becomes more depolarized (in the upward direction in this figure) as rate increases, especially in the secondary range of firing (from spike 6 to 10).

Reprinted by permission from Schwindt and Calvin 1972.

due to accommodation at the initial segment (see figure 1.10). This allows activation of a PIC (an inward current that turns on and is self-sustaining and long lasting) that increases firing frequency at the same level of injected current and thus increases the current–frequency slope.

Maximum sustainable firing frequencies of motoneurons are difficult to measure. Although firing frequencies at the beginning of an intense injection of current can be very high and can exceed the frequencies necessary for tetanization of their innervated muscle fibers, these high frequencies fall off very rapidly.

We might suspect that the most efficient system for gradation of whole-muscle force up to the maximum would be one in which the first recruited motoneuron is still increasing in firing frequency at the time when the last unit is recruited. The best model for this is one in which the slopes of increased firing frequency as a function of injected current are approximately equal for all motoneurons in the pool (Kernell 1984). There is some evidence that this is indeed the case (Kernell 1979; Button et al. 2006). However, there are mechanisms whereby all motoneuron firing frequencies appear to meet at their respective maximal levels during a maximal voluntary contraction (MVC) by braking the firing of some of the early recruited units and accelerating the firing of units recruited later on.

LATE ADAPTATION

Adaptation is the process by which firing frequency decreases with sustained constant-intensity excitation. Late adaptation takes tens of seconds and even minutes to occur, as opposed to the very rapid adaptation that occurs at the beginning of a burst. Figure 1.11 shows examples of early and late adaptation (Sawczuk, Powers, and Binder 1995).

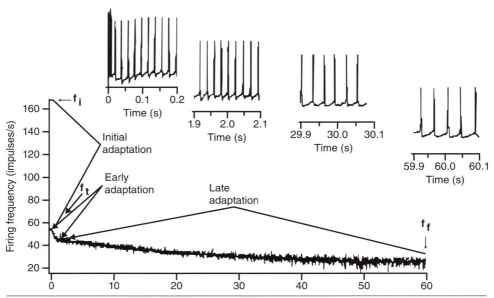

Figure 1.11 Motoneuron adaptation during repetitive discharge of rat hypoglossal motoneurons in vitro in response to sustained intracellular current injection. Three phases of frequency decrease, which correspond to initial, early, and late adaptation, have been identified.

Reprinted by permission from Sawczuk, Powers, and Binder 1995.

IS LATE ADAPTATION
FATIGUE OF THE MOTONEURON?

Motoneuron firing rates decrease when the motoneurons are stimulated with constant excitation from either intracellular or extracellular sources. The phenomenon is not fatigue of the motoneuron in the expected sense—the firing rate increases during adaptation if excitation strength is increased. Thus, if late adaptation occurs in vivo as it does in reduced preparations, an individual generating a constant-load, maintained contraction must gradually increase excitation to the motor pool to compensate for late adaptation and to increase the recruitment of motor units to maintain the force. Indeed, surface electromyography (EMG) reveals a gradual increase in motor unit excitation under these circumstances; however, the importance of motoneuronal late adaptation versus muscle fiber fatigue in these responses is not known. We must also consider that during a voluntary contraction excitatory influences from many sources, including the increased release of excitation modulators (such as epinephrine or serotonin) into the region of the motoneuron pool, may counteract late adaptation. Interestingly enough, although motoneurons demonstrate a wide variation in their amount of late adaptation, the changes in this property that might occur with training and disuse are not currently known.

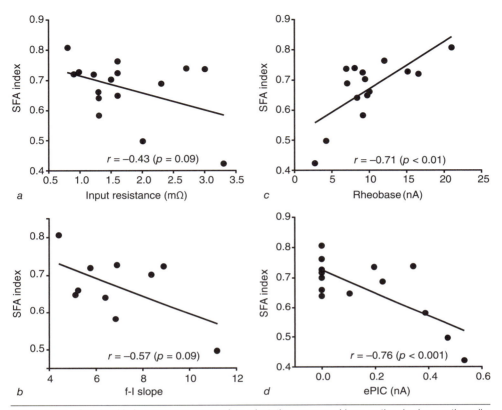

Figure 1.12   Relationship between motoneuron late adaptation, expressed here on the abscissa as the spike frequency adaptation (SFA) index, and *(a) Rin, (b)* rheobase, *(c)* the slope of the frequency–current relationship, and *(d)* the estimated amplitude of the PIC. The higher the SFA index, the more late adaptation present.

Reprinted by permission from Button et al. 2007.

Late adaptation varies among the motor unit types; it is more developed in motoneurons that are less excitable and that innervate fast muscle units (Spielmann et al. 1993; Button et al. 2007; see figure 1.12). The mechanisms for late adaptation are not known, but most likely there are several mechanisms. One candidate may be a gradually increasing outward current that is mediated by GKCa. By this mechanism, calcium entering the cell during rhythmic action potential generation activates an outward potassium conductance, which increases the amplitude and duration of the outward potassium current following the spike. This outward current would increase the time taken to depolarize the membrane back to the threshold for the next spike, resulting in a gradual increase in the intervals between spikes. The apparent systematic difference between fast and slow motoneurons (see figure 1.12) in late adaptation to sustained excitation may indicate different rates of calcium entry and accumulation during rhythmic firing, differing capacities to buffer accumulating calcium, or different spatial relationships between calcium accumulation and channel sites in the motoneuron's membrane.

Another mechanism may include a gradual inactivation of the sodium channels responsible for spike generation. Evidence for this phenomenon as a possible contributor to late adaptation is the change in the shape of the spike and the change in the voltage threshold (that voltage at which sodium channels open to produce the action potential) during late adaptation. Of particular interest is that late adaptation is evident not only during sustained excitation but also during intermittent excitation, such as during rhythmic exercise (Spielmann et al. 1993). The phenomenon of late adaptation is particularly interesting because it relates to the types of contractions that are involved in exercise. The process of motoneuronal adaptation may be an important component, not necessarily in limiting the force produced, but certainly in increasing the excitatory effort required to maintain motoneuron firing frequencies and thus sustain a specific level of neuromuscular effort.

MOTONEURON PICS

Up to now, we have considered motoneurons as linear input–output transducers, in which a given stimulus results in a predictable response. In fact, motoneurons, like many other neuron types, can behave in a sleepy or in a supercharged state. Membrane bistability is a phenomenon by which motoneurons change between these two levels when given specific stimuli to do so. Bistability refers to a shift between two states of

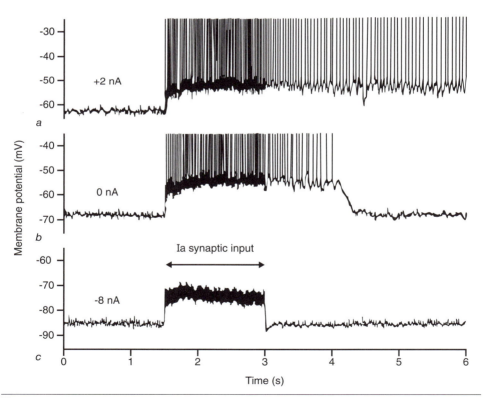

Figure 1.13 Example of bistability in a motoneuron in a decerebrate cat. *(a)* and *(b)* High-frequency, low-amplitude tendon vibration of the triceps surae for 1.5 s evoked firing that continued after the stimulation ceased. *(c)* At hyperpolarized membrane potential, bistability was not present.

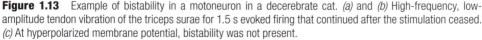

excitability (hence the term *bistability)* via the development of long-lasting plateau potentials that outlast the brief periods of excitation that generate them. An example of membrane bistability is shown in figure 1.13. This change from one level of excitability to another is caused at least in part by activation of voltage-sensitive, slowly inactivating, inward sodium and calcium conductances (Heckman et al. 2008). These conductances are present primarily in the proximal dendrites and to some extent in the soma. Excitatory persistent inward currents (PICs, so-named because they last a relatively long time), which underlie bistability, can be generated by direct intracellular or synaptic excitation. PICs last for up to several seconds after the excitation has been terminated, due to the slow inactivation kinetics of the channels involved. A PIC manifests as a sustained decrease (i.e., shift toward threshold) in the membrane potential or, if the cell is already firing, as an increase in firing rate that can outlast the original stimulation by several seconds. A PIC can be reversed by an inhibitory influence, such as a hyperpolarizing pulse. It can be considered as a type of warm-up phenomenon for motoneurons (Bennett et al. 1998). Nonvolatile anesthetics (such as barbiturates) used during in situ experiments inhibit the generation of plateau potentials, and thus PICs are seen only in experiments where these anesthetics are not used, such as in decerebrate animals or with tissue slices in vitro (Guertin and Hounsgaard 1999).

Several substances uncover plateau potentials in motoneurons; these include serotonin (5-HT), norepinephrine, thyrotropin-releasing hormone (TRH), angiotensin II, substance P, oxytocin, cholecystokinin, and somatostatin (Hultborn and Kiehn 1992). The most intensely studied modulator of plateau potentials in motoneurons is 5-HT. This modulator appears to allow the expression of plateau potentials in motoneurons by enhancing the inward rectifier (sag) current and by reducing the GKCa, which is the outward current responsible for the AHP (see figure 1.14). The various modulators listed earlier probably function by several different mechanisms, all of which uncover the bistability phenomenon. Many of these modulators, most significantly 5-HT and norepinephrine, are released by neurons whose cell bodies are in the brain stem and whose terminals are in the region of the motoneurons. These centers responsible for releasing these monoamines are activated during the shift from rest to locomotor activity, and thus they serve to supercharge motoneurons into action when a person decides to change from a resting to a moving state.

It has been shown that low-threshold motoneurons (those that innervate type S and FR muscle units) demonstrate more marked bistability (show self-sustaining firing for longer durations) than higher-threshold motoneurons demonstrate (Lee and Heckman 1998b). This suggests a functional importance for this property in postural and longer-lasting tonic and rhythmic locomotor activities. Indeed, when monoamines in the spinal cords of rats were depleted using specific neurotoxins, the EMG pattern of the postural soleus muscle changed from a tonic to a more phasic one (Kiehn et al. 1996).

As can be imagined, detecting motoneuron bistability in an awake, voluntarily moving animal or human is difficult, since the excitation intensity of motoneurons is not known or controllable. Eken and Kiehn (1989) examined EMG records of single motor units in rat soleus and found behavior resembling bistability in the presence of excitatory and inhibitory stimulation (of afferents and skin, respectively). Gorassini, Bennett, and Yang (1998) provided evidence that this phenomenon may be of physiological significance in humans performing voluntary contractions (figure 1.15).

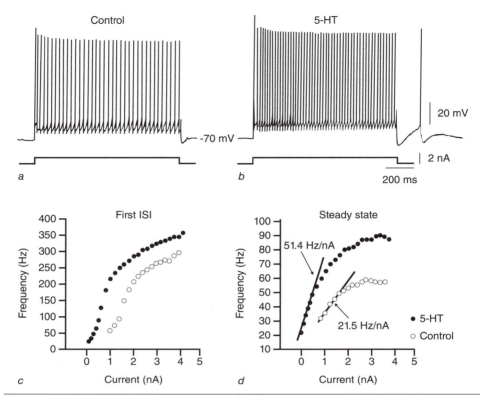

Figure 1.14   Effect of 5-HT on guinea pig trigeminal motoneurons in brain stem slices in vitro. *(a)* versus *(b)* 5-HT increased firing frequency for a given current. *(c)* and *(d)* 5-HT increased the frequency–current slope.

Reprinted by permission from Hsiao et al. 1997.

In their experiments, they asked their subjects to perform low-force, constant-effort, isometric contractions of the tibialis anterior and monitored the firing rate of a low-threshold control unit. When the firing rate of this control unit stabilized at a constant frequency, vibration of the muscle tendon resulted in the recruitment of a second test unit that maintained its firing after the vibration was terminated (the vibration lasted 0.5-1.5 seconds). This occurred in spite of no change in the firing rate of the control unit after vibration, suggesting that descending drive to the motoneuronal pool was the same both before and after the vibration. Sustained activity of the second unit occurred in about 50% of the units that were evoked by the vibration and lasted from a few seconds to a few minutes after cessation of the vibration. These researchers also found that repeated vibrations resulted in an increased number of spikes of the newly recruited unit. In most cases, the sustained firing remained after vibration, even when the subjects were instructed to decrease their effort slightly, during which time the firing rate of the control unit actually decreased. This approach (comparing a test unit that expresses bistable properties with a control unit that does not) has continued to provide some evidence that this phenomenon may have a significant physiological role (Gorassini et al. 2002). Thus, even during effort requiring a constant-level drive of motoneuron pools, the same facilitated firing behavior demonstrated in reduced animal preparations may be at work in humans. The implications for a

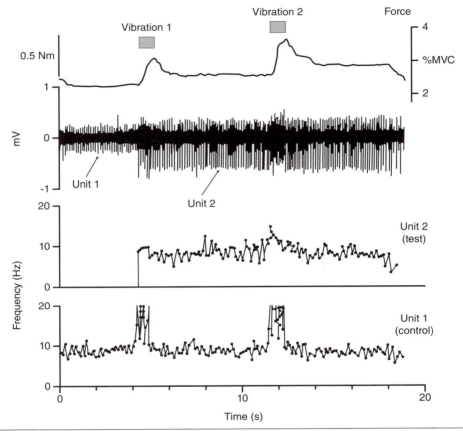

Figure 1.15   Evidence of motoneuron bistability in humans during sustained firing of a tibialis anterior motor unit recruited by brief vibration. The subject was instructed to maintain constant dorsiflexion force while firing of unit 1 was monitored. Brief vibration of the tendon recruited a second unit (unit 2) that continued to fire after the vibration. This occurred without an increase in firing rate of the control unit.

Reprinted by permission from Gorassini, Bennet, and Yang 1998.

possible reversal of the recruitment order of motor units during voluntary effort due to this phenomenon are not clear yet. However, if bistability is organized in a manner similar to *Rin,* such that bistability decreases in the order S > FR > FF, the order of recruitment might remain unchanged if all motoneurons in the pool are given the same plateau-promoting stimulus. More information than this is lacking at present.

SUMMARY

The properties of alpha-motoneurons vary systematically, such that the excitability of innervating motoneurons within motor units varies in the order S > FR > FF. This generates a blueprint for recruitment during voluntary movement, and motor units are recruited in the order of S followed by FR followed by FF as force demand and voluntary effort increase. Ranges of firing frequencies of motoneurons also match the contractile speed of the innervated muscle fibers, such that maximal motor unit force occurs near maximal firing rate of the innervating motoneuron. The tendency for motoneurons to demonstrate late adaptation, or a decrease in firing rate during constant-level excitation, is also matched to the fatigue resistance of the muscle fibers, such that FR muscle units are innervated by motoneurons that show relatively less late adaptation when challenged with constant excitation. The capacity of motoneurons to express self-sustaining currents may also differ among the motor unit types, and this phenomenon may be involved in supercharging these motor units during voluntary movements.

CHAPTER 2

Motor Unit Recruitment During Different Types of Movements

In this chapter, we will discuss how motor unit recruitment can be estimated in human experiments. While the size principle seems to determine motor unit recruitment in most cases, recruitment patterns can change depending on the properties of the task. While the changes occurring during sustained contractions suggest the influence of late adaptation, muscle property changes, and possible motor unit rotation, the exact motor unit recruitment patterns that occur during complex activities such as running are very difficult to measure and are therefore not fully known.

Now that we understand the patterns in which motor units might theoretically be recruited during movements, are motor units actually recruited in ways that we might predict? Is recruitment always based on the size principle, or are there exceptions? What happens to recruitment as humans fatigue? These are issues that we will address in this and following chapters.

MEASURING HUMAN MOTOR UNIT RECRUITMENT

There are several techniques used to determine the recruitment of muscle fibers and motor units during voluntary movement, and each has its advantages and limitations. Here, I provide a brief explanation of the techniques used to investigate motor unit recruitment and then discuss what we know about recruitment during different types of voluntary tasks.

Most of our knowledge in this area comes via the use of fine-wire electrodes that are inserted into the muscle in question to record the activities of the muscle fibers in the electrode's immediate vicinity. In some cases, an attempt is made to reduce the movement of the electrode by hooking it into the muscle or under the skin and by reducing the movement of the contraction. The motor unit can be identified by its characteristic waveform, which can be scrutinized throughout the experimental procedure to ensure that it, and thus the position of the electrode, has not changed. Such studies examine the force threshold (i.e., the percentage of whole-muscle maximal force at which the unit first appears) and firing frequency of motor units under various conditions (such as recruitment of motor units when muscle is used in abduction versus flexion movements or concentric versus eccentric contractions). Techniques for implanting these electrodes have improved over the years, so that reasonable high-force contractions can be examined. However, the type of motor unit is usually not known; only the relative whole-muscle force at which it is first recruited is known. A less invasive extension of this technique is to derive mathematically the waveforms of the individual motor unit potentials contributing to a surface EMG (Mambrito and De Luca 1983).

A permutation of this technique involves passing a fine tungsten electrode through the muscle during a constant-force isometric contraction and recording the firing frequencies of muscle fibers near the electrode. In this technique, the primary concern is sampling as many firing fibers as possible. The result is a sample of firing motor units (unfortunately, not including units that were firing but have ceased to do so) for each contraction condition. This sample allows the investigator to calculate the mean and variation in firing frequency for the activated motor units.

The technique of spike-triggered averaging has been used in conjunction with intramuscular motor unit recording to generate information about the order in which units are recruited during simple contractions. In this technique (MilnerBrown, Stein, and Yemm 1973a, 1973b), EMG impulses of single motor units are used to trigger a signal averager that sums the whole-muscle force immediately following the spike. With enough spikes, the characteristics of the time-locked isometric twitch that follows the spike in question eventually emerge from the force signal. This has been our primary means of identifying the contractile properties (twitch speed and strength) of human muscle units as a function of force threshold during low-force, isometric contractions. The initial experiments suggested, as expected, that motor units are recruited from weakest to strongest and from slowest to fastest (see figure 2.1).

It is possible to estimate motor unit recruitment patterns by examining the spectral properties of the myoelectric signals from surface EMG electrodes. For example, the activity of fast and slow motor units results in myoelectric spectra that have similar spectral power but are high- and low-frequency spectra, respectively (Wakeling 2004). Wakeling and colleagues used this technique to investigate time-varying shifts in motor unit recruitment patterns in human leg muscles during walking and running.

Biochemical indexes of recruitment, in which a biochemical marker of recruitment is measured in muscle fibers from biopsies taken immediately following exercise, have also been used to study recruitment. Recruitment by this method is difficult to determine unequivocally, since different fiber types use substrates at different rates to produce the same amount of external work. In addition, biochemical changes (such as the decrease in glycogen content, which is the most common) may take long to be expressed and may be dependent on oxygen delivery. It has been demonstrated, none-

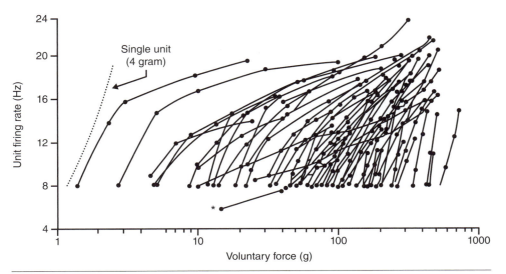

Figure 2.1   Firing rates of 60 motor units in extensor digitorum communis as a function of voluntary isometric contraction strength. The asterisk indicates a unit that is behaving unusually.

Reprinted by permission from Monster and Chan 1977.

theless, that the change in the phosphocreatine/creatine (PCr/Cr) ratio can be quite dramatic after only seven brief isometric maximal voluntary contractions (MVCs) (Beltman, Sargeant, Haan, et al. 2004). As we will see, these indexes have proven useful in estimating the recruitment of muscle fiber types in other forms of activity.

INFLUENCE OF TASK

We have already learned the importance of the size principle in determining the recruitment order of motor units. However, many muscles take part in various movements for which the order of recruitment might be different (consider abduction and flexion, for example). Many muscles have been shown to consist of distinct anatomical compartments that can be identified by (1) separate intramuscular motor and sensory nerve branches that subserve a distinct, regionalized subvolume of muscle fibers; (2) Ia excitatory postsynaptic potential (EPSP) partitioning, in which the strength of connections to motoneurons from Ia afferents from the same neuromuscular compartment is 1.6 to 2.3 times stronger than that from other compartments in the same muscle (Stuart, Hamm, and Vanden Noven 1988); and (3) EMG evidence during various movements indicating that different compartments of a muscle are used differently, depending on the task (Wickham and Brown 1998; Hensbergen and Kernell 1992).

Consistent with the idea of muscle compartmentalization is the concept of task groups of motor units, whereby motor units in a muscle are recruited differently depending on the task that they must perform. The concept of task groups was originally proposed by Loeb (1987) as a functional compartmentalization of the motor apparatus. This functional compartmentalization does not necessarily correspond to the anatomical segregation at the muscle, nerve branch, or motor nucleus level. The idea of task groups of motor units does not necessarily indicate the presence of the muscle compartmentalization described in the previous paragraph.

Wakeling (2004) has shown some evidence that there are time-varying shifts in motor unit recruitment patterns in human leg muscles during walking and running. These shifts vary systematically within each step cycle. The idea is that although the recruitment of motor units is based on the size principle, alterations during the step cycle related to things such as muscle spindle excitation can change this basic recruitment pattern in a systematic way during the step cycle. These alterations differ among muscles and probably among individuals.

While keeping the concept of task specificity in mind, we can also see general patterns of motor unit recruitment based on the size principle in operation during movements, especially simple movements. Following is a discussion of motor unit recruitment during several types of voluntary movements, with emphasis on the degree to which orderly and reproducible recruitment occurs across different movement types. While increased force demands are satisfied at the motor unit level by the recruitment of higher numbers of motor units and the increase in firing rates of those units already recruited (rate coding), the contributions of recruitment and rate coding to force increase can vary considerably among muscles. For example, rate coding is more important in adductor pollicis, where no significant recruitment occurs after 30% of MVC, while recruitment of motor units in biceps brachii is still evident at 88% of MVC (Kukulka and Clamann 1981). In some examples, we will also see how the pattern of motor unit recruitment within a muscle changes as the muscle performs different tasks.

SLOW-RAMP ISOMETRIC CONTRACTIONS

Among the first results obtained from using spike-triggered averaging to extract the twitch characteristics of units as they were being recruited was the finding that units are recruited generally from smallest to largest twitch force and from slowest to fastest twitch contraction time (see figure 2.2). This recruitment order has been

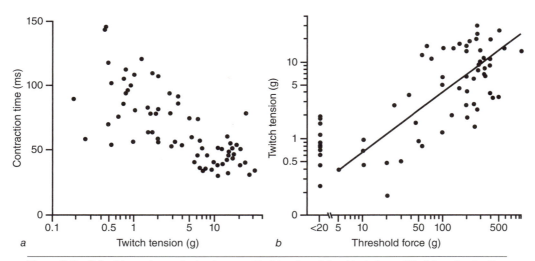

Figure 2.2   Spike-triggered averaging has been used to determine relationships between (a) twitch contraction time and twitch force and (b) twitch force and recruitment threshold in motor units of the human first dorsal interosseus.

Reprinted by permission from Stephens and Usherwood 1977.

demonstrated by this technique in a variety of muscles, especially the muscles of the hand and wrist and the lower leg.

Experiments in which motor unit firing rates have been examined as a function of isometric contraction up to MVC offer insights into the complex way in which recruitment and rate coding are used to augment muscle force. In any given muscle containing a number of motor units of differing strengths and excitabilities, force increase during a slow ramp up to MVC begins with the recruitment of low-threshold units, followed by their subsequent rate coding and the recruitment of additional units. This dual process of recruitment and rate coding proceeds until MVC, at which point all units are recruited at or very near the firing frequencies that result in maximal tetanic force. An example of how units behave during increasing isometric muscle force generation has already been considered (see figure 2.1).

The system is most likely designed to avoid the situation in which the low-threshold units reach maximal firing rates before the highest threshold units are maximally recruited. Such a situation would result in the inefficient generation of surplus action potentials on the part of the low-threshold units and perhaps even in the inactivation of their spike-generating mechanisms. The system is designed to avoid this inefficiency in a number of ways. First, thresholds of motoneurons are spaced more closely in low-threshold units, and this spacing increases as the threshold of the motor units increases (Bakels and Kernell 1994). In this way, recruitment is more important than rate coding for the early recruited units, with rate coding becoming more important as force increases. Second, there may be a synaptic bias in the motor system by which one system facilitates rate coding during smaller forces and another system becomes more significant with higher-threshold units as force increases (Heckman and Binder 1993). An example of this is shown in figure 2.3. The systems that might produce this leveling off of rate coding of low-threshold units as force increases are not known. It has been suggested that PICs in different types of motoneurons might contribute to this nonlinear behavior.

De Luca, Foley, and Erim (1996) used their technical approach of extracting signals of single motor units from the whole-muscle EMG signal to suggest that during ramp-and-hold isometric contractions motor unit firing rates demonstrate an onion skin phenomenon (see figure 2.4). According to this idea, at any particular submaximal force the motor units that are recruited later fire at higher initial frequencies but always fire at lower frequencies than those of previously recruited units. The results emanating from the laboratory of De Luca (De Luca and Erim 1994) indicated that motor unit firing rates converge to similar values near MVC. In several ways, their results were similar to those of Monster and Chan (1977; see figure 2.1) in that low-threshold units demonstrated a plateau phenomenon after their initial rate coding slope. The results differed, however, in that De Luca and colleagues found very little evidence of crossover of motor unit firing rates near MVC. Due to technical difficulties, we cannot say with any certainty what really happens to firing rates at extremely high forces. But there seems to be a consensus that they probably become more similar than they are at submaximal forces.

Interestingly, derecruitment of motor units during a slow decrease in isometric force may involve a different strategy than that used for recruitment. Romaiguère, Vedel, and Pagni (1993) had subjects perform slow-ramp isometric contractions and relaxations of the wrist extensor muscles. They reported that the derecruitment thresholds of motor units were, on average, about 25% lower than the recruitment

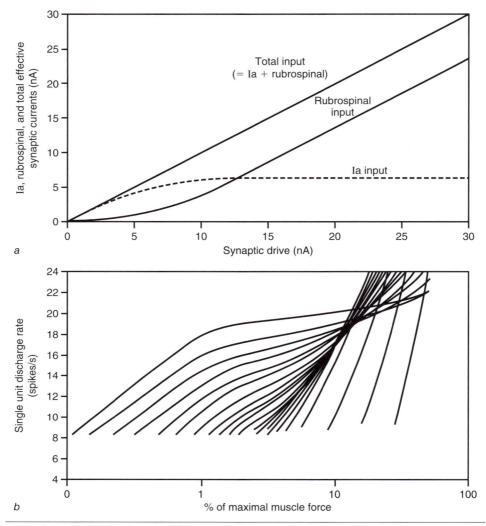

Figure 2.3   *(a)* Example of a systematic motoneuron input scheme that could slow the firing rate increase of low-threshold units as force increases. *(b)* The resultant firing rates in a simulated motoneuron pool. The model in *(a)* shows two inputs to the motoneuron pool: one that has more influence on low-threshold units (Ia input) and one that has more influence on high-threshold units (rubrospinal input). Compare *(b)* with actual recorded motor unit firing rates presented in figure 2.1

Reprinted by permission from Heckman and Binder 1993.

thresholds were. This finding implied that more motor units were active during the relaxation ramp than were active during the contraction ramp at the same force level. Romaiguère and colleagues also reported that the pattern of change in firing frequency differed between the ascending and the descending ramp. Thus, more motor units firing at lower frequency were used to generate the decreasing force than were used for the same increasing force (see the bottom of figure 1 and figure 5B in Romaiguère, Vedel, and Pagni 1993). This phenomenon might involve the activation of PICs during the derecruitment phase.

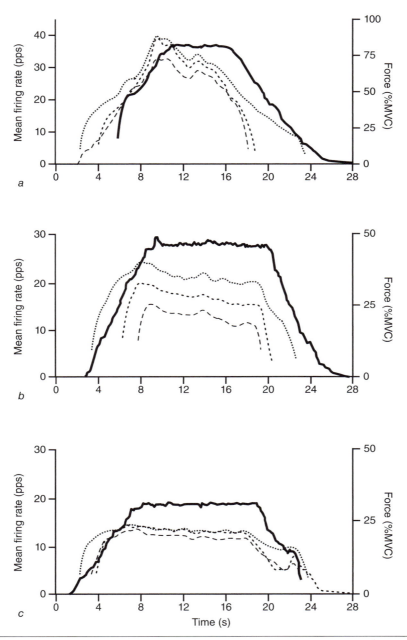

Figure 2.4 Examples of firing rates of motor units in tibialis anterior during isometric dorsiflexion of the ankle at *(a)* 80%, *(b)* 50%, and *(c)* 30% of MVC. The force scale is at right, the firing rate scale is at left, and the force record (showing plateau) is the thick line. The other three lines in *(a)*, *(b)*, and *(c)* show mean firing rates of detected motor units. Note how firing rates decrease throughout the constant-force interval at all force levels.

Reprinted from De Luca, Foley, and Erim 1996.

There is some evidence (see figure 2.5) that motor units that are close to one another in excitation threshold can be rotated in order to offset fatigue during low-threshold, sustained isometric contractions (Bawa et al. 2006). Mechanisms may include, once again, the involvement of PICs that could increase or decrease the excitability and threshold of motoneurons as the contraction progresses and the inactivation of sodium and calcium currents with time.

MAINTAINED ISOMETRIC CONTRACTIONS

Isometric contractions that are maintained for more than a few seconds are among the easiest types of contractions for investigating motor unit recruitment. Generally speaking, contractions of this kind obey the size principle, with recruitment proceeding in the order of S, FR, and FF as force generation increases. Beltman and colleagues (Beltman et al. 2004a) inferred recruitment by examining PCr/Cr ratios in biopsied muscle fibers from the quadriceps immediately following isometric contractions. The subjects performed seven isometric contractions at 39%, 72%, and 87% of MVC.

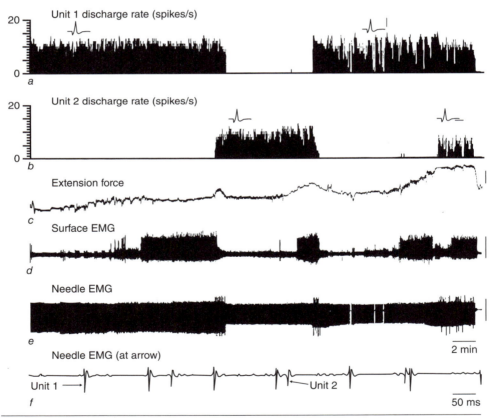

Figure 2.5 Demonstration of motor unit rotation during a sustained contraction of extensor carpi radialis. An active motor unit *(a)* stops firing as another *(b)* starts firing, followed by additional reversals. The scale in *(a)* and *(b)* is instantaneous firing frequency. Also shown are the extension force *(c)*, surface EMG *(d)*, needle electrode EMG *(e)*, and EMG signatures *(f)* of the two units displayed in *(a)* and *(b)*.

Reprinted by permission from Bawa et al. 2006.

MOTOR UNIT ROTATION—ISOLATED PHENOMENON OR GENERAL STRATEGY?

There has been some controversy whether motor unit rotation, in which a recruited unit stops firing while another unit starts firing, may be isolated to only a few muscles, perhaps primarily postural muscles. Results from Dr. Parveen Bawa's laboratory at Simon Fraser University at British Columbia, Canada, demonstrated that motor unit rotation may be more common than previously thought. Bawa and her team (Bawa and Murnaghan 2009) investigated this phenomenon using indwelling EMG electrodes in seven muscles, including the abductor digiti minimi, first dorsal interosseus, extensor digitorum communis, flexor and extensor carpi radialis, tibialis anterior, and soleus. When subjects were asked to maintain the firing of a low-threshold unit (unit 1) for long durations, a second unit (unit 2) frequently started firing, and when subjects were asked to decrease effort to recruit only one unit, often unit 2 maintained firing while unit 1 dropped out. Unit 1 would often reappear, with unit 2 then dropping out. Rotations of motor units also occurred when subjects were asked to rhythmically modulate firing rates in a pseudosinusoidal pattern. The authors proposed that motor units drop out due to an increase in the thresholds for motoneuron firing resulting from inactivation of sodium and calcium channels. Thresholds for motor units that drop in may decrease due to activation of PICs. The acceptance of motor unit rotation as a common occurrence means that we must rethink our interpretations of the neuromuscular fatigue process and consider sites of chronic adaptation (motoneuronal ion channels or PICs) that may influence performance during long-duration efforts.

Their results suggested, as expected from the size principle, a hierarchical recruitment of Type I, IIa, and IIax fibers as intensity increased, with Type IIax fibers showing significant decline on PCr/Cr only after contractions at 87% MVC.

Constant-force isometric contractions, when maintained for more than a few seconds, involve interesting changes in motor unit behaviors (see figure 2.4). De Luca, Foley, and Erim (1996) showed that motor unit firing rates in human muscles (first dorsal interosseus and tibialis anterior) appear to decrease gradually during the maintenance (up to 15 seconds) of constant force output, without evidence of the recruitment of additional motor units. How can muscle force be maintained if motor unit frequencies are declining and no new units are being recruited? De Luca, Foley, and Erim proposed that the declines in firing frequency partially compensate for the tension potentiation (a temporary increase in twitch force due to a preceding high-force contraction, probably attributable to phosphorylation of myosin light chains; see chapter 1) that occurs in fast motor units at low to moderate stimulation frequencies. In fact, according to these authors, several things occurring during this type of contraction can influence the voluntary generation of muscle force in various ways: twitch potentiation and staircase (the gradual increase in force at a constant stimulation frequency), afferent signals from muscle that could either increase or decrease motoneuronal firing frequency, and intrinsic membrane properties such as late adaptation (whereby a constant transsynaptic activation of the motoneuron results in a gradually decreasing firing frequency).

ISOMETRIC CONTRACTIONS IN VARIOUS DIRECTIONS

We previously mentioned that task may alter the recruitment pattern of motor units. Thomas, Ross, and Calancie (1987) examined the recruitment of motor units in abductor pollicis brevis during isometric contractions in two directions: abduction and opposition of the thumb. Using spike-triggered averaging, they found that the threshold force and twitch amplitudes were positively correlated in both movements, as predicted by the size principle of orderly recruitment. In addition, thresholds of motor units for both movements were very similar, and the same units were recruited in both tasks, suggesting that the recruitment of motor units was virtually identical for these two isometric movements (see figure 2.6). Similar results have been presented for motor unit recruitment in the extensor carpi radialis during wrist extension and radial deviation, two movements in which this muscle is recruited extensively (Riek and Bawa 1992). However, extensor digitorum communis, which is primarily a finger extensor but also takes part in wrist extension, demonstrates different task groups of motor units. This complex muscle functions as an extensor for all four fingers, via separate tendons, as well as a wrist extensor, and it exemplifies a muscle within which motor units are recruited as task groups according to the desired task. However, as with simpler muscle–joint systems, recruitment within each task seems to proceed according to the size principle. Likewise, in flexor carpi ulnaris, a wrist flexor, recruitment of motor units was similar for isometric wrist extension, anisometric wrist extension, ulnar deviation, and cocontraction of extensors and flexors (Jones, Bawa, and McMillan 1993).

Jones and colleagues (1994) recorded the recruitment of first dorsal interosseus units during three tasks that were near isometric: abduction, a rotation task (loosening a threaded knob), and a pincer task (pressing between the thumb and index finger). The same units were recruited in all three tasks, and the order of recruitment was always the same.

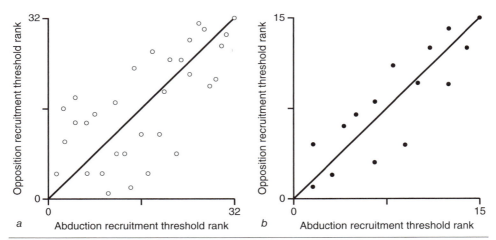

Figure 2.6 Recruitment threshold rank of motor units of abductor pollicis brevis during isometric contractions in two directions: abduction and opposition. Data from two subjects are shown. Recruitment order is similar for the two tasks.

Reprinted by permission from Thomas, Ross, and Calancie 1987.

Thus, motor units that take part in several different types of movements around a joint seem to be recruited for each movement according to the size principle. This seems to hold true whether the same units within the muscle are used for the various movements.

ISOMETRIC CONTRACTIONS VERSUS MOVEMENTS

It is not surprising that motor units are recruited differently during contractions that involve movements but generate the same relative force as a corresponding static contraction. Thickbroom and colleagues (1999) showed that the functional magnetic resonance imaging (fMRI) signal from the sensorimotor cortex was considerably greater during a rhythmic, dynamic finger flexion task than it was when the task involved the same level of static force generation. The technique of fMRI, which measures changes in cerebral metabolism relative to oxygen supply, is unfortunately not suited for localization of the increased activity or identification of its form (increased number of cells involved versus increased frequency of those activated). This result does suggest, nonetheless, that signals arriving at motoneurons are different for static and dynamic contractions.

Tax and colleagues (1989) examined the recruitment of biceps brachii motor units during isometric and slow isotonic (change in elbow joint angle of 1.5°-3° per second) contractions. Their results showed a substantially lower force threshold for motor units during slow isotonic contractions than during isometric contractions (refer to figure 3 in Tax et al. 1989). In addition, minimal firing frequency of the motor units was lowest during isotonic extensions and highest during isotonic flexions. Thus, motor units appear to be activated differently during these three different types of task. Tax and colleagues (1990) continued their studies of this phenomenon to show that the different recruitment was due not to differing peripheral signals in the isotonic and isometric conditions but to differences in central activation. They added an experimental condition in which subjects exerted a constant or slowly increasing torque against an immovable torque motor and experienced the same change in elbow angle as they would during active isotonic contraction. The finding that the recruitment thresholds and minimal firing rates were identical in this condition and the isometric contraction condition led these researchers to conclude that central, and not peripheral, components are involved in determining the specificity of these motor unit responses.

The finding of Tax and coworkers (1989) that recruitment thresholds are reduced when contractions involve joint movements has been substantiated and extended by Kossev and Christova (1998). These investigators showed that recruitment thresholds of motor units in biceps brachii were smaller during isovelocity concentric contractions than they were during isometric contractions. The data are particularly interesting in that these researchers were able to record motor units recruited at up to 54% of MVC because of the stability of their recording electrodes. Their results showed a reduction in torque threshold of motor units during concentric movements. This reduction became more significant as the speed of the concentric contraction increased and was more pronounced for the lower-threshold units.

Sogaard and colleagues (1998) showed that firing rates are higher during dynamic (30° per second) versus static wrist flexions involving the same torque generation. In addition, although mean motor unit firing rates were higher during an isometric contraction corresponding to 60% of MVC than they were during one at 30% of MVC, firing rates were not different when these two torque levels were generated during a dynamic contraction. Thus, not only the recruitment pattern is different for dynamic and static contractions at the same force level—the recruitment strategy used to increase force is also different under these two conditions.

Van Bolhuis, Medendorp, and Gielen (1997) addressed this research using a rather novel approach. They had subjects perform horizontal elbow flexions against small loads, during which they generated forces in a sinusoidal manner at various frequencies and under isometric and isotonic conditions. The researchers found that the behavior of the same motor units in brachialis and biceps muscles varied depending on the type of contraction (isometric or isotonic) and the speed of the sinusoidal movements. They found that the phase lead of motor units (peak in burst rate as a function of peak in torque) was very reproducible among motor units during isometric contractions but varied much more among units during flexion–extension sinusoidal movements of about the same force. In addition, they found that some units showed higher firing rates when the movement changed from isometric to isotonic, whereas other units decreased in firing rate (figure 2.7). These two findings indicate that various motor units are activated differently in isometric and movement tasks. Interestingly, these investigators found that some units had a phase lead of 180% or more during flexion–extension activity, which suggests that some flexor units are active during the extension phase. This supports the results of others that suggest that motor units are activated differently during lengthening and shortening types of contractions. These motor unit studies support whole-muscle studies demonstrating that EMGs for concentric and eccentric movement tasks are higher than they are for isometric tasks (Van Bolhuis and Gielen 1997).

Do these observations indicate that the order of recruitment of motor units can be changed depending on the isotonic or isometric nature of the task? In this case, not necessarily. While the lower force threshold of the motor units might indicate that other units now have a higher threshold (since the total force at which these thresholds are being compared is the same), which would indicate a reversal of order of recruitment for at least some units, there is evidence that a decrease in the minimal firing frequency also occurs. Thus, in the study of Tax and colleagues (1989), more motor units were activated during a slow isotonic contraction, but each generated slightly less force because of lower firing frequencies.

Thomas, Ross, and Calancie (1987) addressed this question of recruitment order by investigating the recruitment of first dorsal interosseus and abductor pollicis brevis motor units during the repetitive opening and closing of scissors and compared the order of recruitment to that seen in these same muscles during ramp isometric contractions. In their limited data set, they found that the recruitment according to increasing motor unit size that was seen with isometric contractions was largely preserved during the repetitive opening and closing of scissors.

There is strong evidence that motor units are recruited differently during shortening contractions as opposed to isometric contractions, even when the movement is slow. The force at which recruitment occurs is lower and the frequency of firing is higher for motor units during concentric contractions. In addition, an increase in torque appears to rely more on increasing firing frequency during an isometric contraction compared with a concentric contraction. It also appears that each motor unit may have a specific contributory role during isometric and concentric contractions, and there is significant variability among units as to their function in isometric versus

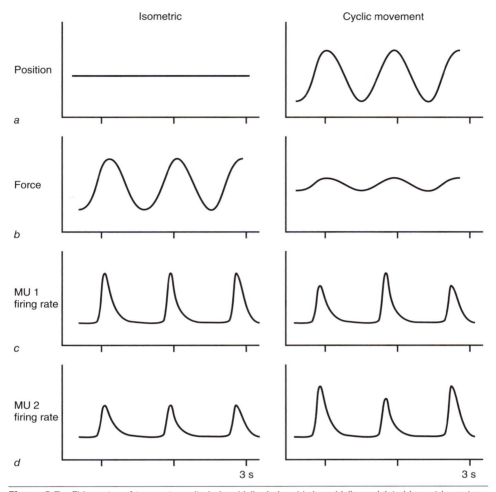

Figure 2.7 Firing rates of two motor units in brachialis during *(a)* sinusoidally modulated isometric contractions and *(b)* sinusoidal movements. Note the lack of consistent change in firing rates of the two units between isometric and movement conditions. Firing rates *(c)* decrease for unit 1 and *(d)* increase for unit 2 when changing from isometric to cyclic movement contractions.

Based on Van Bolhuis, Medendorp, and Gielen 1997.

anisometric contractions. Many of these differences are due to variations in central programming for each of these tasks.

During ballistic isometric contractions, the order of recruitment of motor units seen during slow-ramp contractions is preserved. However, as in the slower anisometric contractions mentioned earlier, force thresholds are lower and initial firing frequencies are much higher during ballistic contractions (Desmedt and Godaux 1977, 1978).

Contractions involving concentric movements often include an initial high-frequency double discharge (doublet) of the motor unit, followed sometimes by a longer interpulse interval, during the initial force increase preceding the movement (Kossev and Christova 1998). The obvious advantage of this phenomenon would be to increase force rapidly in order to assist in overcoming the inertia of the load. The source of this phenomenon is, however, not known. As previously pointed out, it may be more pronounced in motoneurons displaying a pronounced delayed depolarization (Kudina and Alexeeva 1992; Kudina and Churikova 1990). Interestingly, this phenomenon also appears to be subject to adaptive changes with ballistic training (see chapter 10).

LENGTHENING CONTRACTIONS

Lengthening contractions are used in everyday activities. Muscles lengthen when the force applied exceeds the force produced by the muscle, resulting in braking and storage of elastic energy. It is not entirely unexpected, then, that motor units are recruited differently during shortening versus lengthening contractions. This is evident in the example shown in figure 2 of Del Valle and Thomas (2005). During the same submaximal load, and often at maximal load, muscle activation, as estimated using EMG, is less for the eccentric phase than it is for the concentric phase of the contraction (Moritani, Muramatsu, and Muro 1988; Nakazawa et al. 1993; Del Valle and Thomas 2005).

Nardone, Romano, and Schieppati (1989) reasoned that when a person performs lengthening contractions in which the muscle yields to the load, there would be an advantage to preferentially recruiting fast-twitch units, since these units have faster relaxation times and therefore would provide better control over the movement trajectory. Furthermore, these researchers hypothesized that, given the relatively large size of fast units, it would be advantageous to recruit them at low frequency and also to decrease the background force level by derecruitment of slow units. This hypothesis implies a recruitment order different than that seen during isometric contractions of slowly increasing force. In their study, the researchers examined the firing patterns of motor units in soleus and medial and lateral gastrocnemius during plantar flexion and dorsiflexion (the latter achieved by gradually lowering a load on the plantar flexors to create a lengthening, or eccentric, contraction). They found that units could be classified as S (active during shortening), L (active during lengthening), or S + L. L units (15% of all units recorded in soleus and 50% of units in gastrocnemius) often were not recruited at all during shortening movements, but when they were, they were recruited only at relatively high forces or during very rapid contractions. L units also had relatively large spike amplitudes and were difficult to keep activated, and their axons had relatively high conduction velocities and relatively low voltage thresholds to electrical stimulation. The authors contended that these L units were therefore high-threshold units. Their conclusion was that, since it certainly does not require greater force to lower a load than to maintain a load, and since some units are replaced by others when a shortening contraction is changed to a lengthening contraction, motor unit recruitment is different for shortening and lengthening contractions. The contention that the order of motor unit recruitment changes with lengthening versus shortening contractions has been challenged under highly controlled experimental conditions (Pasquet et al. 2006; Stotz and Bawa 2001). Differences in the literature may reflect responses of different muscles or muscle groups. In any case, what is a clear and consistent finding is that recruitment of motor units is not identical at the same level of force generation during shortening versus lengthening contractions.

What are the mechanisms responsible for altered recruitment patterns during lengthening contractions? One reason for the lower activation of motor units during lengthening contractions could be inhibition of motoneurons. The idea that inhibitory influences on alpha-motoneurons increase during eccentric contractions is supported by the finding of Westing, Seger, and Thorstensson (1990). They demonstrated that torque could be increased by 21% to 24% during eccentric MVCs by superimposing electrical stimulation during the contraction. Abbruzzese and colleagues (1994) proposed that spinal cord circuits are altered in their excitability during eccentric movement. They based their conclusion on the findings that magnetic brain stimulation (which stimulates cortical neurons transsynaptically) and electrical brain stimulation (which stimulates cortical axons) both showed a decreased motor-evoked potential (MEP) in elbow flexors during eccentric contractions and that the H-reflex (reflecting excitability of the motor pool) was also decreased. This latter finding has been confirmed (Duclay and Martin 2005). It was once thought that enhanced presynaptic inhibition of motoneurons during lengthening contractions might explain the experimental results; this hypothesis has been investigated specifically and found not to be the case (Petersen et al. 2007).

Apparently not all motor units are affected to the same extent during lengthening contractions, and this results in changes in their behaviors. For example, in a study of first dorsal interosseus motor units, Howell and coworkers (1995) found instances where motor units that normally had high force thresholds under isometric conditions were recruited during concentric–eccentric contractions of low torque and during the eccentric phase of contractions (as Nardone, Romano, and Schieppati found for plantar flexors in 1989; also see figure 2.8).

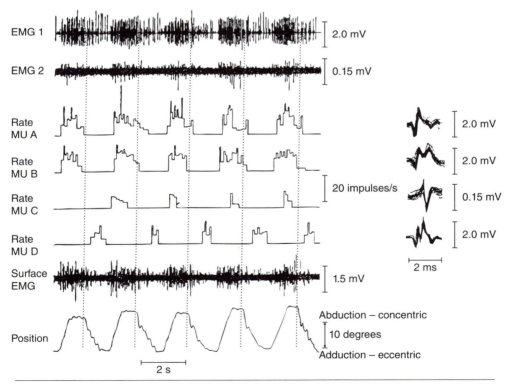

Figure 2.8 Activity of four motor units (MU A to MU D) during repeated abduction–adduction movements against a constant torque load. Note the differences among units in the phase of the contraction during which they are active.

Reprinted by permission from Howell et al. 1995.

The mechanism may involve a combination of increased descending enhancement of recurrent inhibition from large to small motoneurons and presynaptic inhibition to all motoneurons, which would affect small motoneurons more than large motoneurons. However, the activation of specific task groups of motor units may also be involved. Lengthening contractions evoke greater and longer anticipatory responses and more intense cortical activity over a larger cortical area (Fang et al. 2004). Experiments using transcranial magnetic stimulation (TMS) show clearly that the central nervous system has a different activation strategy for lengthening versus shortening contractions, perhaps to counter the higher discharge of spindles that occurs during lengthening contractions (Sekiguchi et al. 2001; Sekiguchi et al. 2003). The timing of inputs to motoneurons is more synchronous during lengthening contractions (Semmler et al. 2002).

Beltman and colleagues have used PCr/Cr ratios measured in single fibers from biopsies to estimate recruitment during short-term contractile tasks (Beltman, Sargeant, Van Mechelen, et al. 2004). Their results showed that fibers are recruited in the normal hierarchical pattern during an eccentric contraction but that recruitment is less intense (i.e., the PCr/Cr ratios were less perturbed).

What we can take away from all of this concerning motor unit recruitment during lengthening contractions is the following: (1) Recruitment of motor units is different, (2) the motor unit pool is less activated, (3) the cortex uses a different strategy of motor unit recruitment, and (4) the order of recruitment of motor unit types is generally preserved (with some exceptions in the literature). We will discuss lengthening contractions a bit later when considering the training (chapters 8 to 10) and fatigue processes (chapter 5).

COCONTRACTION OF AGONISTS AND ANTAGONISTS

Coactivation of agonists and antagonists is worth considering when discussing recruitment, since it may under certain circumstances limit the full expression of torque around a joint as well as involve an added metabolic cost. Coactivation is the simultaneous contraction of antagonists and agonists and is a common phenomenon that serves to stabilize a joint, distribute pressure across the joint, decrease the strain on ligaments during forceful contractions, and increase damping. Coactivation increases with increased rate of agonist shortening (Weir et al. 1998). Coactivation is under central control via presynaptic inhibitory mechanisms (Nielsen and Kagamihara 1993; Hansen et al. 2002), and it may play a role during fatigue and in the training process.

UNILATERAL VERSUS BILATERAL CONTRACTIONS

Reports of the bilateral deficit phenomenon, which involves a slight loss in the maximal voluntary force of a muscle or muscle group that occurs when the contralateral muscle or muscle group contracts at the same time, have appeared in the research literature since 1961 (Vandervoort, Sale, and Moroz 1984; Schantz et al. 1989; Howard and Enoka 1991; Bobbert et al. 2006; Jakobi and Chilibeck 2001). Bilateral deficit has been reported for a number of movements, including knee extension, hand grip, plantar flexion, wrist flexion, elbow flexion, finger flexion, combined hip and knee extension, and jumping, and has been observed in isometric and dynamic movements (Jakobi and Chilibeck 2001). It has been difficult to demonstrate consistently and unequivocally that this deficit results in a concomitant decrease in EMG during the

> ## NEUROMODULATION OF MOTONEURON
> ## _____ EXCITABILITY DEPENDS ON JOINT ANGLE _____
>
> As mentioned in this chapter, there are several modulators (such as epinephrine and 5-HT) that are released in the spinal cord in the vicinity of motoneurons and that can increase motoneuron excitability by facilitating the generation of excitatory self-sustaining currents, or PICs. The release of neuromodulators from brain stem neurons is rather diffuse, and thus it might influence the excitability of motoneurons in a rather random manner. Heckman and colleagues (Hyngstrom et al. 2007) at Northwestern University have demonstrated quite elegantly that dendritic fields may be more excited or less excited, even when endogenous levels of neuromodulators are constant, depending on the reciprocal inhibitory input to the motoneuron from antagonists. In their experiment, in which they measured PICs electrophysiologically in decerebrate cat preparations, they showed that PICs generated in dendrites of an extensor motoneuron were larger when the ankle was flexed and smaller when the ankle was extended. This modulation was due primarily to the influence of reciprocal inhibition, since cutting these pathways removed the effects specific to the joint angle. The authors pointed out the importance of this mechanism for coordinated voluntary movements. They also suggested that these mechanisms may be subject to conditioning effects, whereby voluntary movements might be changed acutely or chronically through repeated movements involving specific changes in joint angle. Could this be one of the benefits of warming up before a performance?

bilateral contraction, probably because surface EMG is not sensitive enough to detect such a small difference in activation (Howard and Enoka 1991). Bilateral deficit is probably not a limit in the amount of muscle tissue that the central nervous system can activate, since it does not occur with other than homologous muscles on opposite sides of the body (Howard and Enoka 1991), and the degree of deficit is not increased if arm exercise is added during the bilateral task (Schantz et al. 1989). Interestingly, not all subjects demonstrate a bilateral deficit; some demonstrate no difference or even a bilateral facilitation (Howard and Enoka 1991). Since electrical stimulation of the contralateral muscle increases the performance of the ipsilateral muscle during a maximal contraction (Howard and Enoka 1991), there clearly are interlimb signals that modulate the expression of maximal force during bilateral efforts. Not all muscle groups are subject to this phenomenon; Herbert and Gandevia (1996), for example, were unable to demonstrate a bilateral difference with thumb adductor muscles. The mechanisms by which bilateral deficit occurs are not known at present. A fine review of the literature, up to 2001, concerning this phenomenon is available (Jakobi and Chilibeck 2001).

RHYTHMIC COMPLEX CONTRACTIONS

Our primary information for recruitment during rhythmic complex contractions, such as running and cycling, comes from glycogen depletion studies. Vollestad and colleagues (Vollestad, Tabata, and Medbo 1992; Vollestad and Blom 1985; Vollestad,

Vaage, and Hermansen 1984) had subjects perform on a bicycle ergometer at different speeds for various durations and examined the disappearance of glycogen in single vastus lateralis fibers from sequential biopsies. Their results generally confirmed a recruitment scheme consistent with the size principle. For example, during ergometer cycling corresponding to 75% of maximal oxygen consumption ($\dot{V}O_2$max), there was virtually no glycogen loss from Type IIab and IIb fibers, corresponding to 10% to 30% of the muscle cross section, during the first 20 minutes of exercise. During this same time, Type I and IIa fibers showed extensive glycogen loss. As exercise continued, Type I and IIa fibers became depleted, and involvement of Type IIab and IIb fibers became more significant. At exhaustion, after 140 minutes of exercise, Types I and IIa were severely depleted of glycogen, while a considerable proportion of Type IIab and IIb fibers were not yet depleted. This finding is consistent with the idea that the highest-threshold units require the most effort to be recruited and cannot be recruited continuously, the latter partly because of their propensity for late adaptation.

These same investigators also demonstrated that the number of fibers involved increases with intensity, in the order I > IIa > IIab > IIb. The rate of glycogen depletion also increases as a function of exercise intensity.

Finally, these investigators showed quite convincingly that all muscle fibers are recruited at a supramaximal intensity (up to 200% of cycling intensity corresponding to $\dot{V}O_2$max). Glycogen depletion rates in the different fiber types indicated that, contrary to the belief of many, Type I fibers are capable, when called upon, of using significant amounts of glycogen anaerobically in order to take part in high-intensity tasks. The increase in rates of glycogen breakdown in all fiber types as intensity increases suggests that there is no significant derecruitment of Type I fibers during high-intensity, primarily anaerobic tasks. Thus, even for such tasks, where the decreased anaerobic potential and lower power might render these fibers inappropriate, the size principle is, for the most part, still in force.

Of course, the limits of using glycogen depletion to infer fiber recruitment must be understood when examining this literature. The loss of glycogen from fibers during exercise depends on, among other things, the level of oxygen delivery to the working muscle, the initial glycogen levels, and the activities of the enzymes of energy metabolism, all of which vary among individual fibers. In addition, the rate of glycogen loss tells us virtually nothing about the rate of energy turnover or of work performance of the fiber. More recently, investigators have found that by using individual fiber PCr/Cr ratios in muscle biopsy material, they can detect evidence of recruitment much sooner (after 1 minute of running) than they can when using glycogen content. This technique has been used to demonstrate that both Type I and Type II fibers are recruited within 1 minute of ergometer exercise at 75% of $\dot{V}O_2$max, at a muscle force corresponding to 38% of maximal dynamic force (Altenburg et al. 2007).

MAXIMAL VOLUNTARY CONTRACTIONS

Since firing rates of motor units during MVCs appear to be sufficient to evoke maximal tetanic forces of the innervated muscle units in most muscles studied, it appears that maximal activation is at least theoretically possible through voluntary effort. However, under a variety of conditions, maximal expression of muscle forces may not occur, even though maximal efforts are called for. How can we know when MVCs are indeed maximal?

Twitch interpolation is a technique in which a muscle's nerve is stimulated while the subject is performing an MVC. The idea is that if the muscle has been voluntarily recruited to maximum, additional force should not be observed when the motor nerve is stimulated. This technique, first applied to the adductor pollicis, has now been used extensively to test just how maximal voluntary contractions are in a variety of muscles and muscle groups, with varying results. An increased force resulting from stimulation during an MVC implies that there is a limit to the expression of maximal contraction that is at or above the level of the motoneuron.

This approach has been modified to incorporate TMS, such that incomplete activation of muscles during an MVC can imply a failure at or above the level of output from the motor cortex (Todd et al. 2004). This technique has been used very successfully in pinpointing potential sites of fatigue.

The degree of voluntary activation can be estimated by plotting the added force resulting from electrical stimulation (ordinate) against the absolute force (abscissa) at a variety of force levels. If the relationship is linear, which many have shown it to be, theoretical maximal force capability should be where the extrapolated line meets the abscissa. Dowling and colleagues (1994) suggested, from their results using elbow flexors, that the relationship of interpolated twitch and percentage of voluntary force is best fitted by an exponential function. This would mean that subjects are never able to activate their elbow flexors 100% but that the small residual twitch response becomes too small to measure at near-maximal effort. Whatever relationship is used, the number derived from these procedures is difficult to interpret quantitatively. For example, what does 95% activation mean? Is one motor unit not recruited? Are 10 motor units not generating maximal firing frequencies?

It may be that many repetitions are necessary in order to permit subjects to generate a true maximal contraction. In a systematic study of the intra- and interindividual variability in degree of activation, Allen, Gandevia, and McKenzie (1995) applied the twitch interpolation technique to elbow flexors of five subjects tested 10 times per day on 5 different days. Their results, summarized in their figure 4, showed that most subjects, if given enough trials, demonstrated maximality of voluntary contractions. In addition, subjects showing high variability on any given day were also highly variable on other days, and subjects showing low variability on any given day always showed low variability. These results underline the importance of taking several measurements in order to determine whether maximal contractions are possible in any given subject. Apparently, at least for elbow flexors, maximal activation is possible in most subjects but does not occur with all contractions.

It has been suggested that a high-frequency train of stimuli, as opposed to the single or double pulses normally used, is more sensitive in determining central failure to activate muscles maximally. For example, Kent-Braun and Le Blanc (1996) showed that a 50-hertz, 500-millisecond stimulus applied during an MVC detected 33% of control subjects who could not fully activate their dorsiflexors, whereas double stimuli detected only 4.8% of subjects and single stimuli detected no cases. In spite of this, the mean ratio for control subjects was still 0.96, indicating that these subjects were capable of maximally stimulating their dorsiflexors voluntarily.

SUMMARY

The study of motor unit recruitment during more complex types of movements, such as those involving changing force levels, shortening and lengthening, and unilateral and bilateral efforts, shows that recruitment patterns exhibit a certain degree of flexibility. Recruitment patterns during more complex and prolonged motor tasks, such as running and cycling, are less evident due to technical considerations. Motor unit rotation may be a strategy used by the neuromuscular system to deal with phenomena such as motoneuron late adaptation and muscle contractile changes, and it may also be used to offset fatigue.

Finally, it appears that although humans are capable of recruiting nearly all of the maximal force capability of muscles, there is a significant inter- and intraindividual variation in this capability. This might be significant in determining an athlete's capacity to perform maximal efforts repeatedly in competition.

Muscle Blood Flow and Metabolism

Cardiovascular dynamics involved in blood flow regulation and muscle metabolic changes with exercise are among the topics that are covered most completely in introductory undergraduate texts on exercise physiology. I assume that these basic tenets are familiar to the reader, and thus in this chapter I restrict discussion to more recent and ongoing developments pertaining to these two important areas of neuromuscular exercise physiology.

MUSCLE BLOOD FLOW

During exercise, blood flow increases in working muscle as a function of workload, and the relationship is quite linear (Saltin et al. 1998). Muscle perfusion during exercise for a sedentary individual is about 250 milliliters per minute in 100 grams of tissue, and it can rise to 400 milliliters per minute in trained athletes (Saltin et al. 1998; Saltin 2007). Mechanical and biochemical mechanisms cause this increased muscle blood flow. However, the blood available for exercising muscles is not limitless—exercising at a high intensity with 15 kilograms of muscle (about 50% of the total muscle mass of a 75-kilogram man) can exceed the heart's capacity to supply blood (see figure 3.1; Calbet et al. 2004). Athletes such as cross-country skiers, who use their arms and legs at relatively high intensities for prolonged durations, experience this limit in blood availability. Therefore, there must be in place mechanisms that increase blood flow when there is metabolic demand as well as mechanisms that cause active muscles to undergo some level of vasoconstriction during exercise (Rowell 1997). During high blood flow, sympathetically mediated vasoconstriction is necessary to maintain peripheral resistance in order to maintain blood pressure (Saltin 2007). When metabolic rates and blood flow demands are high, such as when the arms and legs are working simultaneously, noradrenaline spillover from the legs helps to counteract the strong local vasodilator signals and helps to maintain central pressure.

This section discusses the general physiological mechanisms that maintain blood pressure when muscle demands for oxygen delivery are high and also the local

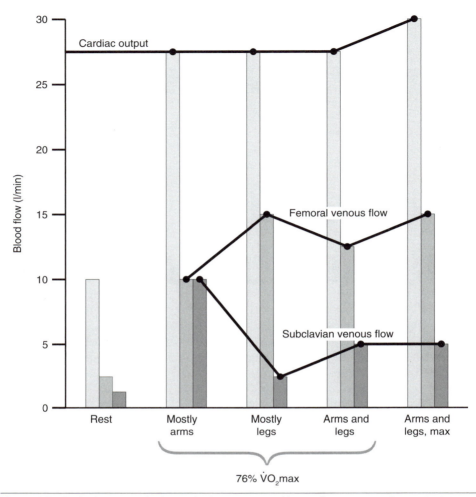

Figure 3.1 During heavy exercise by a trained cross-country skier, cardiac output remains constant at near-maximal and maximal exercise, but blood flow to the legs is compromised when the arms are also working (femoral venous flow is lower for the arms and legs condition than for the mostly legs condition). The light-colored bar, medium-colored bar, and dark bar are total cardiac output, blood flow to legs, and blood flow to arms, respectively.

Data from Calbet et al. 2004.

mechanisms that match blood flow to metabolic demand. As we shall see, there are many proposed mechanisms controlling exercise blood flow, with perhaps a significant degree of redundancy, so that finding the role played by individual mechanisms has not been possible to date.

Oxygen Delivery as the Regulated Parameter

What is the regulated variable in blood flow regulation? It appears to be oxygen delivery, and not blood flow per se, at least from studies in which arterial oxygen content has been decreased. Flow adjusts accordingly to achieve the appropriate oxygen delivery to the working muscle (Marsh and Ellerby 2006). The regulation of blood flow by oxygen delivery appears to be restricted to the oxygen bound to hemoglobin, since experiments have shown a close relationship between blood flow and oxygen bound to hemoglobin, independent of partial pressure of oxygen in blood (Gonzalez-Alonso et al. 2001). The relationship between blood flow and oxygen demand is also supported by the observation that blood flow appears to be a linear function of oxygen consumption, independent of the fiber type composition of the muscle (Marsh and Ellerby 2006). In addition, blood flow is related more to the metabolic rate of the contractile activity than to the work performed (Hamann et al. 2004, 2005).

There are several ways in which blood flow may be regulated by the oxygen content of the blood and, more specifically, the saturation level of hemoglobin. Hemoglobin may, via conformational and redox transitions, stimulate the release of vasodilator substances such as nitric oxide (NO), S-nitrosylated thiols, nitrite, and adenosine triphosphate (ATP). Such a mechanism would help explain the close relationship between venous hemoglobin saturation and venous blood flow (Calbet et al. 2007; Gonzalez-Alonso et al. 2001).

Functional Sympatholysis

A rise in sympathetic nerve activity with exercise causes vasoconstriction in non-muscular regions and also in inactive and active muscles. Mechanisms are in place to override active muscle vasoconstriction in an attempt to match blood flow to metabolic demand while maintaining blood pressure. One way this occurs is via functional sympatholysis, which is an inhibition of the sympathetically mediated vasoconstrictor response in active muscle as long as blood pressure is maintained. This is probably due to substances produced in the contracting muscle that inhibit alpha-1 or alpha-2 adrenergic receptor activation by the noradrenalin (NA) released from the nerve terminals. NO may be the blocker, although conflicting results have been reported (Dinenno and Joyner 2003; Saltin 2007; Tschakovsky and Joyner 2008; see later discussion). Adenosine may also play this role. It has also been demonstrated that the receptors for endothelin-1 (ET-1), endothelin receptor type A (ETA) and type B (ETB), may also play a role in muscle blood flow regulation during exercise. Like the blunting of adrenoreceptor sensitivity that occurs with exercise, responsiveness to the powerful vasoconstrictor ET-1 is also diminished during exercise by currently unknown metabolic mechanisms (Wray et al. 2008).

Mechanical Events

Investigators have demonstrated in elegant experiments that mechanical events alone can explain at least the initial rapid increase in muscle blood flow. Using the hamster cremaster microvascular preparation, Mihok and Murrant (2008) demonstrated that single electrically evoked contractions of 3 to 5 fibers (lasting 250-500 milliseconds) produced significant dilations in the arterioles overlapping these fibers. Clifford and colleagues (2006) carried this further to measure vasodilatation in isolated feed arteries of rat soleus muscles in response to externally applied pressure pulses (see figure 3.2). Their results also showed that mechanical stimuli alone result in vasodilatation. These researchers also denuded arterioles of the epithelium and found that the response was reduced but not eliminated. This finding demonstrated that the mechanical response of vasculature is mediated by both epithelial-dependent and epithelial-independent mechanisms.

CAN MUSCLE MASSAGE AFTER EXERCISE ACCELERATE RECOVERY?

Muscle blood flow can be increased in resting muscles by muscle compression, which evokes a rapid vasodilator response. Therefore, if accumulation of substances in exercising muscle, such as lactate and hydrogen ion, is assumed to contribute to fatigue, is it not reasonable to assume that massage of the muscle immediately following exercise maintains elevated blood flow and thus accelerates the removal of such substances and thus facilitates recovery? This is the question that Dr. Michael Tschakovsky and colleagues at Queen's University at Kingston, Ontario, addressed experimentally (Wiltshire et al. 2010). Subjects maintained a handgrip task at 40% of MVC for 2 minutes, after which they performed either passive rest or 10 minutes of rhythmic forearm contractions at 10% of MVC or were treated to forearm massage (effleurage and petrissage). The researchers found that massage substantially reduced forearm blood flow and removal of lactate primarily due to the suppression of flow during the massage strokes. There was also a reduction in flow with active recovery, although in this situation some of the lactate was removed and metabolized by the working muscle fibers. On the basis of these results, the researchers recommended that sport massage be avoided in situations of acute bouts of repeated exercise.

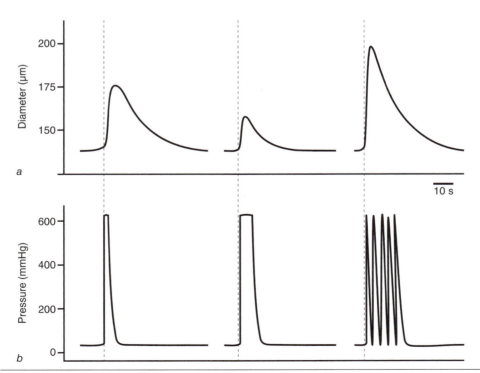

Figure 3.2 Effects of externally applied pressure pulses (*b,* bottom trace) on the diameter of rat soleus feed artery in vitro (*a,* top trace).

Data from Chifford et al. 2006.

Tschakovsky and colleagues (2004) measured changes in forearm blood flow resulting from single 1-second isometric contractions at intensities ranging from 5% to 70% of MVC. Their results, which are illustrated in figure 3.3, demonstrated that there is a rapidly acting vasodilation that is proportional to muscle activation and is

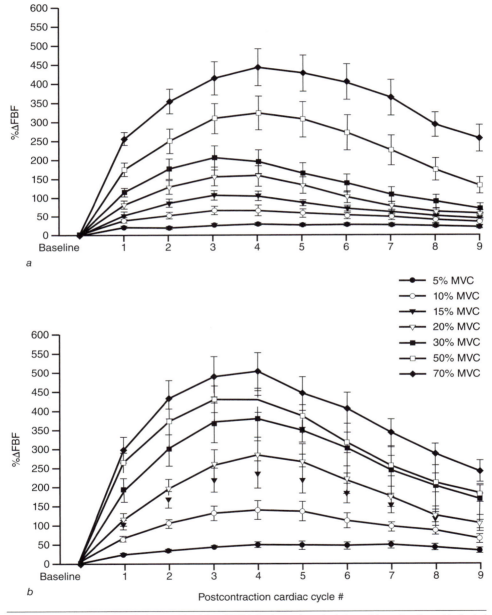

a

b

Postcontraction cardiac cycle #

Figure 3.3 Percent change in forearm blood flow (FBF) following various single isometric contractions at indicated percentages of forearm MVC (handgrip). Flows are measured beat by beat (for each postcontraction cardiac cycle) following the contractions. *(a)* Arm above heart and *(b)* arm below heart.

influenced by local arterial driving pressure (since flow increases were greater when the arm was below heart level).

Kirby and colleagues (2007) demonstrated the importance of extravascular pressure on the initial increases in muscle blood flow. In their studies, they compared immediate increases in blood flow resulting from pressure induced by inflating pressure cuffs around the muscle with voluntary and electrically evoked contractions. Their results demonstrated quite clearly the importance of mechanical influences on the rapid vasodilatation that occurs at the beginning of contraction. While the immediate increase (within two cardiac cycles) appears to be totally attributable to mechanical factors, other factors are involved in determining the peak flow response to contractile activity (Kirby et al. 2007). Generally, the results from these experiments demonstrated quite convincingly that mechanical factors are most important at the beginning of contractile activity, giving time for the release, accumulation, and diffusion of other substances that then play major roles in sustaining blood flow and matching it to metabolic demand.

There is little evidence that muscle pump plays a significant role as a mechanical factor in controlling muscle blood flow during exercise. While muscle pump does play a role, its contribution is modest in determining muscle blood flow, and it is not required for sustaining venous return, central venous pressure, stroke volume, or cardiac output during exercise (Gonzalez-Alonso et al. 2008; Tschakovsky et al. 1996).

Locally Produced Metabolites

Various locally produced substances may contribute to the maintenance of vasodilatation during exercise. These include NO, adenosine, prostaglandins and arachidonic acid, potassium, and ATP released from red blood cells.

Nitric Oxide

NO is synthesized from l-arginine in the presence of oxygen and a number of cofactors in a reaction that is catalyzed by the enzyme nitric oxide synthase, or NOS (Tschakovsky and Joyner 2008). NOS is present in the endothelium (eNOS) and in the muscle (neural NOS, or nNOS). NO may be released from the vasodilator nerves, the vascular endothelium, or the muscle fiber itself. The latter may result from stimulation of NOS via shear stress, mechanical stress, or ATP release from red blood cells. NO may also originate from red blood cells (Clifford and Hellsten 2004). Deoxygenated hemoglobin is also capable of reducing nitrite in the blood to NO, and this phenomenon provides an interesting link between oxygen demand and blood flow (Clifford and Hellsten 2004).

The evidence regarding the role played by NO in the control of exercise blood flow is far from clear. The literature has shown that NOS inhibitors have a minimal effect on skeletal muscle blood flow. The effect seems to be the most pronounced when NO production is blocked during—as opposed to before—exercise. NO and prostaglandin synthesis inhibitors reduce blood flow only modestly when given during rhythmic forearm handgrip exercise (Schrage et al. 2004). NO is apparently not involved in blood flow changes that occur during transition from one intensity of exercise to another (Saunders et al. 2005). NOS inhibition does return blood flow to baseline levels faster after exercise (Joyner and Wilkins 2007).

Recently, Hocking and collaborators (2008) showed evidence of an interesting mechanism for the involvement of NO in blood flow regulation. They showed that fibronectin, an extracellular fibrillar matrix protein, may transduce signals from contracting muscle fibers to local blood vessels via an NO mechanism. More specifically, their results from experiments using in vitro hamster cremaster preparations suggested that muscle contractions alter the conformation of fibronectin proteins, exposing an active site that interacts with receptors on smooth muscle or skeletal muscle cells. This interaction subsequently activates an NO-mediated signaling pathway that leads to vasodilation.

Adenosine

Adenosine is probably responsible for 20% to 40% of the maintained phase of exercise hyperemia (Marshall 2007). The adenosine may originate from adenosine monophosphate (AMP) released by muscle fibers that is converted to adenosine through dephosphorylation by ecto-5'-nucleotidase bound to the sarcolemma (Marshall 2007; Clifford and Hellsten 2004). During exercise the concentration of this enzyme may increase via translocation from the cytosol. Adenosine is also released from endothelial cells and generated from ATP released from endothelial cells, nerve terminals, or red blood cells and produces vasodilation through stimulation of A_1 adenosine receptors. Adenosine causes vasodilation by acting on A_{2A} adenosine receptors on the extraluminal surface of the arterial smooth muscle.

In lower hind-limb muscles of anesthetized cats, blocking the A_{2A} adenosine receptor significantly reduces blood flow resulting from electrical stimulation (Poucher 1996). Results from human experiments are less clear. Adenosine does, however, appear to be important in determining blood flow heterogeneity in human muscles during exercise. Blood flow heterogeneity in a large muscle such as the quadriceps normally decreases, as bulk flow increases, with increasing workload. When adenosine receptors are blocked, blood flow in the exercising leg remains more heterogeneous, although bulk flow is not influenced. Thus, adenosine might play a role in making blood flow more homogeneous in different parts of the muscle as workload increases (Heinonen et al. 2007).

To confound things, there appear to be adenosine responders and nonresponders (Clifford 2007; Joyner and Wilkins 2007). The basis for this variability is currently unclear, and this variability may be significant in influencing exercise performance. More research is needed on this issue.

Prostaglandins

Prostaglandins are produced by factors that increase the calcium level in endothelial cells and by mechanical stress (Clifford and Hellsten 2004). They exert their vasodilator effect via binding to receptors on smooth muscle; this results in decreased smooth muscle calcium levels. The contribution of prostaglandins to exercise blood flow is modest at best (Joyner and Wilkins 2007). Prostaglandins are not involved in the changes in blood flow that occur during transitions from one exercise intensity to another (Saunders et al. 2005). At steady-state exercise, prostaglandin synthesis inhibition results in a transient and modest decrease in blood flow, suggesting that prostaglandin synthesis exerts redundant control of exercise hyperemia.

Potassium

Potassium is released from muscle fibers during contractile activity and accumulates in the interstitial space, where it could influence blood flow by interacting with the vasculature. In dog muscles contracting in situ, the increase in blood flow at the onset of contractions is eliminated if vascular smooth muscle hyperpolarization is prevented by potassium infusion (Hamann et al. 2004a). Potassium released from muscle fibers may be involved, but this is currently controversial (Clifford 2007) in spite of the observation that the infusion of potassium at rates that might reproduce levels seen during exercise increases muscle blood flow (Juel 2007). Increased interstitial potassium might also potentiate the effects of other vasodilators (Juel 2007). Although the mechanism of how elevated potassium might increase blood flow is not known, it may involve activation of smooth muscle inward-rectifier potassium channels (K_{ir}), since the effects of potassium infusion on blood flow are absent when this channel is blocked with barium (Clifford and Hellsten 2004).

ATP

As previously mentioned, hemoglobin may act as a sensor for the oxygenation state of the muscle and may help control blood flow by release of ATP in proportion to the release of oxygen (Rosenheimer et al. 2004). Other possible sources of ATP include endothelial cells and motor and sympathetic nerve terminals (Clifford and Hellsten 2004). ATP may also make a small contribution to exercise hyperemia by binding to the purinergic P2γ receptors located on the vascular endothelial cells. Such binding could result in the release of endothelium-derived hyperpolarization factors (EDHFs), NO, prostaglandins, and perhaps other substances (Clifford and Hellsten 2004; Rosenheimer et al. 2004; Saltin 2007; Marshall 2007). ATP infusion blunts sympathetic vasoconstriction induced by augmenting alpha-adrenergic receptor stimulation with tyramine in resting as well as exercising muscle (Rosenheimer et al. 2004).

EDHF

It is thought that agonist-induced increases in intracellular calcium levels in endothelial cells may lead to hyperpolarization of these cells, followed by hyperpolarization of smooth muscle cells. Mechanisms may include gap junctions or diffusion of a substance such as EDHF from endothelial cells to smooth muscle cells, resulting in their relaxation and vasodilation (Clifford and Hellsten 2004).

Summary of Blood Flow Control During Exercise

The exact mechanisms and combinations of mechanisms that are responsible for control of muscle blood flow during exercise remain elusive. Clearly, mechanical factors are involved, at least during the initial phase of the increase in blood flow, and metabolic factors fine-tune the blood flow to meet metabolic demand while maintaining blood pressure. Results using blockers or agonists of single pathways are equivocal and often negative (e.g., blocking NO production generally has no effect on exercise blood flow). The current thought is that metabolic mechanisms of blood flow regulation during exercise are redundant, with some mechanisms covering for others when they are blocked. There is also evidence that blood flow might be muscle specific and even individual specific. Clearly more experimental work must be done in this area.

MUSCLE METABOLISM

Recent discoveries on the factors controlling muscle metabolism and how it relates to muscle adaptation have changed our ways of thinking in these areas. A sensor molecule that detects muscle energy state, AMP-activated protein kinase (AMPK), can regulate the conservation of cell ATP and has been implicated in modulating pathways that are involved in training-induced muscle adaptations. With the growth of the inactivity and type 2 diabetes epidemic, more attention has turned to the metabolic pathways that lead to insulin sensitivity and disruptions in fat metabolism and transport across membranes. These recent issues are discussed in this section.

AMPK

It would be extremely advantageous and efficient if muscle possessed a metabostat—a regulator like a thermostat that could control energy flux instead of temperature. Researchers are proposing that the muscle enzyme AMPK is one such metabostat. AMPK is a kinase that is activated under conditions of metabolic stress, such as exercise, when the ratio AMP to ATP rises. The overall effect of AMPK activation is activation of catabolic pathways that generate ATP and inhibition of pathways that consume ATP. Two important upstream kinases are involved in AMPK activation. One is the tumor-suppressing protein kinase serine/threonine kinase 11 (LKB1) and the other comprises the calcium/calmodulin-dependent protein kinase

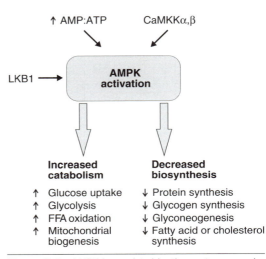

Figure 3.4 AMPK is regulated by the upstream regulators CaMKK and LKB1, which both phosphorylate AMPK. Phosphorylation by LKB1 but not CaMKK is stimulated by the binding of AMP to AMPK. Also included are processes of energy metabolism regulated by AMPK.

kinases CaMKKα and CaMKKβ. Phosphorylation of AMPK by LKB1 is facilitated by the binding of AMP to AMPK (see figure 3.4).

AMPK activation is sensitive to circulating hormones such as leptin, adiponectin, tumor necrosis factor-alpha (TNFα), and ciliary neurotrophic factor (CNTF). However, activation of AMPK occurs in isolated muscle preparations, and thus the importance of circulating factors in this metabolic regulatory mechanism is limited (Jorgensen et al. 2007).

The responses to AMPK activation are various and have the effect of conserving ATP. For example, biosynthetic pathways that use ATP are downregulated by AMPK activation. AMPK activation inhibits muscle glycogen synthesis via phosphorylation of glycogen synthase. AMPK also inhibits protein synthesis by inhibition of mRNA translation via the mammalian target of rapamycin (mTOR) pathway and also by inhibition of peptide elongation via elongation factor 2 kinase—this action is consistent with the tumor-suppressing effect of the upstream kinase LKB1.

AMPK activation, besides inhibiting the biosynthetic pathways mentioned in the previous paragraph, also stimulates catabolic pathways that generate ATP. For example, AMPK activation stimulates glucose uptake via translocation of glucose transporter type 4 (GLUT4) to the membrane. The most likely mechanism for the AMPK stimulation of glucose uptake is via inhibition of a protein called *Akt substrate of 160 kDa (AS160)* that normally prevents the translocation of GLUT4 to the plasma membrane (Turcotte and Fisher 2008). AMPK also phosphorylates the beta-isoform of acetyl-coenzyme A (CoA) carboxylase (ACC), resulting in a reduction in malonyl-CoA levels. This relieves the inhibition of fatty acid transport into mitochondria via the carnitine carrier system. Since malonyl-CoA inhibits carnitine palmitoyltransferase I (CPT1) on the outer surface of the inner mitochondrial membrane, reduced levels of malonyl-CoA permit entry of more fatty acyl-CoAs into mitochondria (Watt et al. 2006; Hardie and Sakamoto 2006). Thus, inhibition of malonyl-CoA results in increased fatty acid oxidation and more ATP. Activation of AMPK also induces the translocation of the fatty acid transporters CD36 (also known as *FAT)* and plasma membrane fatty acid binding protein (FABPpm) from intracellular sites to the cell surface and thereby upregulates fatty acid transport (Nickerson et al. 2007).

A major change in energy status is not required for AMPK activation. For example, there is evidence that increased fatty acid availability increases AMPK by allosteric modulation of the AMPK molecule, making it a better substrate for the upstream protein kinase LKB1. This happens with no change in cellular energy status. Such a mechanism would increase fat oxidation in the presence of increased fatty acid levels and thus would help reduce intramyocellular fat storage (Watt et al. 2006).

Figure 3.4 summarizes the known effects of AMPK activation in muscle (Hardie et al. 2006). As we shall see in the chapter on aerobic endurance training (chapter 6), AMPK is also involved in mitochondriogenesis that occurs with adaptation to training through increased expression of peroxisome proliferator-activated receptor gamma coactivator 1-alpha (PGC-1α). Interestingly, the training state seems to determine the activation of AMPK in response to acute exercise. Endurance-trained and strength-trained athletes demonstrate enhanced responses when performing an unfamiliar activity and blunted responses when performing the activity in which they are trained. This emphasizes the concept of AMPK signaling for the level of metabolic stress, which should be higher in an unfamiliar activity (Coffey et al. 2006).

Exercise, Insulin, and Glucose Uptake

Knowledge of how glucose uptake by skeletal muscles is regulated has increased significantly in the past 5 years, fueled by the concerns surrounding the increased incidence of insulin resistance and type 2 diabetes in the population. I have therefore summarized what we know at present about these mechanisms and about what goes wrong in the insulin-resistant and type 2 diabetic state.

Initially the contribution of glucose to exercise energy metabolism is minor, but it increases substantially as muscle glycogen stores are diminished, reaching up to 35% of leg metabolism and 100% of carbohydrate metabolism (Wojtaszewski and Richter 2006). Mechanisms that promote the entry of glucose into muscle cells in order to present this important substrate for oxidation provide an intriguing, complex, and continually evolving story.

How Insulin Promotes Glucose Uptake

Figure 3.5, which is taken from the fine review article by Wojtaszewski and Richter (2006), is a summary of our current knowledge of the control of glucose uptake into muscle fibers. In summary, the activated insulin receptor phosphorylates the insulin receptor substrate (IRS) and some other substrates that bind to other target proteins when phosphorylated. One of these target proteins is the p85 subunit of phosphatidylinositol 3-kinase (PI3K), which activates its p110 catalytic subunit and results in a rise of phosphatidylinositol 3-phosphate (PIP3) within the membrane. Increased

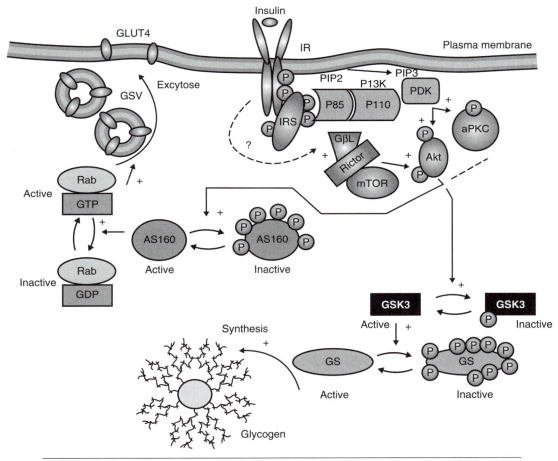

Figure 3.5 Events from insulin binding to its receptor to the activation of the effector proteins GLUT4 and glycogen synthase (GS). aPKC, atypical protein kinase C; GSV, GLUT4 storage vesicle; IRS insulin receptor substrate; p85 and p100 are subunits of phosphatidylinositol 3-kinase; PDK phosphoinositol-dependent protein kinase; Akt is a serine-threonine kinase; Rictor-mTOR kinase complex; GSK3 glycogen synthetase kinase; AS160 substrate for Akt which links IR signaling and GLUT4 trafficking; Rab is a G-protein.

_____ AMPK AS A TARGET FOR TYPE 2 DIABETES _____

It would seem that activating AMPK pharmacologically is an obvious strategy to control type 2 diabetes. However, according to Dr. Bei Zhang and colleagues (Zhang, Zhou, and Li 2009) at Merck Research Laboratories in New Jersey, there are a number of issues to resolve before this can occur. First of all, AMPK can exist in 12 different complexes that differ among different tissues. This raises the question of whether an all-purpose activator is feasible or whether activators should be directed toward specific AMPK complexes. AMPK complexes may also differ across species, which renders it difficult to extrapolate results from animal models to humans. AMPK activation might also have unforeseen and undesirable effects in tissues such as brain, adipose tissue, and pancreas, and so researchers think that tissue-selective AMPK activation may be necessary. In addition, issues such as dose, continuous versus intermittent activation, and efficacy when compared with other agents such as metformin must be resolved before a viable pharmacological intervention using AMPK activators can become a reality.

PIP3 promotes GLUT4 translocation from storage vesicles to the plasma membrane and thus facilitates glucose entry.

Increased PIP3 also activates Akt, a serine/threonine protein kinase, and atypical protein kinase C (aPKC). Akt activation may require additional phosphorylation from other sources, which are shown in figure 3.5 as the rapamycin-insensitive companion of mTOR (Rictor) and mTOR complex. Akt is activated and recruited to the plasma membrane in conjunction with phosphatidylinositol-dependent protein kinase-1 (PDK1). Akt and aPKC play facilitatory roles in activating glycogen synthase through glycogen synthase kinase 3 (GSK-3) and in activating AS160, a protein essential for translocation of GLUT4 to the plasma membrane. In fact, AS160 is considered to be the essential link between insulin receptor signaling and GLUT4 trafficking. The actual mechanism for the AS160 promotion of GLUT4 translocation is thought to be through GTP-Rab, since AS160 contains a GTPase-activating protein domain. It is worth noting that AS160 is also activated during contractile activity through the AMPK system discussed earlier in this chapter (Wojtaszewski and Richter 2006).

How Exercise Promotes Glucose Uptake

During exercise, insulin-independent mechanisms contribute to glucose uptake. This is evident in the increased sensitivity to glucose that occurs in muscles of exercising subjects with type 2 diabetes. Increased calcium during contractions enhances glucose uptake through calmodulin kinase and typical and atypical PKC. As discussed previously, AMPK activation also enhances glucose uptake through increased GLUT4 translocation to the plasma membrane. Dynamic, nondamaging types of exercise performed at a moderate intensity increase insulin sensitivity both during exercise and during postexercise. What are the mechanisms by which this occurs? Results are currently controversial. One explanation is that there are two forms of IRS, one activated more by exercise (IRS2) and one activated more by insulin (IRS1), with crossover such that downstream signaling mechanisms to a given insulin stimulus are enhanced. Another possible mechanism involves an enhanced sensitivity of aPKC to a given PIP3 stimulus (Wojtaszewski and Richter 2006). A third mechanism may include AMPK activation enhancing the insulin signaling response to a given stimulus—substances that mimic AMPK action enhance insulin sensitivity (Wojtaszewski and Richter 2006).

Insulin Resistance

Increased insulin resistance is the phenomenon in which the body loses its ability to clear a glucose load from the circulation in response to insulin. Since skeletal muscle makes up such a large proportion of the body's tissues, problems in this regulation generally originate in increased resistance in this tissue (Turcotte and Fisher 2008).

A main factor involved in insulin resistance seems to be a disconnect between fatty acid uptake and oxidation in muscle cells (Turcotte and Fisher 2008). Fatty acid uptake is partially determined by circulating plasma free fatty acid (FFA) levels. In individuals who are obese and diabetic, there is a permanent relocation of fatty acid transporters to the plasma membrane without increased expression. This allows for a greater level of fatty acid uptake for any given level of circulating fatty acids (Nickerson et al. 2007). Since fat uptake significantly exceeds fat oxidation, there is an accumulation of intracellular lipid intermediates other than triglycerides, such as diacylglycerols (DAGs), ceramides, and long-chain fatty acyl-CoAs. This buildup may interfere with normal insulin signaling. Intramuscular DAGs act as a ligand for several serine kinases that phosphorylate serine sites on the insulin receptor and IRS1. Phosphorylation at these sites reduces downstream insulin signaling and results in reduced GLUT4 translocation. Ceramides appear to block insulin effects by preventing insulin-evoked phosphorylation of Akt and GSK (Thyfault 2008).

There is some speculation that the number of fatty acid transporters in mitochondria and plasma membrane may have a strong genetic component. This would predispose certain individuals to increased insulin resistance and type 2 diabetes.

A reasonable explanation for the development of insulin resistance that relates to neuromuscular activity concerns muscle oxidative capacity. Perhaps decreased activity, coupled with a tendency to eat more food (and fat) than is necessary for fuel, may result in increased uptake but not oxidation of fats. In individuals who are obese or diabetic, mitochondrial oxidation and electron transport are deficient compared with these same processes in people who are not obese or diabetic (Turcotte and Fisher 2008). The main benefit of exercise is reducing this disconnect by increasing the capacity for mitochondrial FFA oxidation.

Role of Chronic Exercise

Chronic exercise reduces insulin resistance. A major way that this occurs is via increased GLUT4 expression. Exercise at 73% of maximal aerobic power (MAP) for 1 hour results in increased gene expression (more than double) of GLUT4 that lasts for at least 3 hours after cessation of the exercise (Kraniou et al. 2007). This stimulation of gene expression eventually results in increased protein; performing up to 16 sessions of high-intensity exercise for 6 minutes (90% of $\dot{V}O_2$max, once per hour) increases GLUT4 and the monocarboxylate transporter 4 (MCT4, lactate transporter) in human muscles (Green et al. 2008).

GLUT4 expression is controlled to a significant extent by the transcriptional coactivator PGC-1, which plays a major role in the overall response of muscles to aerobic

IMPAIRED INSULIN-MEDIATED CAPILLARY RECRUITMENT IN TYPE 2 DIABETES

In individuals without type 2 diabetes, insulin has a capillary recruitment function, mediated via endothelial cells, that directs blood preferentially toward those parts of the muscle that are metabolically most active. This is demonstrated by the fact that glucose uptake by muscle is higher per unit blood flow than expected under conditions of uniform capillary perfusion. Dr. Scheede-Bergdahl and colleagues (2009) at the University of Copenhagen demonstrated that this insulin-mediated control of forearm muscle blood flow is impaired in individuals with type 2 diabetes. In their experiment, they showed that non-insulin-mediated increases in forearm blood flow and glucose uptake in response to infusion of the vasodilators acetylcholine (ACh), sodium nitroprusside, and adenosine were similar. This demonstrated that bulk muscle flow capacity was not impaired with type 2 diabetes. However, the ability for vasodilation in response to prolonged mild hyperinsulinemia was impaired in subjects with type 2 diabetes. These results suggested that patients with type 2 diabetes have defective insulin-mediated capillary recruitment. Such a deficit could prove significant in limiting performance in events where control of blood flow to different parts of the muscle or body becomes critical, such as exercising with arms and legs in a hot environment.

endurance training. Cultured muscle cells normally do not express GLUT4 or PGC-1; however, when an adenovirus vector was used to allow expression of PGC-1 protein in L6 myotube cultures, GLUT4 mRNA levels changed from undetectable levels to levels normally found in muscle tissue in vivo (Michael et al. 2001). This was accompanied by increases in basal and insulin-stimulated glucose transport. The results of these experiments also strongly suggest that the effect of PGC-1 in vivo in stimulating GLUT4 expression is achieved via binding to and coactivating the muscle-specific transcription factor myocyte-specific enhancer factor 2 (MEF2; Michael et al. 2001).

Peroxisome proliferator-activated receptors (PPARs) regulate lipid utilization and storage and regulate metabolic substrate utilization during conditions such as fasting and exercise. Agonists specific to the predominant isoform in muscle, PPARΔ, enhance insulin sensitivity by increasing lipid oxidation, thereby lowering plasma FFA levels and generating a negative feedback on the insulin signaling cascade such that glucose uptake is increased and plasma levels reduced (Kramer et al. 2007). PPARs are implicated in the coordinated training responses of muscle to aerobic endurance training.

Protein-Assisted Fatty Acid Entry Into Muscle Cells

Although FFAs are capable of entering the muscle cell by passive diffusion, evidence is mounting that the actions of fatty acid transporters in the plasma membrane exert considerable control over this process. Fatty acid uptake into muscle fibers is increased within minutes of exercise initiation. This is accomplished through activation of pathways that translocate transporters from intracellular stores to the cell membrane.

We have already mentioned the transport of FFAs into the muscle cell, the way in which AMPK participates in this phenomenon, and the proliferation in the transporters in insulin resistance. The principal players in fatty acid transport into muscle are the plasma membrane transporters themselves, CD36, the plasma membrane fatty acid binding proteins (FABPpm), the cytoplasmic fatty acid binding proteins (FABPc), the fatty acid transport proteins (FATPs), and the acyl-CoA binding protein (ACBP; Nickerson et al. 2007; Pelsers et al. 2008). Transporters are involved not only in FFA entry into the cell but also in FFA transport to the mitochondria and transport across the mitochondrial membrane. The latter function is accomplished by a cooperation between CD36 and CPT1, an important regulator of FFA entry into mitochondria (Nickerson et al. 2007).

The importance of this transport system was evidenced by the finding that in CD36 knockout mice FFA uptake into muscles is deficient (see figure 3.6; Bonen et al. 2007). The importance of this system was also evident in experiments with human subjects. Oxidation rates of palmitate in mitochondria isolated from human muscles correlated closely with CD36 content following 120 minutes of cycle exercise (Holloway et al. 2006). In addition, a specific inhibitor of CD36 added to these mitochondrial preparations reduced palmitate oxidation significantly (Holloway et al.

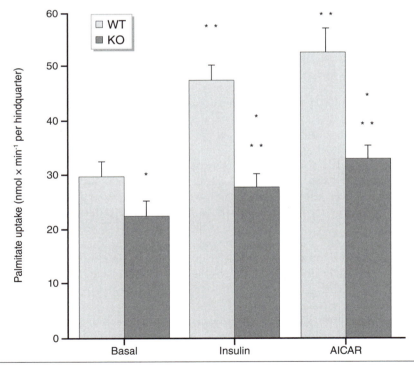

Figure 3.6   In CD36 knockout mice, palmitate uptake by perfused hind-limb muscles is deficient under basal conditions and when insulin and aminoimidazole carboxamide ribonucleotide (AICAR), the pharmacological activator of AMPK, are included in the perfusate. WT, wild-type mice; KO, knockout mice.

Reprinted by permission from Bonen et al. 2007.

2006). It has recently been shown that a high-fat diet, acute exercise, and short-term training rapidly upregulate membrane CD36 content by translocation and—after 60 minutes—by increased protein synthesis (Pelsers et al. 2008). Total expression of CD36 can be rapidly upregulated (within days) by a high-fat diet. On the other hand, FABPpm protein content may be upregulated after at least 4 weeks of a high-fat diet or aerobic endurance training.

We have seen the important role these FFA transporters play in insulin resistance, and there is some evidence that exercise may reverse this increased insulin resistance through specific signaling pathways. The field of FFA transporters is relatively young, and clearly more information will be beneficial for treating conditions such as type 2 diabetes and obesity.

Summary of Muscle Metabolism Issues

The principal issues under investigation at this time relate to the control of substrate entry and use during exercise and in conditions where inactivity and disease (such as type 2 diabetes) are involved. Molecular biological technology is moving us toward a better understanding of the systems used to transport substrates into the muscle cell and mitochondria as well as how these systems work, how they are altered in disease, and how they are affected by acute and chronic exercise.

SUMMARY

Muscle blood flow is controlled to maintain adequate oxygen delivery. The state of hemoglobin saturation may participate in this control by stimulating the release of vasodilator substances. Mechanisms are in place to override the active muscle vasoconstriction that occurs during exercise so that blood flow is adequate for energy demands while blood pressure is maintained. This is accomplished by functional sympatholysis, which is an inhibition of the sympathetically mediated vasoconstrictor response in active muscle that occurs as long as blood pressure is maintained. Functional sympatholysis is probably due to substances produced in the contracting muscle that inhibit receptors involved in muscle vasoconstriction. Mechanical events during contractions are also important in increasing muscle blood flow; rapid dilation of arterioles has been shown in in vitro preparations of animal arterioles, and rapid increases in muscle blood flow in human subjects can be induced by using rapidly inflating pressure cuffs to increase extravascular pressure. Release of vasodilatory substances, such as NO, adenosine, prostaglandins, potassium, ATP, and EDHF, is responsible for more long-term control of muscle blood flow during exercise. There appears to be considerable redundancy in the control of muscle blood flow, such that attempts to determine the significance of any one mechanism via blockade experiments have proven difficult.

The muscle enzyme AMPK is a muscle energy sensor, or metabostat, that is activated under conditions of metabolic stress, such as exercise. The response to AMPK activation is activation of catabolic pathways that generate ATP and inhibition of pathways that consume ATP. AMPK activation helps conserve ATP during exercise and is implicated in pathways leading to chronic muscle adaptations to aerobic endurance exercise, such as mitochondriogenesis. Insulin and exercise increase muscle glucose uptake by distinct but overlapping metabolic signaling pathways. Insulin resistance occurs in muscles of subjects who are inactive or obese or who have type

2 diabetes. Resistance is due to the accumulation of fatty acid intermediates in the muscle cell cytoplasm that interfere with the insulin signaling cascade. Fatty acids are transported into the muscle cell and mitochondria by fatty acid transporters. The fatty acid transport system may become defective in insulin sensitivity, and exercise may counter insulin resistance in part by improving the availability and function of these transporter systems.

Peripheral Factors in Neuromuscular Fatigue

It is naive to think that neuromuscular fatigue resulting from exercise can be explained by a single mechanism. Although fatigue during certain kinds of effort involves failure of the contractile apparatus, some types of fatigue cannot be explained by a decrease in the force-generating capacity of the muscle tissue alone. Some evidence suggests that systems providing input to the muscle fiber may be involved. During maximally sustained or intermittent contractions, for example, stimulation of the motoneuron with a supramaximal voltage may or may not evoke an increased force; muscle force increased by an external voltage implies that either motivation was lacking or fatigue was occurring at a more central location (so-called *central fatigue*). This central contribution to neuromuscular fatigue can differ among muscles and muscle groups and can vary depending on body temperature or the intensity and type of contraction generated by the muscle or muscle group. Even during submaximal contractile activity, either sustained or intermittent, there are signs of changes in the nervous system as exercise progresses that might influence the efficacy or pattern with which the contractile machinery is activated.

While contractile activity of a muscle stimulated supramaximally by an external voltage source gives an objective measure of fatigue (i.e., the decline in maximal force response in spite of continued supramaximal stimulus), the notion of fatigue in the exercising organism can include an increased effort necessary to maintain a submaximal contractile force even as the maintained force level remains the same. Thus, as depicted in figure 4.1, the individual keeps exercising at the same performance level while experiencing an increase in the excitation of the motor pool necessary to maintain the performance and a simultaneous decrease in the maximal capacity of the contractile system. Whether there are limits to the length of time during which this

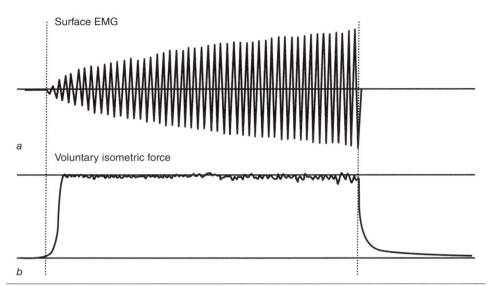

Figure 4.1 *(a)* EMG and *(b)* force of first dorsal interosseus of a subject asked to maintain a target abduction force of 35% of MVC for as long as possible (210 s in this case). Note the gradual increase in EMG. MVC decreased to about 70% at fatigue.

Data from Fuglevand et al. 1993.

increased effort can be sustained by central and peripheral nervous system structures is a topic of intense research.

In this chapter we will discuss the fatigue factors that are found in the periphery, while in the next chapter we will focus on factors occurring in the spinal cord and above. I have purposely shied away from referring to *central fatigue* and *peripheral fatigue,* since some peripherally generated fatigue signals result in modulation of central command, in which case there is difficulty in specifying central versus peripheral fatigue.

INTRAMUSCULAR FACTORS AND MUSCLE FORCE

Figuring out what intracellular mechanisms are responsible for fatigue in exercising humans or other animals is difficult. The measurement of metabolites and ions in fatigued muscles is problematic from a number of perspectives. Biopsies taken at the point of fatigue may still change up to the time of freezing. In addition, biopsy materials or even whole animal muscle samples are highly compartmentalized, and we do not know in what compartment (interstitial versus intracellular, fast versus slow, perinuclear versus subsarcolemmal, peri-Z-line versus peri-A-band, mitochondrial versus sarcoplasmic) the metabolites being assayed are located. Decrements in neuromuscular performance often involve central as well as peripheral components. For this reason, tools such as electrical stimulation of the motor axons innervating the muscle are used to assess the effects of fatigue directly on the muscle properties. When stimulated, fatigued muscles demonstrate slower and lower amplitude twitches, lower maximal force capacity, and a decreased maximal rate of tension development and decline (Fitts 2008). The extent of change in each of these properties can differ depending on the protocol of exercise or the electrical stimulation used to produce the fatigue.

Many factors determining muscle force, contractile speed, and power may be influenced by biochemical events that occur during high-intensity, repetitive contractile activities that lead to fatigue. Interference with the transition states of the actomyosin cross bridges from weakly bound, low-force states to strongly bound, high-force states that occur during contraction may develop as products such as calcium, hydrogen ion, inorganic phosphate, and reactive oxygen and nitrogen species (ROS and RNS) change during contractile activity (figure 4.2). Changes in these metabolites may influence the kinetics of change among actomyosin binding states, which would slow down contractile speed; may influence the number of cross bridges in the high-force configuration at any given time; or may influence the force-generating capability of individual cross bridges. Since power is the product of force and velocity, changes in force-generating capacity and the shape of the force–velocity curve resulting from fatigue can result in rather drastic decreases in power. Fatigue increases the curvature of the force–velocity relationship, thereby exacerbating the loss of power (Jones et al. 2006). Interestingly, changes in the concentrations of ATP and adenosine diphosphate (ADP) probably play a minor role, if any, in the fatigue process.

In addition, the force–calcium relationship may change in fatiguing muscles such that more calcium is required to attain a given force. The amplitude of the calcium transient may decrease for a number of reasons: The action potential that activates the voltage sensor may be attenuated because of potassium accumulation in the transverse tubules (T-tubules), the fall in intracellular ATP and the rise in magnesium may influence the calcium channel opening, and inorganic phosphate may enter the sarcoplasmic reticulum and precipitate with calcium (Allen et al. 2008a, 2008b). More calcium becomes trapped in the cytoplasmic compartment, with the added disadvantage of desensitizing the calcium-binding sites on contractile proteins. While a catastrophic scenario involving all or even some of these events is feasible, we currently have very little evidence of this scenario occurring at in vivo temperatures, since most experiments have been conducted on muscle fibers in vitro at relatively low temperatures (Fitts 2008).

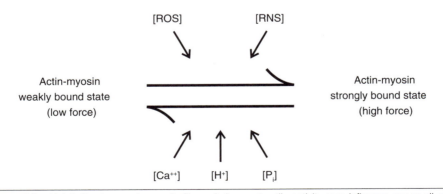

Figure 4.2   Metabolic changes in muscle fibers during contractile activity may influence any or all of the steps involved in transitioning from weakly bound to strongly bound actomyosin cross bridges, thereby causing a decreased force production.

Attempts have been made to examine muscle fiber fatigue in reduced preparations at temperatures closer to body temperature. For example, Karatzaferi and colleagues (2008) showed that at 30 °C muscle fiber velocity is decreased, as is seen during fatigue, when hydrogen ion, phosphorylation of the myosin regulatory light chain, and inorganic phosphate are all increased within physiologically relevant ranges. In addition, Knuth and colleagues (2006) reported that the effects of reduced pH (6.2, which is substantial) on unloaded velocity of shortening and peak power was more significant at 30 °C than at 15 °C and that the effect of this reduced pH on maximal force generation was minimal.

Although in the past it was thought that lactic acid production and the resultant acidification of the interior milieu of the muscle cell impede contraction and lead to fatigue, this idea has been challenged. In fact, physiologically relevant increases in lactate and hydrogen ion have little detrimental effect on skinned muscle fibers in vitro (skinning the muscle fibers allows us to set the intracellular concentrations of molecules, since there is no longer a semipermeable barrier or transporter system that can become limiting), and their accumulation may even have a protective effect during fatigue (Cairns 2006). The effects of plasma lactate and hydrogen ion concentration during exercise may therefore be due to their effects on the nervous system.

Ionic disturbances (other than those involving hydrogen ion) during contractile activity may also be involved in fatigue, mainly through effects on muscle fiber excitability. Failure to maintain resting concentrations of the principal ions, sodium and potassium, results in changes in ion driving force and ultimately membrane potential. Inactivation of sodium channels ensues (McKenna et al. 2008). The ionic candidate most likely to play a major role in fatigue is potassium, which may accumulate in the interstitial space and reduce membrane excitability (McKenna et al. 2008; Juel et al. 2000; Cairns and Lindinger 2008). The critical interstitial potassium concentration at which muscle tetanic force is affected in fresh muscle in vitro (figure 4.3) is well within the levels that have been recorded in muscles of exercising humans (10-13 millimolar; Juel et al. 2000; see figure 4.4). As might be expected, intracellular sodium rises during contractile activity, and this also leads to fatigue. However, the relationship between this increase in intracellular sodium and fiber function is less clear, since it is more difficult to manipulate intracellular ion content of fresh muscle fibers than it is to alter interstitial ion content, as has been done with potassium. Although other ions show exercise-related changes in concentration gradients across the sarcolemma (refer to table 1 in Cairns and Lindinger 2008), current thought is that these changes influence fatigue by modulating the effects of potassium change (Cairns and Lindinger 2008). A major player in the alteration of intracellular versus extracellular sodium and potassium during exercise is the membrane-located sodium-potassium ATPase (Na^+/K^+ ATPase), which pumps sodium out of and potassium into the cell. This ATPase helps maintain resting membrane potential by pumping out more sodium ions (3) and pumping in fewer potassium ions (2), making the pump electrogenic. Exercise has been shown to decrease maximal Na^+/K^+ ATPase activity, although data from human and animal experiments are not consistent. Training, on the other hand, increases the content of the alpha-subunits of Na^+/K^+ ATPase, and this increase is related to a slower accumulation of interstitial potassium and a longer time to fatigue (Nielsen et al. 2004). A schematic summary of the major mechanisms thought to cause muscle fatigue is presented in figure 4.5.

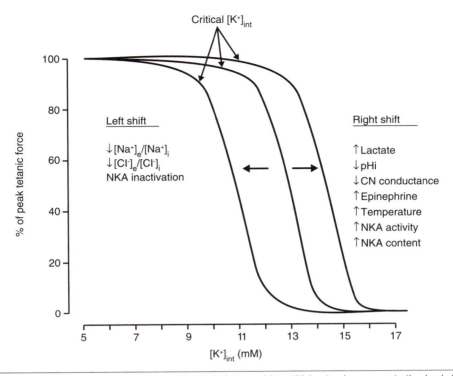

Figure 4.3 The relationship between peak tetanic force and interstitial potassium concentration in skeletal muscle. Critical interstitial potassium corresponds to that concentration at which peak tetanic force starts to decrease abruptly.

Reprinted by permission from McKenna, Bangsbo, and Renaud 2008.

INVOLVEMENT OF STRUCTURES OTHER THAN MUSCLE

Proposed biochemical mechanisms of fatigue in muscle are the subject of several thorough reviews (Fitts 1994; Westerblad et al. 1998; Allen and Westerblad 2001; Allen 2004; Cairns and Lindinger 2008). Concentrating on components other than muscle fiber contractile elements provides the opportunity to focus on less frequently considered phenomena that have been proposed as possible limiting components during exercise and that can be considered alongside biochemical changes. In fact, there is growing awareness that the biochemical events occurring in fatiguing muscles interact with the nervous system to modulate the voluntary fatigue process in a variety of ways and through a variety of mechanisms. Looking at these other mechanisms also allows us to consider neuromuscular fatigue as more than merely an end point beyond which continued function is not possible—we can view it as a condition that develops gradually as exercise continues.

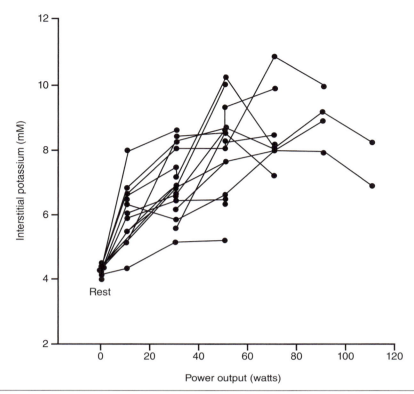

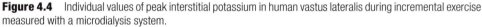

Figure 4.4   Individual values of peak interstitial potassium in human vastus lateralis during incremental exercise measured with a microdialysis system.

Reprinted by permission from Juel et al. 2000.

Role of Neuromuscular Transmission Failure

Neuromuscular transmission failure is failure of a nervous impulse to be translated into an impulse on the sarcolemma immediately beneath the motoneuron terminal. Failure at this level is notoriously difficult to determine, even in anesthetized animal preparations, and especially in the voluntarily exercising human. In many studies of fatigue, neuromuscular transmission failure is eliminated as a contributing factor by examination of the M-wave. The M-wave is the surface EMG response to a single shock (usually of an intensity high enough to evoke a maximal force response) at the level of the axon of the peripheral nerve. The observation of an unchanged M-wave (i.e., no decrease in amplitude or major change in shape) at a time when neuromuscular fatigue is present suggests that electrical signals are still capable of crossing the neuromuscular junction.

Caution must be exercised in trying to implicate neuromuscular transmission failure on the basis of measurements that represent not only neuromuscular transmission but also propagation of the sarcolemmal action potential from the motor end plate to the recording electrode, which is in fact what the M-wave represents. For example, M-waves might change at fatigue not because of neuromuscular failure but because of changes in muscle membrane excitability and propagation speed from the junction to

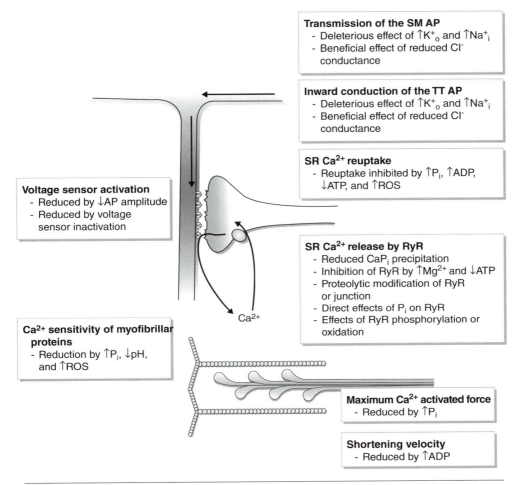

Figure 4.5 The major mechanisms involved in muscle fatigue. SM, sarcomembrane, also called *sarcolemma*; AP, action potential; TT, T-tubules; SR, sarcoplasmic reticulum; RyR, ryanodine receptor.

Reprinted by permission from Allen, Lamb, and Westerblad 2008b.

the recording electrode. Decreases in M-wave amplitude during sustained contractions can occur in the absence of decreases in evoked force (Zijdewind, Zwarts, and Kernell 1999). M-waves can also be potentiated when a muscle changes from rest to contractile activity (Hicks et al. 1989). Thus, generally speaking, M-wave measurement after fatigue seems useful if no changes occur compared with measurements taken at rest, but results are difficult to interpret if changes occur in M-wave amplitude or shape.

Neuromuscular transmission failure can be induced, especially in reduced animal preparations but also in vivo in humans, if the system is pushed at high enough frequencies for long enough durations or if factors required for optimal functioning are in limited supply. Later, I discuss the most pertinent evidence for neuromuscular transmission fatigue when frequencies and durations of excitation are within a reasonably physiological range.

Probably the best evidence to date for the possibility of neuromuscular transmission involvement in fatigue emanates from studies in which muscle forces in response to direct and indirect stimulation in situ or in vitro were compared before and after fatigue induced at relatively physiological stimulation rates. Pagala, Namba, and Grob (1984) found significant differences between direct and indirect stimulation in rat extensor digitorum longus, diaphragm, and soleus after fatigue induced by stimulation at 30 hertz in trains lasting 500 milliseconds every 2.5 seconds for 3 to 5 minutes. This approach has been confirmed, especially for diaphragm, with frequencies as low as 10 hertz, which is well within the physiological range (Van Lunteren and Moyer 1996; Aldrich et al. 1986; Johnson and Sieck 1993; see figure 4.6). Researchers have shown neuromuscular fatigue in intact rabbit (Aldrich 1987) and sheep (Bazzy and Donnelly 1993) diaphragm in response to loaded breathing by demonstrating that phrenic nerve electrical activity (ENG) continues to increase at a time when diaphragm EMG has plateaued. The most recent and systematic evidence of this kind has emanated from studies of diaphragm.

As we will see in a subsequent chapter (chapter 7), many components of neuromuscular transmission adapt to improve transmission and offset fatigue. While this in itself is not proof that neuromuscular transmission is one of the weak links in exercise, it definitely is strong anecdotal evidence of the importance of neuromuscular transmission in the fatigue process.

If we accept that the research to date constitutes evidence that neuromuscular transmission failure can occur under relatively physiological conditions, then we should consider the possible mechanisms behind this failure. There are three main possibilities: neurotransmitter depletion, postsynaptic membrane failure, and failure to propagate the action potential into axon branches (branch block failure).

Neurotransmitter Depletion

Continued stimulation of the neuromuscular unit results in a gradual decrease in the amplitude of end plate potentials (EPPs) measured in the muscle fiber just under the nerve terminal. This is due to lowered quantal content (number of vesicles), smaller quantal size (ACh per vesicle), or both (Wu and Betz 1998; Dorlöchter et al. 1991; Reid, Slater, and Bewick 1999). Decreases in ACh release with fatigue have generally been observed in studies using unphysiological patterns of electrical stimulation, and therefore their implications for voluntary fatigue are unknown. The safety factor for neuromuscular transmission is probably in the neighborhood of 4 (i.e., a change in EPP of approximately 10 millivolts generates an action potential in the muscle fiber, while the EPP is approximately 40 millivolts, or 4 times higher than necessary), implying that a rather drastic reduction in the EPP to 25% of its original value is required for failure to occur due to lack of transmitter.

Van Lunteren and Moyer (1996) demonstrated that decreased neurotransmitter release may contribute to fatigue. They found that fatigue induced in rat diaphragm in vitro via intermittent stimulation at 20 impulses per second (much like breathing) resulted in a decrease in maximal force of the diaphragm that was more pronounced when the muscle was stimulated by its motor nerve than it was when the muscle was stimulated directly. This finding suggested that fatigue was caused at least partially by a decrease in neurotransmitter. These researchers confirmed this hypothesis by conducting another experiment in which 3,4-diaminopyridine was added to the bath. This substance increases neurotransmitter release at the neuromuscular junction,

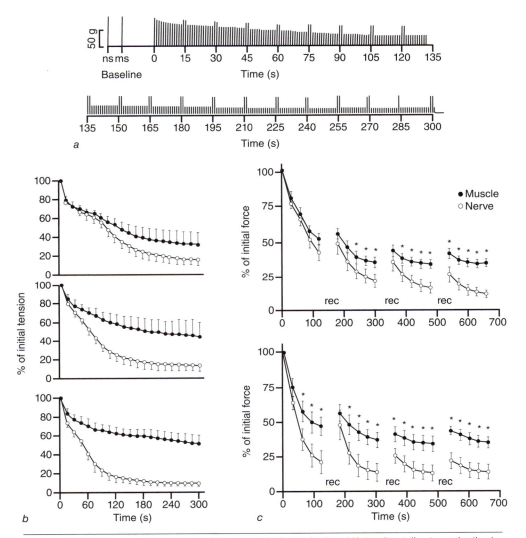

Figure 4.6   Neuromuscular transmission failure in rat diaphragm in vitro. *(a)* Intermittent direct muscle stimulation during phrenic nerve stimulation reveals an increasing contribution of neuromuscular transmission failure as stimulation continues. *(b)* Difference between direct (closed circles) and indirect (open circles) stimulation at 20, 40, and 70 impulses per second (top to bottom). *(c)* Direct and indirect stimulation at 10 Hz, intermittent protocol.

Reprinted by permission from Kuei, Shadmehr, and Sieck 1990.

most likely via an increase in the voltage-dependent calcium conductance in the nerve terminals (Van Lunteren and Moyer 1996). They found that the 3,4-diaminopyridine decreased the force difference between direct and indirect stimulation in fatiguing rat diaphragm. These researchers also found that the tetanic contractions maintained their force plateau better in the presence of 3,4-diaminopyridine. This report lends support to the hypothesis that neurotransmission failure plays a role in fatigue under reasonably realistic stimulation conditions (20-hertz intermittent stimulation for 1.5 minutes). It also suggests that calcium dynamics at the nerve terminal may limit neuromuscular transmission. However, until we can find a way to measure this in human experiments, under conditions of voluntary contractile activity, we cannot be sure of its importance as a fatigue mechanism.

Postsynaptic Membrane Failure

Part of the problem in determining the role of presynaptic mechanisms (discussed in the preceding sections) in neuromuscular junction fatigue is that in reduced preparations the measurements (such as EPP amplitude, used to estimate quantal content) are usually taken from under the postsynaptic membrane, which might also be influenced by continued activation. Thesleff (1959) demonstrated this quite effectively by adding pulses of ACh onto end plates in vitro at various times after stimulation of their motoneurons. He showed, for example, that the voltage response of the end plate to a given pulse of ACh was reduced immediately following a tetanic stimulation of the nerve 20 times per second for 150 milliseconds, and this reduced sensitivity lasted several 10ths of a second. This reduction was higher with higher frequencies and was similar to the reduction in the EPP amplitude observed during the train of electrical impulses. Thus the author concluded that a major proportion of the EPP decrease during trains of stimuli was due to a decrease in end plate sensitivity and not to neurotransmitter depletion.

It is now recognized that desensitization of ACh receptors is at least in part to blame for this end plate desensitization (Magleby and Pallotta 1981; Giniatullin et al. 1986). Prolonged exposure of the receptor to ACh appears to convert it first into a desensitizable state and then into a desensitized state, characterized by a smaller and slower time course associated with ACh binding (Magleby and Pallotta 1981; Giniatullin et al. 1986). Prolonged here is relative; at the frog neuromuscular junction, desensitization was demonstrated with two pulses of ACh separated by 10 milliseconds, which was insufficient time for the ACh from the first pulse to be completely removed from the synaptic cleft before the arrival of the second pulse (Magleby and Pallotta 1981). Desensitization is very easily demonstrated in neuromuscular junctions that have been treated with anticholinesterases, which do not allow the normally rapid removal of ACh from the cleft. The importance of receptor desensitization in the fatigue process, as of other processes at the neuromuscular junction, has not been established unequivocally.

Failure of Axon Branches to Pass on Action Potentials

Axon branch block failure occurs when the action potential generated in the axon is not propagated into all of the branches extending to the muscle fibers. In fact, there is a significant decrease in the excitability of the smaller axon branches relative to the larger branches, so that the safety factor (the amount of excitation produced compared with that which is necessary) for action potential generation decreases as the caliber of the axon decreases (Krnjevic and Miledi 1959). Thus, any variable that affects this safety factor during contractile activity could result in increased blocking at these branch points and thus a failure to excite a proportion of the muscle fibers in the motor unit. Krnjevic and Miledi (1959), in their classic studies of neuromuscular junction responses to fatigue, found evidence of branch block failure in in vitro rat diaphragm preparations stimulated at 20 to 50 hertz for several minutes. They found that this blocking was not altered by substantial variations in potassium, calcium, or hydrogen ion in the bath and concluded that the level of oxygen played a role. Evidence of branch block failure has been shown in cat muscles subjected to maintained, rather high frequencies (80 hertz) of stimulation (Sandercock et al. 1985; Clamann and Robinson 1985). Since we know that high motoneuron firing frequencies are not maintained during voluntary exercise, branch blockade probably plays a minor role, if any, during fatigue.

Muscle Wisdom

During both submaximal and maximal efforts, there is abundant evidence that the way in which muscles are activated begins changing within a few seconds following the commencement of activity. During maximal effort, this change in activation expresses itself most vividly as a decline in EMG activity due to a reduction in the frequency of firing of the recruited motor units (figure 4.7). Interestingly enough, this decreased

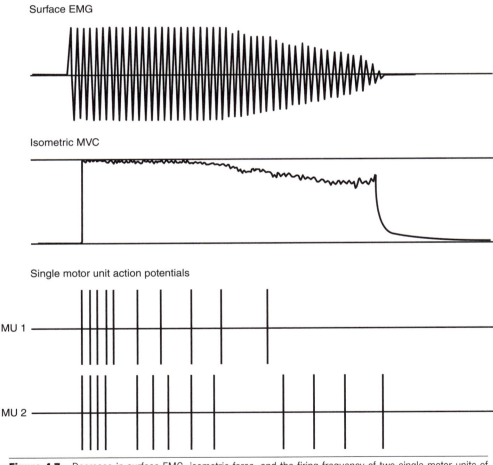

Figure 4.7 Decrease in surface EMG, isometric force, and the firing frequency of two single motor units of adductor pollicis during a maximal voluntary effort lasting 1 min.

Data from Marsden, Meadows and Merton, 1983.

firing frequency also occurs in some motor units during submaximal contractions sustained to fatigue, even at a time when additional motor units are being recruited to maintain the force (Garland, Griffin, and Ivanova 1997). There is considerable speculation as to whether this decrease in firing frequency is a sign of failure somewhere in one or more of the mechanisms involved in increasing and maintaining the excitation of motoneurons or whether it is in fact a sensible response to changes in the contractile properties of the fatiguing muscle. A simple test of the sensibility of this mechanism is to determine whether the decreasing activation of muscle results in a failure to express the full muscle contractile force available at the time. This has been tested during fatigue using supramaximal electrical stimulation, but results have been mixed (BiglandRitchie et al. 1983; Woods, Furbush, and Bigland-Ritchie 1987; Thomas, Woods, and Bigland-Ritchie 1989).

This gradual decline in muscle activation during fatigue, exemplified by the decrease in the firing frequency of motor units during rhythmic or sustained effort, has been given the moniker *muscle wisdom* (Marsden, Meadows, and Merton 1983). It seems wise for the neuromuscular system to decrease motor unit firing rates over time during continued excitation, since muscle fibers are slowing in contractile speed and thus require lower frequencies to maintain maximal force (BiglandRitchie and Woods 1984). Lower frequencies for the motoneuron mean less action potential generation per unit time and less stimulus for late adaptation, which is more pronounced at higher frequencies (Granit, Kernell, and Shortess 1963b), and less stress on the neuromuscular junction and the sarcolemma propagation mechanisms. In fact, several investigators have demonstrated that the optimal pattern for sustained force generation using electrical stimulation is a constantly decreasing frequency of stimulation (BiglandRitchie, Jones, and Woods 1979; see figure 4.8). This finding lends support to the concept of muscle wisdom.

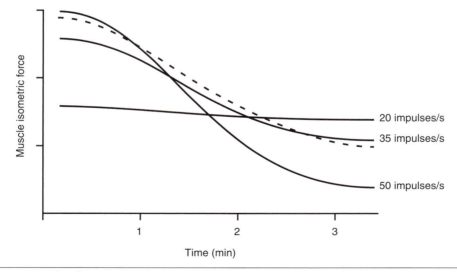

Figure 4.8 Dorsiflexion tension of the big toe evoked by electrical stimulation of the peroneal nerve at 50, 35, and 20 Hz during 1 min. The dotted line shows the optimal force–time envelope that results from a continuous decrease in the stimulation rate from 50 to 20 Hz. Calibration bar = 10 N.

Data from L. Grimby, Hannerz, and Hedman, 1981.

What are the mechanisms underlying this decline in muscle activation during fatigue? Is this decline truly muscle wisdom, or is it a failure of physiological systems? How can we determine from where muscle wisdom is dispensed? There are several candidate mechanisms working either individually or in combination that could explain the decreased muscle activation that occurs with fatigue. They include (1) decreased excitability of the motoneurons, (2) decreased excitatory influence from peripheral sources, (3) increased inhibitory influences on motoneurons, and (4) decreased excitation of motoneurons from supraspinal sources. In this chapter, we will consider only the third possibility, since it refers to the effects of altered metabolic by-products in fatiguing muscles as an inhibitory influence on the motor system. We will consider the remaining three mechanisms in the next chapter.

Inhibition of Motoneurons During Sustained Contractions

Increased inhibition of motoneurons could contribute to the decreased motoneuronal excitability and decreased firing rates observed during fatigue at submaximal and maximal levels. Undeniably, the most popular hypothesis for inhibition as a mechanism during fatigue involves the stimulation of afferents whose muscle endings are sensitive to metabolic by-products of fatiguing contractions that ultimately have an inhibitory influence on the motoneurons (see the sidebar below). There is a considerable amount of information on the effects of metabolic substances on muscle receptors. These effects are discussed in a later section in this chapter summarizing evidence from animal experimentation.

THE FATIGUE REFLEX

Bigland-Ritchie and colleagues (BiglandRitchie et al. 1986; Woods, Furbush, and Bigland-Ritchie 1987) conducted ingenious experiments that strongly pointed toward an inhibitory influence on motoneuron excitability emanating from fatiguing muscles. The design of their experiment and an example of their results are presented in figure 4.9. These investigators found that the reduced firing rates that occurred with a sustained MVC did not recover when the muscle was kept ischemic. This finding suggests there is a reflex by which information is transmitted via afferents from the fatigued muscle to the spinal cord and that has an inhibitory effect on motoneuron excitability. After all, how else would the motoneuron cell body in the ventral horn of the spinal cord, where firing rates are determined, know what is happening at the metabolic level of the muscle? Garland, Garner, and McComas (1988) obtained similar results in soleus muscle stimulated to fatigue under ischemic conditions. This finding demonstrated that the decline in muscle activation does not require voluntary muscle activation and supported the hypothesis of a reflex origin for the decrease in firing rates. It has since been demonstrated that the decrease in EMG during MVC occurs to the same degree even if the large afferents are blocked using compression during the ischemic fatigue (such that the H-reflex is almost abolished, thus implicating the Ia afferents). This finding suggests that smaller afferents, not large afferents, are probably involved in this fatigue reflex.

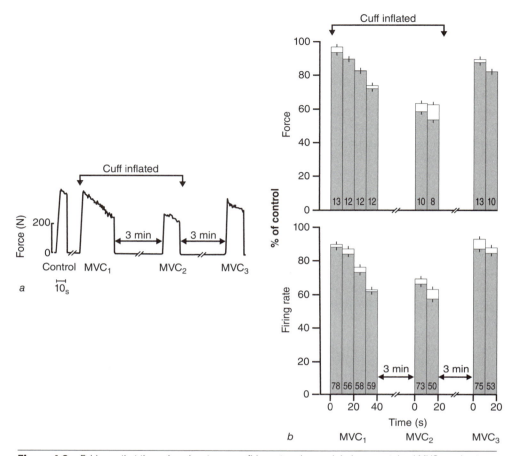

Figure 4.9 Evidence that the reduced motoneuron firing rates observed during a sustained MVC are due to an inhibitory influence on motoneurons from muscle afferents. *(a)* The experimental design. Quadriceps and adductor pollicis were examined. Subjects performed three MVCs to fatigue, separated by 3 min of rest. From the beginning of the first to the end of the second MVC, ischemia was induced by a tourniquet. *(b)* Example of the results obtained from these experiments (quadriceps). Motor unit firing rates did not recover until blood flow was restored (during the second bout of rest), implying that the fatigued muscles influenced the motoneuron firing rates.

Reprinted by permission from Woods et al. 1987.

The role of Group III and IV afferent stimulation in motoneuronal excitability may be quite different for extensors versus flexors. Butler and colleagues (2003) demonstrated that the recovery of biceps brachii motoneuronal excitability following cessation of an MVC, as measured by the amplitude of the MEP (which is the electrical response recorded on the muscle surface in response to cortical stimulation) in response to cervicomedullary stimulation, is not influenced by whether the muscle is kept ischemic after the MVC. Thus, stimulation of Group III and IV afferents is not directly related to decreased motoneuronal excitability—at least for this muscle. In the extensor (triceps brachii), however, maintenance of ischemia following the fatiguing MVC does prolong the recovery of motoneuronal excitability. Thus, extensors may require greater levels of cortical drive during fatigue to override this inhibition (Martin et al. 2006).

Using this same model of stimulating muscle under ischemic conditions, Sacco, Newberry, and colleagues (1997) found that EMG was depressed during an MVC in medial gastrocnemius after fatigue of the lateral gastrocnemius. Thus they suggested that this inhibitory reflex also affects close synergists.

The decrease in motoneuron excitability found by Duchateau and Hainaut (1993) after fatigue in intrinsic hand muscles was too slow to be attributed to motoneuron adaptation or to decreased spindle support, both of which exert their most significant effect early in the contraction; the decrease in motoneuron excitability persisted as long as ischemia was maintained. By elimination, these researchers concluded that increased inhibition involves a reflex originating in the muscle, but like other investigators in this area, they were unable to pinpoint the exact source of the inhibition.

RESEARCH FROM ANIMAL EXPERIMENTS

This section reviews research in which anesthetized or decerebrate animals were used to investigate the previously discussed fatigue mechanisms. Fatigue of a muscle or muscle group was produced by electrical stimulation of the motoneuron or, in the case of one research group, by simulation of exercise via stimulation of the mesencephalic locomotor region (MLR) in the cat.

Behavior of Spindle and Golgi Tendon Organ Afferents

It does not appear that a major change in the sensitivity of the spindle and Golgi tendon organ receptors is involved in the fatigue response. On the contrary, it appears that Ia and II spindle afferents may actually increase in dynamic sensitivity following fatigue (Nelson and Hutton 1985). Golgi tendon organs, on the other hand, may experience a diminished sensitivity with fatigue that probably acts to increase excitation of homologous and synergistic motoneurons via a reduction in autogenic inhibition (Hutton and Nelson 1986).

When this type of experiment (one using an anesthetized animal preparation) is conducted with fusimotor neurons left intact, information about the sensitivity of the fusimotor–spindle system is gained. Christakos and Windhorst (1986) stimulated single cat medial gastrocnemius motor units to fatigue via single ventral root filaments and recorded the discharge from isolated muscle spindle afferents during the stimulation. They found an increase in spindle gain as fatigue progressed; simply put, afferent firing rates did not decrease as much as force did.

The most recent work in this area (Hayward, Wesselmann, and Rymer 1991) demonstrated that spindle and Golgi tendon organ afferents show no major change in sensitivity with fatigue. In the researchers' anesthetized cats, the motoneuron was not severed, so fusimotor neurons were left intact. After fatigue of triceps surae via electrical stimulation (at a voltage below the threshold for stimulation of fusimotor neurons), spindle afferents showed no consistent change in spontaneous firing or in response to stretch. In addition, responses of Golgi tendon organs did not noticeably change with fatigue and were always related to force production.

Spinal Cord Inhibition

Hayward, Breitbach, and Rymer (1988) fatigued the medial gastrocnemius of decerebrate cats and found evidence of increased inhibition of the synergistic soleus muscle (as measured by a decrease in soleus force induced by a crossed extensor reflex) during fatigue of the gastrocnemius. However, these and other researchers (Lafleur et al. 1992; Hayward, Breitbach, and Rymer 1988; Zytnicki et al. 1990) found that this inhibitory influence subsided very quickly with continued stimulation during evoked contractile activity in medial gastrocnemius because of the gradually reduced efficacy of transmission of inhibitory inputs to motoneurons innervating this muscle and its synergists. Declining inhibition was not due to a decrease in the firing of Golgi tendon organs. The researchers proposed that the declining inhibition was due to presynaptic Ib inhibition via Ib fibers, which filtered information from Ib fibers during the contraction to help maintain optimal excitation of the motor pool.

The role of the inhibitory Renshaw cells in the fatigue process is uncertain. Windhorst and Kokkoroyiannis (1991) presented evidence that Renshaw cell inhibition may be particularly strong during the first few seconds of repetitive motoneuron firing and could thus contribute to the rapid decline in firing that occurs at this time.

Excitation of Group III and IV Afferents

Group III afferents respond mainly to mechanical stimuli, while Group IV afferents are primarily nociceptors but also include mechanoreceptors. Group III and IV afferents exert an inhibitory influence on motoneurons innervating the muscle of origin of the afferents and their synergists (Windhorst et al. 1997; Kniffki, Schomburg, and Steffens 1981). For this reason, and because of their slow and longlasting response (compared with that of spindles, for example), they have been targeted as the possible source of the fatigue reflex, by which muscle-related changes during fatigue diminish the excitation of motoneurons during maximal sustained contractions (as discussed earlier in this chapter).

Receptors of Group IV afferents increase their firing rates in the presence of a number of substances that might increase in a fatiguing muscle, including bradykinin, potassium chloride, lactic acid, 5-HT, and arachidonic acid (Djupsjöbacka, Johansson, Bergenheim, and Wenngren 1995; Djupsjöbacka, Johansson, Bergenheim, and Sjölander 1995; Sinoway et al. 1993; Windhorst et al. 1997; Djupsjöbacka, Johansson, and Bergenheim 1994; Rotto and Kaufman 1988; Kaufman et al. 1983; Kniffki, Schomburg, and Steffens 1981). During locomotion generated by electrical stimulation of the MLR in the decerebrate cat, firing of Group III and IV afferents is reduced when a cyclooxygenase inhibitor (arachidonic acid products) is administered (Hayes et al. 2006). The mechanically sensitive Group III afferent receptors can increase their mechanical sensitivity when exposed to some of these substances, and this finding implies that these afferents become more sensitive during fatigue (Rotto and Kaufman 1988; Kaufman et al. 1983; Hayward, Wesselmann, and Rymer 1991; Sinoway et al. 1993). Other substances released during contractile activity that have no effect on the firing of these afferents when injected into a resting muscle might potentiate receptor responses during contractile activity (Rotto and Kaufman 1988).

Hayward, Wesselmann, and Rymer (1991) recorded responses from nonspindle Group II and III afferents in triceps surae of anesthetized cats, examining their sensitivity to contraction, stretch, and surface pressure before and after fatigue produced by muscle stimulation. The main findings were an increased sensitivity to mechanical stimulation and an increased spontaneous discharge rate.

In 1994, Pickar, Hill, and Kaufman recorded responses from Group III afferents to stimulation of the MLR in decerebrate cats walking on a treadmill. These investigators reported an overall increase in the firing of these receptors, even at the very low forces generated in triceps surae during this locomotor task, attesting to the mechanical sensitivity of these receptors. They later substantiated their findings by demonstrating that Group III afferents are also excited during this rather low level of activity. These experiments were particularly important because they used a model that closely mimics voluntary locomotion (as opposed to electrical stimulation of the muscle) to demonstrate that these afferents transmit information regarding the contractile and metabolic state of the muscle not just during fatiguing activity but also during relatively mild activity. Apparently, these mechanisms are also implicated in fatigue of diaphragm—Hill (2000) has shown that Group IV afferents from fatigued rat diaphragms show increased firing rates compared with those of fresh diaphragms.

Fusimotor Neuron (Gamma-Motoneuron) Responses

Normally, flexor alpha-motoneurons are excited and extensor alpha-motoneurons are inhibited by the activity of high-threshold (Groups III and IV) afferents (Appelberg et al. 1983). The situation is somewhat less clear for gamma-motoneurons. Appelberg and colleagues (1983) examined the effects of stimulating Group III afferents on the firing of gamma-motoneurons and found that excitatory effects outweighed inhibitory effects.

Djupsjöbacka, Johansson, Bergenheim, and Sjölander (1995) and Appelberg and colleagues (1983) measured the effects of injecting lactic acid, arachidonic acid, and potassium chloride into the muscle circulation on the firing of muscle spindle afferents. Their hypothesis was that the effects of these substances on the firing level of the fusimotor neuron would be evident from the discharge of the spindle afferents when the muscle was subjected to sinusoidal stretches. The researchers found increases in spindle sensitivity and modulation (range in frequency in response to stretch) even when the injection was in a contralateral muscle. No such effect was evident when the muscle's nerves were cut. Thus, stimulation of chemosensitive afferents increased the excitability of fusimotor neurons.

A similar approach was undertaken by Ljubisavljevic and colleagues (Ljubisavljevic, Jovanovic, and Anastasijevic 1994; Ljubisavljevic et al. 1995), except that they directly examined the effects of fatiguing muscle stimulation on spontaneous firing of fusimotor neurons. Their various treatments (fatigue of contralateral muscles, ischemia following fatigue, denervation, and so on) demonstrated that the effects on fusimotor discharge (increased frequency) were mediated by Group III and IV afferents from the fatigued muscle and were evident in nonfatigued ipsilateral and contralateral muscles. In addition, they found an early increase in firing, which they attributed to excitation by mechanoreceptors, and a late and longer-lasting increase in firing, which they attributed to the chemosensitive receptors. They concluded that this mechanism is a compensatory attempt to optimize the firing rates of the spindle afferents at a time when spindle support is declining during fatigue.

DEMONSTRATION OF A CONTRIBUTION OF FUSIMOTOR NEURONS TO FATIGUE

Pedersen and colleagues (1998) took a unique approach to investigating the role of spindle afferents and fusimotor neurons in fatigue. They fatigued the medial gastrocnemius of anesthetized rabbits via electrical stimulation of the motor nerve and examined the sensitivity of ensembles of spindle afferents (3 to 10 afferents) from the lateral gastrocnemius to stretches of different amplitudes. Analyzing groups of afferents allowed these researchers to derive an index of separation, or an overall idea of how well different afferents simultaneously convey distinct information about the stimulus. Statistical analysis showed that ensembles of muscle spindle afferents of lateral gastrocnemius lost the ability to discriminate the different stretch paradigms with fatigue of the synergist (the medial gastrocnemius); in a way, the afferents became more similar in their responses to stretch. This effect did not occur when the medial gastrocnemius nerve was cut before the muscle was stimulated to fatigue; this finding suggested that the effect was transmitted to the lateral gastrocnemius via afferents from the medial gastrocnemius. By examining the effects of stretch on the modulation and frequency of firing of single afferents, the researchers found that fusimotor drives also became more similar (less discriminatory) after fatigue. Their results suggested that events in the fatigued medial gastrocnemius, probably communicated to the spinal cord via Group IV afferents, influenced fusimotor neurons such that the normal variability of responses to stretch among these cells was reduced. As a result, important information regarding muscle length was lost during and immediately following fatigue. Such changes with fatigue would have significant effects on neuromuscular coordination, something that anyone who has exercised for more than 15 minutes has probably experienced—especially anyone who is not used to performing that exercise regularly.

Changes in Intrinsic Motoneuron Properties

Very little is known about fatigue-related changes in intrinsic motoneuron properties. Stimulation of Group III and IV afferents results in changes in intrinsic motoneuron membrane properties that would change their firing behavior during excitation. Specifically, in decerebrate cats the injection of substances that stimulate these afferents (bradykinin, 5-HT, potassium chloride) decreases membrane potential and cell input resistance, decreases amplitude of afterhyperpolarization, and increases synaptic noise (Windhorst et al. 1997). The authors of these studies pointed out that such changes do not necessarily contribute to adaptation during fatigue; rather, these changes indicate a general change in parameters that are significant in cell excitability and rhythmic firing properties.

SUMMARY

Changes in muscle fibers induced by fatiguing exercise might interfere at several levels to allow maximal expression of contractile force. This is difficult to verify in vivo, however, which explains why we cannot arrive at definite conclusions as to the most significant factors involved in fatigue. Ionic changes seem to be the most important influence on the ability of the muscles to continue contracting. At the same time as metabolic muscle changes are occurring, changes are taking place in the excitability of the neuromuscular junction and in axon branch points. There is indirect evidence that metabolic changes in muscles result in inhibitory signals to motoneurons that decrease their excitability and thus add to the fatigue scenario.

Central Factors in Neuromuscular Fatigue

In the last chapter, we considered the peripheral changes that occur with fatiguing exercise that might result in reduced performance. In this chapter, we will consider central components of fatigue—that is, factors occurring in the spinal cord and above. Since the experiments conducted to investigate the central components of fatigue involve voluntary movements (as opposed to the peripheral components described in the last chapter, which have often been investigated in vitro, or with reduced animal preparations), the working definition of neuromuscular fatigue that is used for this chapter is that of Vollestad (1997, 220): "any exercise-induced reduction in the maximal voluntary force or power output." Such a reduction normally occurs during submaximal exercise, even when maintained at a constant force output.

MOTONEURON ACTIVITY DURING SUSTAINED CONTRACTIONS

In general, motoneuron activity gradually decreases during sustained contractions. In experiments using maximal sustained contractions, contractile durations are relatively short (usually less than 2 minutes and no longer than 5 minutes), since maximal effort is required from the beginning of the contraction. Whole-muscle EMG begins falling immediately (Stephens and Taylor 1972), and so do the firing frequencies of the motor units (BiglandRitchie et al. 1983).

Some motor units cease to discharge in spite of continued maximal effort (Peters and Fuglevand 1999). In response to supramaximal stimulation of the muscle's nerve, the M-wave is either unchanged or slightly decreased at fatigue (Jones, Rutherford, and Parker 1989; Stephens and Taylor 1972), but its area invariably increases, which is usually interpreted as a result of decreased conduction velocity of the action potential along the fatiguing muscle fiber membranes (BiglandRitchie, Jones, and Woods 1979).

Interpretation of changes in the M-wave with fatigue is complex (BiglandRitchie, Jones, and Woods 1979; Stephens and Taylor 1972), especially with the recent observation that motor unit action potentials can increase with fatigue in a nonsystematic way across motor units (Chan et al. 1998). Stimulation of the fatigued muscle via its motoneuron with supramaximal electrical stimulation may or may not result in a slightly increased force. This finding signifies that, in many situations, central activation of muscle units may be less than complete, in spite of high motivation levels (Gandevia et al. 1996; Thomas, Woods, and Bigland-Ritchie 1989; BiglandRitchie et al. 1983; Woods, Furbush, and Bigland-Ritchie 1987).

The situation is a bit different when contractions start at a submaximal intensity. When a subject is asked to generate a submaximal contraction until fatigue, muscle EMG increases gradually as additional motor units are recruited to maintain the force (Garland et al. 1994; Bigland-Ritchie, Furbush, and Woods 1986; Löscher, Cresswell, and Thorstensson 1996b; Macefield et al. 1991; Petrofsky and Phillips 1985; Häkkinen and Komi 1983b, 1986), as was shown back in figure 4.1. This increased recruitment is substantiated by an apparent increase in the excitability of the motoneuron pool, since H-reflex amplitude increases as contraction proceeds during a maintained contraction of the triceps surae at 30% of MVC (Löscher, Cresswell, and Thorstensson 1996b; see figure 5.1). Simultaneously, and somewhat paradoxically, the majority of recruited motor units demonstrate a decline in their firing frequencies (Christova and Kossev 1998; Petrofsky and Phillips 1985; Garland, Griffin, and Ivanova 1997; Garland et al. 1994; De Luca, Foley, and Erim 1996), although some may actually demonstrate an increase (Garland et al. 1994), especially those that are recruited later during the task (Garland, Griffin, and Ivanova 1997). Thus, as more units are recruited, at least some, and perhaps most, of those already recruited experience a decrease in firing rate.

De Luca, Foley, and Erim (1996) made some interesting observations of this phenomenon for contractions of the tibialis anterior and first dorsal interosseus maintained at various percentages of MVC. Even at relatively short durations before the

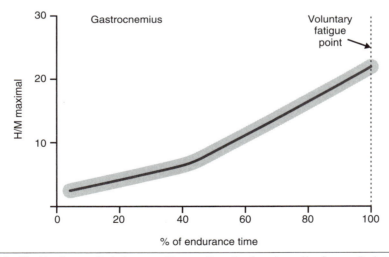

Figure 5.1   The H-reflex amplitude (expressed here as the ratio of maximum H-reflex amplitude to maximum M-wave amplitude) increases in gastrocnemius during a submaximal contraction sustained until fatigue.

Data from Löscher, Cresswell, and Thorstensson 1996b.

onset of significant fatigue (10-20 seconds), motor units began to show a consistent decline in firing rates once the target force was attained. Those recruited at higher thresholds with lower firing rates demonstrated a greater decrease in firing rates than those with lower thresholds and higher firing rates. This was seen even during contractions at levels at which no additional recruitment was found, thus suggesting that force can be maintained early during constant-force contractions in spite of lack of further recruitment and declining firing rates of active motor units. The authors suggested that, at least early during these maintained contractions, lower firing rates were compensated for by contractile changes such as staircase and potentiation, which would tend to increase the force per excitation. However, this effect would last only a minute or so and does not explain the decreasing rates that occur as muscle unit forces drop with fatigue.

As motor unit firing rates decline with fatigue during submaximal contractions, electrical stimulation of the muscle may or may not result in force equal to a voluntarily generated MVC, indicating that the decline in motor unit firing frequency may limit the expression of muscle force to a greater extent than it does during maximal sustained contractions (Löscher, Cresswell, and Thorstensson 1996b; Sacco, Thickbroom, et al. 1997). When the target force can no longer be maintained, not only MVC but also maximum EMG is depressed (Petrofsky and Phillips 1985; Fuglevand et al. 1993; Löscher, Cresswell, and Thorstensson 1996b). This deficit can persist for at least 6 hours following the exercise and probably can last longer (Bentley et al. 2000). The lower the target force and thus the longer the contractile duration, the greater the depressions in maximum EMG and motor unit firing frequencies at the point of fatigue (BiglandRitchie, Furbush, and Woods 1986; Dolmage and Cafarelli 1991; Fuglevand et al. 1993; Sacco, Thickbroom, et al. 1997). The depression of the voluntary EMG at fatigue, when compared with a fresh MVC, is more than that of the M-wave, suggesting that the decrease in EMG is not entirely due to neuromuscular failure, although some neuromuscular failure might be present (Fuglevand et al. 1993; Löscher, Cresswell, and Thorstensson 1996b; Sacco, Thickbroom, et al. 1997). These low firing rates at fatigue do not appear to substantially limit the full expression of available muscle force, however, and thus this phenomenon may constitute a demonstration of muscle wisdom.

If motoneurons decrease in activation during a sustained effort, what is the cause? We have already seen in the previous chapter that fatigue-related metabolites might inhibit motoneurons via afferent activity from the fatiguing muscles. Are there other mechanisms?

Gradually Decreasing Motoneuronal Excitability

When sustained or intermittent current injection is used to excite alpha-motoneurons above their current threshold for rhythmic firing, their firing frequency decreases gradually over several seconds and even minutes. This phenomenon is known as *late adaptation* (see chapter 1 for a discussion of this phenomenon). This intrinsic decrease in motoneuron excitability may possibly play a role in the decreased muscle activation that occurs with sustained or intermittent effort, although this suggestion is difficult to confirm. Late adaptation is not a fatigue of the motoneuron and can be overcome via increased current injection into the motoneuron. A recent review on whether this intrinsic phenomenon of motoneurons plays an important role in voluntary fatiguing contractions is worth consulting (Nordstrom et al. 2007).

Motoneuron excitability is frequently estimated using the H-reflex. This technique involves stimulating the muscle's nerve with an electric shock of a duration and intensity that preferentially excite Ia afferents (which are the largest axons in the peripheral nerve and thus the most excitable). The amplitude of the muscle EMG response that follows the stimulation at monosynaptic latency is used as an estimate of the excitability of the motoneuron pool of the muscle of interest and has been used to estimate changes in motoneuron excitation with fatigue. There are caveats to its use, however (Capaday 1997). For example, H-reflex amplitude changes can occur due to alteration in presynaptic inhibition on Ia terminals, with no change in motoneuron excitability (Nielsen and Kagamihara 1993; Capaday 1997).

Following a maximal sustained effort to fatigue, H-reflex amplitude decreases, suggesting a decrease in motoneuron excitability. Garland and McComas (1990) demonstrated a 50% decrease of the H-reflex amplitude after stimulating soleus muscle at 15 hertz for 10 minutes under ischemic conditions. Duchateau and Hainaut (1993) also found a decreased H-reflex amplitude after both MVC to fatigue and electrical stimulation in first dorsal interosseus and adductor pollicis muscles. Their results were comparable to those of an earlier experiment in which researchers found a similar decrease in the stretch reflex response using a similar fatigue paradigm. This suggests that spindle fatigue is not involved in the decreased H-reflex amplitude (Balestra, Duchateau, and Hainaut 1992). McKay and colleagues (1995) also reported a decrease in H-reflex amplitude immediately following MVC of the ankle dorsiflexors; H-reflex amplitude recovered to control values after 1 minute of rest. Thus, the excitability of the motoneuron pool (which is the cumulative effect of all excitatory and inhibitory influences acting on the motoneurons, any neurohumoral influences, and motoneuron intrinsic properties) decreases with fatigue of the muscle during a maximal effort, whether activation is voluntary or artificial. This decrease is probably not due to late adaptation of the motoneurons involved, since these H-reflex changes persist after fatigue if the muscle is maintained in an ischemic state (Duchateau and Hainaut 1993; Garland and McComas 1990). These findings support the concept that at least part of the decrease in the H-reflex in response to maximal effort is due to a decreased excitatory or increased inhibitory influence emanating from the fatigued muscle itself.

Butler and colleagues (2003) recently provided intriguing evidence that motoneurons may lose excitability during MVCs. These authors stimulated descending motor paths at the cervicomedullary level that contact motoneurons of the elbow flexors and found that the resulting MEP (similar to an M-wave, only stimulating descending tracts above the motoneuron instead of the motoneuron axon) decreased in amplitude during a sustained MVC. This finding indicated reduced motoneuronal excitability. The authors also found that the MEP recovered after the exercise, regardless of whether muscle ischaemia was maintained during the recovery, thus eliminating the primary role of stimulation of Group III and IV muscle afferents in this reduced motoneuronal excitability (see the discussion on increased inhibition of motoneurons later in this chapter).

Muscle afferents include the dynamic and static spindle afferents (Groups Ia and II), afferents arising from Golgi tendon organs (Group Ib), mechanoreceptors (nonspindle Group II and Group III), and nociceptor and metabotrophic receptors (Group III and the unmyelinated Group IV afferents). In view of their excitatory and inhibitory effects on motoneuron excitation, much effort has been expended in attempting to determine whether they are involved in the decreased activation of the neuromuscular system observed during submaximal activation. Some of these ideas were considered in the previous chapter.

There is little evidence that decreased excitability of alpha-motoneurons is a major player in fatigue during sustained submaximal contractions. Löscher, Cresswell, and Thorstensson (1996b) examined H-reflex amplitude at various times during a submaximal (30% of MVC) isometric contraction of plantar flexors maintained to fatigue. Their findings showed a gradual increase in H-reflex amplitude that continued over time until the point of fatigue (see figure 5.1). More recently, Racinais and colleagues (2007) reported evidence that motoneurons become less excitable following 90 minutes of treadmill running. They found that the H-reflex and V-response (H-reflex measured during voluntary contraction) of soleus were both decreased following the exercise.

Decreased Muscle Spindle Excitation

In fresh muscles, spindle activity increases with an increase in voluntary drive, so that new spindles are recruited and those already firing increase their firing rate as the strength of the contraction increases (Bongiovanni and Hagbarth 1990). Since muscle spindle discharge is an excitatory influence on motoneurons, we might expect a decrease in excitation of the motoneurons innervating a fatiguing muscle to accompany a decline in spindle discharge. Several investigators have examined the possible involvement of alterations in spindle afferent discharge in fatigue.

Fusimotor neurons (gamma-motoneurons) innervate intrafusal muscle fibers and cause contraction of their contractile components. Thus, fusimotor neuron discharge constitutes an important component in the response of muscle spindle afferents during movement. Hagbarth and colleagues (1986) showed that tibialis anterior motoneuron firing rates during a maximal effort decrease significantly when nerves are partially blocked with a local anesthetic (gamma-fibers are more sensitive than alpha-fibers are to anesthetics; see figure 5.2). Reduced firing rates in the blocked preparations could

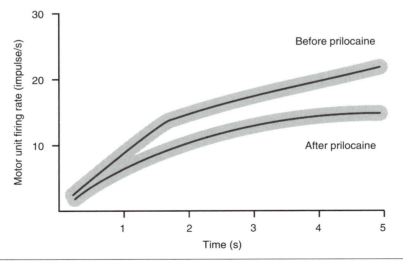

Figure 5.2 Effects of blockade of gamma-motoneurons on motor unit firing in tibialis anterior during MVC: The top line is before blockade, and the bottom line is after partial block of the peroneal nerve via injection of prilocaine around the nerve. Note how firing frequencies are lower after blockade.

Data from Hagbarth et al. 1986.

be increased by stretch of the muscle or by tendon vibration during contraction; this observation suggests that fusimotor neuron activity is important in providing excitatory support to motoneurons during sustained effort. The finding that relatively high vibration frequencies (165 hertz) were particularly effective in countering the effects of nerve block suggests that Type Ia endings, which can be driven up to 200 hertz by tendon vibration (Hagbarth et al. 1986), are more important than Type II endings, which seldom follow vibration frequencies higher than 100 hertz (Gydikov et al. 1976). Furthermore, these authors found that reductions in firing rates were exacerbated by tendon vibration of the antagonist. Subsequent studies (Bongiovanni and Hagbarth 1990) showed that all of the decrease in EMG that occurred with a maximal sustained contraction of the dorsiflexors for 4 minutes could be compensated for by short durations of tendon vibration (figure 5.3).

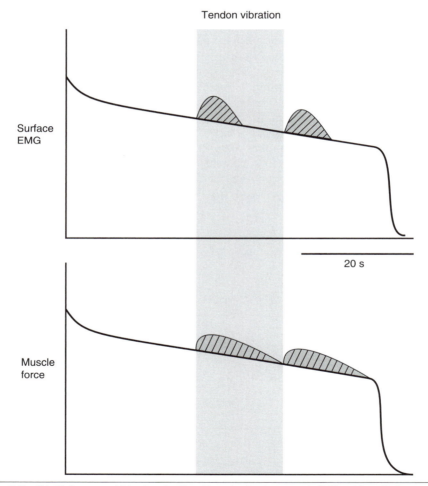

Figure 5.3 Brief (a few seconds) tendon vibration attenuates the decreases in EMG and force that occur during foot dorsiflexor MVC sustained to fatigue. Increases that occur following the cessation of vibration are unexplained.

Data from Bongiovanni and Hagbarth 1990.

Macefield and colleagues (1993) examined the recruitment of dorsiflexor motoneurons deafferentated by pharmacological blockade of the peroneal nerve. They found that firing frequencies at maximal effort were lower (18.6 hertz) when motoneurons were denied afferent excitation than they were when motoneurons were unblocked (28.2 hertz). These findings are consistent with the interpretations of previous investigators. However, they also noted that the firing rates of their deafferentated motoneurons did not drop during the maximal effort as much as intact afferents dropped in experiments with the same muscle and task. This led them to conclude that there are probably at least two mechanisms that limit motoneuron firing frequencies during fatigue: one operating early in the task (declining spindle support) and one operating later (increased inhibitory influence; see figure 5.4).

These results suggest that the fusimotor system provides an important source of motoneuron excitation for ongoing contractile activity during a sustained MVC and that a decrease in the functioning of this system might explain in part the decreased firing rate of motor units observed during fatigue at this intensity. A reduction in the sensitivity of the spindle itself would not explain why H-reflexes are reduced even when afferents are electrically stimulated (Garland and McComas 1990; Duchateau and Hainaut 1993), thereby bypassing the spindle, or why the time course of H-reflex decrease is longer than expected from the rapid decrease in spindle support. A possible effect of fatiguing contractions on fusimotor activation has been proposed (discussed later in this chapter); a decrease of fusimotor activation would decrease spindle activation. Altogether, the current research tells us that a decline in the excitatory influence of muscle spindles during a maximal contraction can explain at least a proportion of the decreased motoneuron activation that occurs as fatigue progresses.

During submaximal sustained contractions, spindle discharge decreases in spite of the increased EMG that results as more motor units are recruited to maintain the

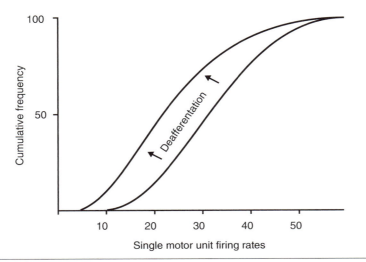

Figure 5.4   Firing behavior of single motor axons in the common peroneal nerve during an MVC in the absence of feedback from the muscle. The latter was achieved via distal pharmacological deafferentation. Notice how firing rates are lower when information from muscle receptors is removed.

Data from Macefield et al. 1993.

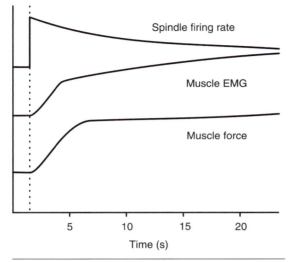

Time (s)

Figure 5.5 Extensor digitorum longus during the first 30 s of a contraction at 60% of MVC. Firing of spindle afferents during constant-load isometric contractions decreases while excitation is still increasing. Tracings are, from top to bottom, firing rate of the spindle ending, muscle EMG, and muscle force.

Based on Macefield 1993.

target force (Macefield et al. 1991; see figure 5.5). Macefield and colleagues (1991) demonstrated this by recording afferent fibers in the peripheral nerve during sustained submaximal contractions of the dorsiflexors. The decrease in firing frequency was quite rapid and occurred even during the initial contraction as force was increasing to the target level. Spindles could be stimulated to higher frequencies by vibration of the tendon and by pressure applied over the spindle receptive field; thus, spindles were not fatiguing. The researchers also found decreased firing frequencies for Golgi tendon organs and cutaneous afferents during the sustained effort and decreased sensitivity to pressure applied to the receptive field (Golgi tendon organ afferents) and to maintained skin stretch (cutaneous afferents). The explanation for these decreases, according to the authors, may include a reduction in fusimotor output to spindles, a gradually reduced responsiveness of the spindles to fusimotor drive (because of axon branch block failure or neuromuscular junction failure at the gamma-motoneuron–intrafusal fiber level), or an adaptation of the fusimotor neuron that might be more pronounced than that of the alpha-motoneurons.

Avela and Komi (1998) asked their subjects to perform maximal leg extension countermovements before and after a marathon competition. They found that the short-latency stretch reflex that occurs during the eccentric phase of the contraction was virtually absent after the marathon. This finding adds support to the contention that prolonged submaximal exercise may decrease spindle support.

These results are similar to those of Hagbarth, Bongiovanni, and Nordin (1995), who examined fatigue effects on finger extensor muscles. They asked their subjects to sustain an isometric contraction at 30% of MVC for as long as possible. During the effort, the experimenters imposed torque pulses and extensor unloading in order to study reflex responses in both extensor (unloading reflex) and flexor (stretch reflex) muscles. They found reflex responses to be less pronounced and to have a longer latency in fatigued muscles. This attests to a reduction in servo control with fatigue.

It does not appear that spindle sensitivity to length change decreases with submaximal fatigue, however, since the EMG response to a standardized tendon tap actually increases following a fatiguing contraction at 50% of MVC (Häkkinen and Komi 1983b; Nelson and Hutton 1985). More recently, Biro and colleagues (2007) confirmed this increased sensitivity of spindle afferents following fatigue and suggested that the increased sensitivity is an attempt to compensate for the declining voluntary drive to motoneurons, which they also found in their fatigued subjects.

Increased Central Inhibition of Motoneurons

Ia input to motoneurons can be regulated via Ia presynaptic inhibition. Presynaptic inhibition involves a depolarization of Ia afferent terminals to motoneurons, brought about by axoaxonal synapses, which reduces the size of the Ia excitatory postsynaptic potential. Presynaptic interneurons are subject to excitatory and inhibitory control from several brain centers. Hultborn, Meunier, Morin, and colleagues (1987) found that presynaptic inhibition influencing a motoneuronal pool decreases at the beginning of a contraction of the implicated muscles and increases in muscles not involved in the contraction. They proposed that at least part of the increase in presynaptic inhibition of the nonrecruited muscle might be due to activation of presynaptic neurons by Group I afferent information coming from the contracting muscle. However, their results indicated at least one supraspinal control mechanism of the presynaptic inhibitory system: the inhibition of motoneurons of the contracting muscle (which are disinhibited upon activation). Most probably, a tonic presynaptic inhibition is in place during the noncontracting state, and a supraspinal center (or centers) selectively removes this inhibition from recruited motoneurons by inhibition of these interneurons (Nielsen and Kagamihara 1993; Hultborn, Meunier, Pierrot-Deseilligny, et al. 1987). A decrease in presynaptic inhibition at the onset of recruitment ensures that sensory information from spindles is received in full fidelity by motoneurons of the activated muscle. There is some evidence, nonetheless, that presynaptic inhibition, which decreases at the beginning of contraction, may increase as the contraction continues, which would involve a gradual reduction in motoneuron excitation (Hultborn, Meunier, Morin, et al. 1987).

A possible role for the presynaptic inhibition system in maximal efforts to fatigue is supplied by the case of cocontraction, or coactivation, of agonists and antagonists. An altered activation of Ia inhibitory interneurons via supraspinal centers may explain the variable degree of agonist and antagonist coactivation that can occur with certain types of tasks. Ia inhibitory neurons appear to be controlled differently during cocontraction than during flexion–extension movements (Nielsen and Kagamihara 1993). During repeated isometric MVCs continued to fatigue, cocontraction of agonist (quadriceps) and antagonist (biceps femoris) muscles increases, as evidenced by a decrease in EMG of the agonist and no change in EMG of the antagonist. Increased coactivation also occurs during fatiguing isokinetic knee extensions at low and high speeds, although coactivation is higher at high speeds, even before fatigue (Weir et al. 1998). One hypothesis is that supraspinal influence reduces the Ia presynaptic inhibition of afferents that convey reciprocal inhibition to antagonists during agonist fatigue, thus increasing coactivation.

Presynaptic inhibition might also explain the phenomenon by which prolonged tendon vibration accentuates the fall in EMG, motor unit firing rates, and force during a sustained MVC in foot dorsiflexors. This is, of course, in contrast to the effects of short durations of tendon vibration on fatiguing muscle (see figure 5.3). Bongiovanni, Hagbarth, and Stjernberg (1990) reported this phenomenon, noting also that this effect developed slowly and persisted for up to 20 seconds after the end of vibration, was accentuated by previous activity, and preferentially affected the ability to drive high-threshold units at high firing rates.

Kukulka, Moore, and Russel (1986) established that reciprocal inhibition, via Renshaw cell activation, might play a role in reducing motoneuron excitability during maximal activation of the soleus muscle. They showed that, following fatigue, the inhibitory effect of a conditioning stimulus (preceding the H-reflex by 10 milliseconds) on the H-reflex of soleus was altered in a way that suggested that reciprocal inhibition was increasing during the effort. Highest-threshold motoneurons contribute several times more to the excitation of Renshaw cells than lowest-threshold motoneurons contribute (Hultborn, Katz, and Mackel 1988; Hultborn et al. 1988). Thus significant reciprocal inhibition is not unexpected when recruiting high-threshold units for these unusually long durations (up to 5 minutes).

Thus, motoneurons may show decreased activation during a sustained maximal contraction due to a number of inhibitory influences. These influences include inhibitory afferents and interneurons as well as inhibition through Ia afferent connections.

Similar mechanisms of motoneuronal inhibition during fatigue have been reported during submaximal contractions. Reciprocal inhibition of the antagonist soleus increased after fatigue of the agonist tibialis anterior produced by a sustained contraction at 30% of MVC or by electrical stimulation (Tsuboi et al. 1995). This was expressed as a depressed H-reflex of the soleus in response to stimulation of the nerve innervating the fatigued tibialis anterior 2 to 4 seconds before.

Aymard and colleagues (1995) fatigued wrist extensors (extensor carpi radialis) in human subjects (via sustained isometric contractions at 50% MVC performed for 1 minute on, 1 minute off) and examined the degree of inhibition that was evoked in an antagonist (flexor carpi radialis, a wrist flexor) and a transjoint synergist (biceps brachii, an elbow flexor) when Group I afferents of the fatigued muscle were stimulated. As in the study of Tsuboi and coworkers (1995), these researchers did this by evoking an H-reflex in the nonfatigued muscles and stimulating the afferents of the fatigued muscles at a latency that would demonstrate the inhibitory effect of the disynaptic inhibitory pathway on H-reflex amplitude. Their results, unlike those of Tsuboi and coworkers (1995), showed a decrease in the inhibitory response of afferent stimulation among synergists and no change between antagonists (i.e., reciprocal inhibition) with fatigue. The best explanation of their results was a selective effect of fatigue on the Ia inhibitory interneurons that govern inhibition between synergists, so that synergist inhibition was reduced.

A centrally mediated common drive between agonist–antagonist pairs results in a variable degree of agonist–antagonist coactivation during voluntary contractions (Psek and Cafarelli 1993; Rothmuller and Cafarelli 1995). The flexibility with which cocontraction can occur suggests that it is modulated by a supraspinal mechanism in which reciprocal inhibitory pathways are suppressed, most likely via modulation of the Ia inhibitory system. Coactivation is particularly evident during strong and fast-displacement contractions (Psek and Cafarelli 1993; Hagbarth, Bongiovanni, and Nordin 1995; Weir et al. 1998). Consistent with this idea is the finding that gradual fatigue of an agonist during a series of submaximal, intermittent, isometric contractions results in an increased agonist EMG as more motor units are recruited to maintain the target force and a gradual and correlated increase in the recruitment of the antagonist (Psek and Cafarelli 1993). During repeated maximal contractions of knee extensors, coactivation gradually increases, even as the EMG of the fatiguing agonist decreases, since the ratio of EMG in agonists to that in antagonists remains relatively constant while the force of the agonists (but not the antagonists, obviously) decreases (Weir et al. 1998). This is an example of a fatigue-induced decrease in an inhibitory system in the spinal cord; Ia reciprocal inhibition of an antagonist decreases. The inhibition may occur via a centrally controlled Ia presynaptic inhibition that releases the inhibitory effect of agonist contraction on its antagonist. Hagbarth, Bongiovanni, and Nordin (1995) feel that at least a part of this decreasing reciprocal inhibition is due to declining Ia afferent inflow into the spinal cord. This idea is supported by the results of Belhaj-Saïf, Fourment, and Maton (1996). They recorded from motor cortical cells in monkeys performing submaximal elbow flexion tasks to fatigue and found a significant number of cells increased in their antagonist facilitatory function.

Such coactivation during rhythmic contractions is expected to attenuate the limb displacement against the load by the agonist as fatigue progresses. This is shown in figure 5.6: An increase in activation of biceps femoris with time tends to oppose the mechanical consequence of activation of the vastus lateralis. This increased coactivation with fatigue could serve as a damping mechanism, however, to reduce the risk of oscillations in the agonist–antagonist pair as fatigue increases (Rothmuller and Cafarelli 1995). Finally, it might serve to limit the time to exhaustion of the agonist and thus reduce major perturbations in metabolism, morphology, or both that might result from more prolonged activation.

Decreasing Excitation From Supraspinal Centers

Our knowledge concerning the possible contribution of decreased excitation of motoneurons from supraspinal sources to neuromuscular fatigue has emanated from studies stimulating the cortex via either transcranial electrical stimulation (TES) or TMS and the recording of the resulting MEP, which is a compound action potential of short latency (Taylor et al. 1996), measured at the muscle. TES excites pyramidal cortical cells at the level of their axons, while TMS stimulates these cells either

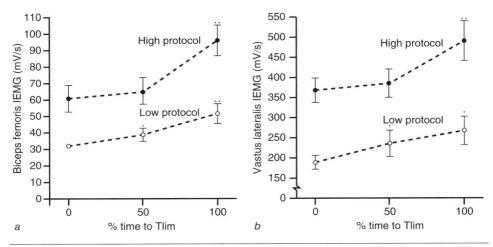

Figure 5.6   Coactivation of agonists and antagonists increases with submaximal contractions maintained to fatigue (100% Tlim). Subjects were asked to generate static, intermittent leg extensions at 30% (LO) or 70% (HI) of MVC until fatigue. *(a)* Biceps femoris integrated EMG increases with time in both protocols. *(b)* Vastus lateralis EMG. The related increases in EMG in agonists and antagonists suggest central control.

Reprinted from Psek and Cafarelli 1993.

TUG-OF-WAR BETWEEN CORTICAL INHIBITION
_____ AND SPINAL EXCITATION? _____

Mills and Thomson (1995) asked subjects to generate a sustained MVC with the first dorsal interosseus muscle and then used TMS to deliver stimuli during the fatigue process. The contraction lasted about 100 seconds, and isometric force decreased to about 25% of the original. The authors reported no change in MEP amplitude (more precisely, they observed a slight increase that was statistically not significant) but did report an increase in the duration of the silent period during fatigue. They concluded that either (1) there was no change in cortical excitability with fatigue or (2) there was change, but the changes in cortical and spinal excitability changed in equal and opposite directions (a possibility that they were not in a position to resolve with their techniques). They attributed the increase in cortical inhibitory influence to peripheral feedback from the fatiguing muscle.

A similar approach was taken by McKay and colleagues (1996), again using fatigue of the ankle dorsiflexors. Their results were similar to those of Mills and Thomson (1995), in that MEP amplitude did not change significantly at fatigue, but the silent period did increase. However, they also reported that the MEP of the contralateral muscle significantly increased, suggesting that cortical excitability may increase during sustained MVCs in an attempt to counter the decreased excitation from the periphery (although this was not measured). Thus, their results might suggest that the lack of major change in MEP of the fatigued muscle during TMS is due to an equal and opposite change in cortical and spinal excitability, as was offered previously as a possible explanation for these results. They also succeeded in demonstrating that the source of change in the silent period was cortical, since the increase in this period was not evident when stimulation was at the level of decussation of the pyramids (near the junction of the medulla and spinal cord).

directly or indirectly through intracortical connections (Taylor et al. 1996; McKay et al. 1996). The basic approach has been to deliver stimuli at submaximal intensity to the cortex in control and voluntary fatigue conditions. A change in the amplitude of the MEP yields information on the general excitability of the cortical cells and spinal cord (increased amplitude = increased excitability), whereas the silent period that immediately follows the stimulation (an interruption of the voluntary EMG lasting approximately 250 milliseconds) yields information on the degree of inhibition present in the cortical area (a longer silent period indicates more inhibitory influence).

Taylor and colleagues (1996) and Gandevia and coworkers (1996) examined fatigue following sustained MVC in the elbow flexors and combined TMS with several other techniques in an attempt to isolate the locus of fatigue. Their results (see figure 5.7) demonstrated an increase in the amplitude of the MEP as well as a longer silent period at fatigue, evidence of a parallel increase in simultaneous excitation and inhibition in the cortex. When ischemia or tendon vibration was administered at the point of fatigue, the recovery of the MEP amplitude or silent period was not influenced, a finding indicating that they were not linked to peripheral events. In addition, similar

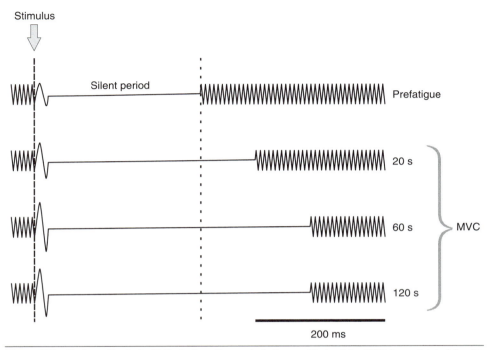

Figure 5.7 During a 2-minute MVC of elbow flexors, both the amplitude of the MEP (immediately following stimulus) and the duration of the silent period evoked by TMS increase.

Data from Taylor et al. 1996.

to previous results, the changes in MEP amplitude were cortical and not spinal, since they were not evident with stimulation of the corticospinal tract at the cervicomedullary junction.

Perhaps more revealing, these studies also demonstrated that cortical stimulation during fatigue resulted in an increased superimposed muscle twitch response, a classic demonstration of central fatigue (figure 5.8). This indicates that, in spite of increasing cortical excitation during the MVC (i.e., increasing MEP amplitude), the cortex was operating at a suboptimal level for maximal force generation. This limitation was not at the level of the motoneuron, since peripheral stimulation also resulted in a larger superimposed contractile response at fatigue. The researchers also noted that when ischemia was imposed at the end of the fatigue protocol, MEP amplitude recovered rapidly, but the superimposed twitch in response to cortical stimulation remained elevated. This means that, in spite of increases during fatigue and the rapid recovery following fatigue, this cortical excitation remained suboptimal for maximal muscle force generation as long as the muscle remained in a fatigued state. The authors proposed that the suboptimal driving of motoneurons was in response to afferent information from muscles, joints, and tendons.

There is a gradual decrease in MEP amplitude after exercise is terminated. Bonato and colleagues (1996) asked subjects to perform abduction–adduction movements of the thumb at a maximum rate for 1 minute and examined the MEP amplitude during

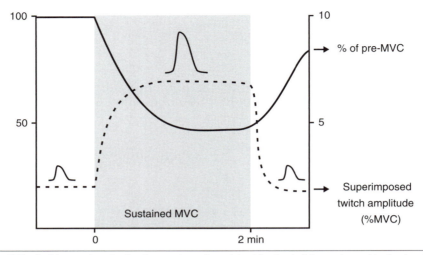

Figure 5.8   In this experiment, stimuli were given alternately at the level of the cortex and brain stem before, during, and after an isometric MVC of the elbow flexors. While the MVC decreased (gray line, left *y*-axis), the amplitude of twitches resulting from cortical stimulation increased (interrupted line, right *y*-axis), indicating involvement of central nervous system mechanisms in fatigue.

Data from Gandevia et al. 1996.

the recovery period. They found that MEP amplitude decreased significantly by 15 to 20 minutes after the exercise for both ipsilateral and contralateral muscles. They also found a contraction of the motor output map of the corresponding cortex and suggested that this was due to inhibitory mechanisms.

More recently, Gandevia and colleagues (1999) showed a possible depression in corticospinal influence on motoneurons that occurs immediately following a 120-second MVC of the elbow flexors. The authors accomplished this by using a stimulation technique that selectively activates these axons (transmastoid electrical stimulation). The time course of the effect is different from the effects of TMS and most likely involves decreased efficacy of corticospinal–motoneuronal synapses.

The results from experiments using cortical and subcortical stimulation during a sustained MVC to fatigue can be summarized as follows: (1) The level of cortical inhibition increases, (2) the level of cortical excitation increases, (3) the cortex appears to drive motoneurons at sufficient intensity to elicit maximal contractile responses in the control condition but perhaps not during fatigue, (4) this central fatigue is related to afferent information from the periphery, (5) decreases in corticospinal stimulation of motoneurons may also occur, and (6) since TMS results in a superimposed contractile response during fatigue, the locus of this central failure is somewhere in the sites driving the cortex. In other words, sites upstream from the motor cortex appear to limit the full expression of contractile force during a sustained MVC.

Changes also occur during submaximal contractions. During a low-level (20% of MVC), maintained, isometric elbow flexion task performed to fatigue, transcortical stimulation reveals a gradually increasing MEP amplitude; this finding indicates an increasing cortex excitability (Sacco, Thickbroom, et al. 1997). Increasing duration of the silent period, indicating increased inhibition, also occurs during the second half of the contractile period, on a different time course from that of the MEP change. The MEP change occurs with TMS (stimulation of pyramidal cells directly and trans-synaptically) and not TES (stimulation of pyramidal cell axons). This observation suggests that inhibition is in the cortex and not the spinal cord.

These results are at odds with those of a previous study on submaximal, sustained contractions of adductor pollicis (Ljubisavljevic et al. 1996). In this study, investigators found a gradual reduction in MEP amplitude and a decrease in silent period as contractions at 60% of MVC continued to the point of fatigue. Interestingly, MEP amplitude and silent period subsequently increased when contractile activity continued past the point of fatigue at a lower force. The biphasic response of MEP amplitude and silent period seen in this study, unlike in the study of Sacco, Thickbroom, and colleagues (1997) cited earlier, may be due to the higher intensity and thus shorter duration of the contractile period, the use of the adductor pollicis (compared with the elbow flexors studied by Sacco, Thickbroom, et al. 1997), or both.

TMS performed following exercise reveals two phenomena: postexercise facilitation (increased MEP amplitude) and depression (decreased amplitude; BrasilNeto, Cohen, and Hallett 1994; McKay et al. 1996; Samii, Wassermann, and Hallett 1997). Samii, Wassermann, and Hallett (1997) found that contraction of extensor carpi radialis at 50% of MVC resulted in a postexercise depression in the MEP only if the exercise was continued until the point of fatigue. For shorter times, at the same intensity, MEP amplitudes were slightly facilitated. These authors concluded that facilitation and depression occurred simultaneously during the contraction.

The experiments of Löscher, Cresswell, and Thorstensson (1996a) were designed to test the idea that central mechanisms, and not muscle afferents, are involved in fatigue during submaximal contractions. They asked subjects to maintain a contraction of the triceps surae at 30% of MVC for as long as possible (about 400 seconds). At the point of fatigue, the subjects stopped contracting and the experimenters continued stimulating the muscle electrically at the 30% torque level at a frequency of 30 hertz for 1 minute. They then asked subjects to generate a maximal contraction. The principal findings were that subjects were capable, after 1 minute of continued stimulation of the fatigued muscle, to resume voluntary torque generation at 30% of MVC for a further 85 seconds, even though the previously fatigued muscle had continued to work via electrical stimulation. The investigators hypothesized that the electrical stimulation after voluntary fatigue, while continuing to fatigue the muscle further, actually allowed the fatigued central mechanisms to rest and thus facilitated the resumption of effort 1 minute later. This evidence seems to speak against using a peripheral source (i.e., effects of muscle state on afferent activity, which would have an inhibitory influence on motoneuron recruitment) to explain the decreased muscle activation observed with fatigue.

During submaximal exercise that is of long enough duration to allow decreased circulating glucose, increased core temperature, or altered circulating hormone levels, central fatigue may contribute more significantly to decreased performance, due to the detrimental effects of these conditions on cortical function (Nybo and Secher 2004). Hyperventilation during strenuous exercise can also blunt the increase in cerebral blood flow, thus hampering oxygen delivery to the brain and contributing to central fatigue (Nybo and Rasmussen 2007).

IS THERE CENTRAL FATIGUE DURING PROLONGED, LOW-FORCE CONTRACTIONS?

Much of the evidence regarding central fatigue at the cortical level has been derived by examining high-force contractions. Sogaard and colleagues (2006) and Smith and colleagues (2007) have used TMS to show that central fatigue occurs even with sustained low-force voluntary contractions of 15% and 5% of MVC, respectively. Their subjects held a submaximal isometric contraction of the biceps brachii for 43 minutes (15%) or 70 minutes (5%). There was evidence of muscle fatigue, in that evoked twitch responses fell. There was also evidence of central fatigue, in that the silent period following TMS increased (which suggests cortical inhibition), as did the superimposed twitch response during periodic MVCs performed during the submaximal task. Thus it appears that during low-intensity, sustained contractions, there is gradual development of an inability to drive the motor cortex optimally during the brief MVCs. The functional implication for how this might influence the low-force contraction is not entirely clear at this point, but it may be related to an enhanced perception of effort and a desire to quit as contraction continues. Perhaps this fatigue mechanism is also at work during intermittent, relatively low-force contractions such as those used for walking, jogging, and swimming. It is worth noting that decreased cortical excitability is present following a marathon run (Ross et al. 2007). The implications of training on this fatigue mechanism have yet to be determined experimentally.

ISOMETRIC VERSUS ANISOMETRIC TASKS

The bulk of the experimental work on fatigue, especially the research necessitating muscle or motor unit recordings during the effort, has used constant-load isometric contractions. The reason for this is obvious: Intramuscular electrode stability during the contractions allows unequivocal identification of the same motor unit over a prolonged duration. Miller and colleagues (1996) investigated changes in firing rates of motor units in the triceps brachii while their subjects performed a slow flexion–extension task to fatigue at a load corresponding to 17% of MVC. The researchers examined the number of spikes per contraction for the identified units recruited at the beginning of the task and found no systematic decrease with fatigue; in fact, this number decreased, increased, or stayed constant for different motor units as fatigue progressed (implying decreased, increased, or constant firing rates as fatigue progressed).

In a more recent study from this same laboratory, Griffin, Ivanova, and Garland (2000) found that motor unit discharge rates decreased when the triceps brachii performed contractions that were isometric but that simulated the torque changes that occur during an anisometric task involving elbow flexion. The implication is that, due to the dynamic nature of this fatigue task, afferent information, blood flow, or both may be different, and thus their effects on motoneuron firing rate during fatigue may also be different. We will have to see more experiments involving dynamic submaximal contractions to fatigue to resolve this issue.

The contributions of central versus peripheral fatigue differ between concentric and isometric fatiguing tasks. For example, when subjects performed three series of 30 maximal voluntary knee extensions, peripheral fatigue occurred before central fatigue was evident (the latter determined via interpolated twitches), while the opposite occurred when contractions were of equal torque but were isometric (Babault et al. 2006).

ROTATION OF MOTOR UNITS?

Given that continued contractile activity (sustained or intermittent) can have such detrimental effects on muscle fiber function, would it not be of some physiological advantage to be able to rotate motor units (recruit additional units to replace previously active units) during prolonged submaximal exercise? Such a strategy would offset fatigue by giving periodic, temporary rests to motor units and thus would improve performance.

There is some evidence that motor unit rotation may occur during contractions maintained at very low intensity. Fallentin, Jorgensen, and Simonsen (1993) demonstrated several examples of motor unit rotation in the biceps brachii muscles of individuals who were asked to maintain an isometric contraction at 10% of MVC for 1.5 to 2 hours. Furthermore, these investigators found that the subjects who were capable of lasting the longest before fatigue were also those who demonstrated motor unit rotation most consistently. Similar results were reported by Tamaki and colleagues (1998), who found, again with contractions at very low intensity (10% of MVC), that there was rotation among the synergists involved in plantar flexion (soleus, medial and lateral gastrocnemius), especially during the latter part of the task to fatigue.

More recent studies on fatigue of the first dorsal interosseus hand muscle revealed that the recruitment of motor units may be quite heterogeneous (as measured from different sites on the surface of the skin over the muscle belly) during a task in which subjects maintain an isometric force of 50% of MVC for as long as possible (Zijdewind, Kernell, and Kukulka 1995). Westgaard and De Luca (1999) demonstrated motor unit substitution in human trapezius muscle during submaximal contractions lasting for 10 minutes. In these studies, several motor units ceased being activated, only to be reactivated later in the session, while the whole-muscle activity and that of other units remained activated at a constant level. Bawa and colleagues (2006) added to this literature by demonstrating rotation of motor units with similar thresholds in muscles of the forearm and distal leg during low-level contractions sustained for more than 30 minutes.

While the concept of rotation of near-threshold motor units seems valid and demonstrated, the mechanisms are less clear. This phenomenon may be made possible by active properties of motoneurons, such as PICs and activation and inactivation of ion currents, that can occur to various degrees among motoneurons during prolonged contractions. Future studies such as these may uncover other neuromuscular strategies like motor unit rotation that we can use, or perhaps learn through training to use, to improve performance.

SUMMARY

There is evidence that failure can occur at a number of sites, from the cortex to the muscle fiber, during exercise. Metabolic changes at the fiber level seem to occur at the same time that changes at other neuromuscular levels occur, such that fatigue is most likely an integrated phenomenon. Neuromuscular fatigue includes a decrease in the activation of the motoneurons, sometimes (but not always) to levels that limit the expression of muscle force. This decrease in net excitation appears to be due to a number of components, including decreased excitatory influence from spindles, increased inhibitory influence from muscle receptors that are sensitive to fatigue substances, and an altered supraspinal signal. The latter appears to involve an increased inhibitory influence as well as an altered signal coming into the primary motor cortex. It is difficult to determine the relative importance of each of these components when comparing submaximal and maximal efforts maintained to fatigue. However, we can venture a guess that muscle receptor activation by fatigue substances is of lesser significance and spindle support is of greater significance during longer efforts to fatigue. The role of supraspinal influences during short-term versus long-term neuromuscular fatigue requires more experimentation, since most of our evidence at present is for short-term efforts.

Muscular Mechanisms in Aerobic Endurance Training

Most of what we know about endurance in general, and about the effects of aerobic endurance training in particular, concerns cardiovascular, respiratory, and muscular adaptations. Information regarding the latter has proliferated especially since the 1960s, due to the evolution of knowledge and techniques in histochemistry, biochemistry, and molecular biology. We are now beginning to have a clear picture of the muscle phenotypic adaptations that characterize aerobic-type endurance training, although the signals that promote the adaptations and the molecular mechanisms by which these adaptations occur remain elusive.

Muscle adapts to stresses such as those imposed by regular exercise by altering its various properties to respond to stress with better performance, less fatigue, and less damage. To do this, muscle changes its protein synthesis profile, upregulating the net production of some proteins and downregulating the production of others. The result is a gradual change in muscle fiber properties.

Muscle fibers possess a myriad of metabolic systems designed to allow for this adaptation. These systems require sensing mechanisms (so the fiber knows that it has undergone an exercise bout), amplification systems (via diverging and often redundant metabolic pathways that result in events such as phosphorylation of other proteins), and effector mechanisms (a resultant change in net protein synthesis). The effector mechanisms involve the nuclei that are present throughout the length of the muscle fiber and include changes in the transcription of some of the 30,000 genes that make up the genome and posttranscriptional mechanisms leading to altered protein synthesis. The purpose of this chapter is to summarize our current knowledge regarding these events and how these events relate to exercising and training the neuromuscular system. With this knowledge, we will be in a better position to understand neuromuscular training responses when discussing them in later chapters.

This chapter emphasizes the adaptations to aerobic-type endurance training that occur in the muscle. The important changes that occur at the neuromuscular junction, at the level of the motoneuron, in the spinal cord, and in the central nervous system that promote the development of neuromuscular endurance are considered in the following chapter.

CHRONIC MUSCLE STIMULATION

Training responses are complex at the level of the whole muscle. One reason for this is the complexity of the exercise being performed during the training—we obtained a glimpse of this complexity in the previous chapters. Which fibers are being recruited to generate what percentage of their maximal force? Are some fibers more overloaded than others are? (Our knowledge from previous chapters tells us that they are.) Do all fibers respond equally to the same degree of stress? Studies with human subjects often measure biochemical muscle responses using the muscle biopsy technique. A biopsy sample taken from vastus lateralis or gastrocnemius is a mixture of fibers of different types and, if taken from trained subjects, of different training states. From the fibers in the sample, we obtain an average of the biochemical or molecular biological property being assayed. This is, of course, assuming that the biopsy technique provides us with a representative sample of the muscle, which is open to argument (see Lexell et al. 1983).

Chronic electrical stimulation of muscles has been used extensively during the past 30 years to demonstrate the limits of adaptability of muscles to increased activity. All indications are that the changes reported with this model may represent the ultimate state of high-intensity and high-volume aerobic training—a state that could never be achieved via voluntary activation. The results from this model have been very instructive with regard to the extent and relative time course of muscle adaptations that occur, against which we can measure changes induced by aerobic-type endurance training. Perhaps more importantly, this model has provided us with information regarding the signals that promote the adaptive changes and regarding the mechanisms of protein metabolism involved (such as gene transcription rates, translation capacity, and post-translational modifications). This information about protein metabolism especially could not always be provided by the complicated model of voluntary exercise, although for the most part, changes noted after endurance exercise are in the same direction as those reported for electrical stimulation.

For these reasons, considerable attention has been paid to chronic electrical stimulation as an extreme model of muscle adaptation to aerobic-type endurance training. Table 6.1 lists adaptations to chronic electrical stimulation; most of these adaptations have been reported in subjects undergoing endurance training or as differences between endurance-trained and untrained subjects, although the magnitude of the adaptations is less than that observed with electrical stimulation, as is expected.

In this chapter, we will discuss the results emanating from studies using chronic neuromuscular electrical stimulation conducted with animals. The changes reported

HOW HARD DO YOUR LEGS WORK
———— DURING AEROBIC ENDURANCE EXERCISE?————

In order to investigate the effects of acute and chronic exercise interventions in humans, we must feel confident that we are exposing all subjects in our experimental cohort to the same exercise stimulus. But how can we normalize a stimulus to different individuals? Do we normalize workload as a percentage of maximal aerobic power (% $\dot{V}O_2$max)? Do we give everyone the same absolute exercise workload? This question was addressed experimentally by the researchers at Manchester Metropolitan University in the United Kingdom (McPhee et al. 2008). They asked subjects to perform a two-leg cycle ergometer test to determine $\dot{V}O_2$max. The two-leg tests were followed on separate days by one-leg tests to determine the $\dot{V}O_2$peak for each leg. The results showed that there was considerable variability among subjects in the relationship between one-leg $\dot{V}O_2$peak and $\dot{V}O_2$max, such that the ratio of these two values ranged from 0.58 to 0.96. Most of this variation among subjects was due to differences in single-leg $\dot{V}O_2$peak and not due to differences in $\dot{V}O_2$max, and this variation was strongly related to leg muscle volume. The investigators demonstrated that for two individuals given an exercise intensity of 70% $\dot{V}O_2$max, one individual would exercise each leg at 36% of the $\dot{V}O_2$peak for the legs, while another subject would exercise at 60%. Thus, when given the same percent of $\dot{V}O_2$max, these two individuals would subject their leg muscles to stresses of widely differing intensities and perhaps of different types. These results may explain some of the variability among subjects in responsiveness to *the same training stimulus* that has been referred to frequently in the research literature. These results also provide strong implications for revisiting guidelines for training that are based on measures of whole-body aerobic power.

with this model are in the same direction as the adaptations that have been reported with aerobic-type endurance training, the most evident of which are changes in calcium regulatory proteins, increased activities of mitochondrial enzymes, and changes in the proportions of fiber types from Type IIb toward Type I. In the analysis of muscle changes induced by electrical stimulation, however, we must keep in mind the following differences between electrical stimulation and whole-body endurance training.

1. Chronic electrical stimulation typically involves stimulation durations of at least 8 hours and often up to 24 hours per day, a duration obviously longer than that of daily endurance training, even for highly trained athletes.

2. Chronic stimulation experiments often do not include rests. Resting is most likely instrumental in determining the final phenotype in response to whole-body endurance training. Resting allows for some recovery from fatigue and thus allows for a higher tension-time index, whereas tension during continuous stimulation falls early during the stimulation and may recover relatively little or not at all (Hicks et al. 1997; Green, Düsterhöft, et al. 1992). In addition, resting might allow changes in protein synthesis that are instrumental in determining the final phenotype. The levels of mRNA most likely decrease during contractile activity, with increased transcription and

translation taking place once this activity has ceased (Cameron-Smith 2002). In this chapter, whenever we refer to altered mRNA levels resulting from muscle contractile activity, we refer to analysis of samples taken from resting muscles.

3. Chronic stimulation does not produce the patterns of impulse activity that a muscle fiber experiences during whole-body endurance training. For example, recruitment, firing frequency, pattern of firing, and number of impulses per burst should be quite variable among motor units within a muscle, depending on the threshold and the type of motor unit. This activity should also change as fatigue occurs. With chronic stimulation, the same pattern is imposed on all motor units, and the pattern does not change with fatigue.

4. Finally, chronic electrical stimulation does not involve voluntary recruitment of motor units. While the significance of this is not fully known at present, it may be that voluntary recruitment involves changes in structure and function at the spinal cord and supraspinal level that are significant contributors to aerobic endurance performance (as we will discuss in the following chapter). Thus, aerobic endurance training might involve muscular changes that are less marked than, and perhaps qualitatively different from, those seen in chronic stimulation, partly because the nervous system becomes trained to alter the way muscle adaptations are translated into better performance.

Rather than summarize the adaptations that have been reported to occur with chronic electrical stimulation, which have been the focus of several exhaustive reviews (Pette and Staron 1997; Pette 1998; Pette and Düsterhöft 1992), I have summarized lessons that we have learned from this technique as they most likely apply to aerobic-type endurance training. This is the focus of the following section. The phenotypic muscle changes that result from chronic electrical stimulation, and also from endurance training, are summarized in table 6.1.

COORDINATION OF MUSCLE PROTEIN SYSTEMS

Many functional properties of the muscle fiber are determined by several proteins working in concert. When these proteins adapt to increased activity, by increasing in content or by changing in isoform, the most efficient adaptive response is one in which all proteins adapt together. Teleologically, it would not be ideal for a fiber to alter its MHC characteristics from fast to slow contracting without also changing its myosin light chains or the functional machinery involved in the regulation of calcium levels by the sarcoplasmic reticulum. This is a big order, since the constituent proteins of a system often vary in size, turnover rate, subunit structure, and three-dimensional complexity, and controlling these proteins requires the control of several steps in the protein synthesis and degradation machinery (see figure 6.1). The coordinated expression of many proteins simultaneously is partially accomplished by the effects of common transcription factors and metabolic signals that promote the expression of several genes.

A clear example of this is the coordinated expression of the proteins associated with Type I fibers, including heavy chains, light chains, and thin filament proteins, that occurs relatively late in the adaptation process (Pette and Staron 1997). Another example involves the coordinated changes in components of the calcium regulatory

Table 6.1 Changes in Muscle Phenotype During Chronic, Low-Frequency Electrical Stimulation

Early	Middle	Late
Increased		
MAPK activation	MHC IIx and IIa	MHC I
Hsp70, Hsp60	Mitochondrial volume density	
AMPKα2	Oxidative enzymes	Troponin T 1s/2s
Total RNA	Sarcoplasmic reticulum calcium uptake Na$^+$/K$^+$ ATPase	Myosin light chains (MLCs)
Hexokinase II	SERCA2a	Troponin I slow
Capillary density	Troponin T 3f/4f	Troponin C slow
Protein synthesis	MLC 1f/3f	MLC 2s/2f
VEGF	Ubiquitin proteasome system PPARα PLB Fatigue resistance Twitch contraction time Twitch half-relaxation time	% of Type I fibers % of hybrid fibers (I/IIa, IIa/IIx)
Decreased		
T-tubular volume	MHC IIb	MLC fast
Sarcoplasmic reticulum calcium uptake	Parvalbumin	MHC IIx/a
Calcium-ATPase	Glycolytic enzymes SERCA1a Calsequestrin DHPR content Ryanodine receptor content Triadin content Troponin T 1f/2f	Oxidative enzymes Troponin T 1f/4f Troponin I fast Troponin C fast Muscle mass Muscle force

Early, middle, and late periods of adaptation refer (approximately) to the time up until week 1, to the time from week 1 to week 6, and to the time after week 6, respectively.

membrane proteins of the sarcoplasmic reticulum and their functionally related proteins, including the dihydropyridine receptor (DHPR), the ryanodine release channel, triadin, sarcalumenin, and calsequestrin, early during chronic stimulation (Ohlendieck et al. 1999). Part of this coordination most likely results from common signals acting to promote the expression of the several genes that code for the proteins of a given system. In other cases, translational and posttranslational modifications may operate to ensure that protein changes take place in a coordinated fashion—later sections describe examples for parvalbumin, sarco/endoplasmic reticulum calcium-ATPase (SERCA) proteins, and MHC IIb.

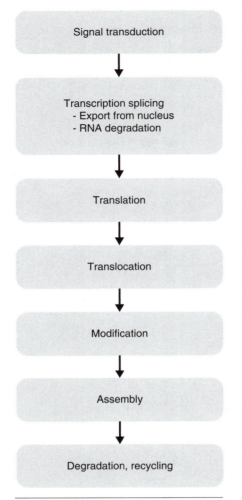

Figure 6.1 Schematic of the control of gene expression. Various extracellular signals initiate changes at the transcriptional, translational, and posttranslational levels, resulting in changes in protein contents.

PRETRANSLATIONAL CONTROL

Pretranslational control is the control of protein synthesis by alteration of the abundance of mRNA. Mechanisms that manipulate mRNA abundance include increased transcription, altered mRNA processing, and increased mRNA stability (Booth et al. 1998; Fluck and Hoppeler 2003).

Changes in the concentrations of specific mRNAs in response to a single bout of exercise can occur after a few contractions or can occur up to several hours after exercise cessation. Each mRNA has a characteristic concentration and time envelope that determines translation of the protein. For most genes that have been studied, we do not know the time course of exercise-induced changes in mRNA; instead, we usually have a single measurement taken at a discrete time point at the end of exercise. It is likely, however, that the most significant change in mRNA concentrations occurs during recovery (Cameron-Smith 2002). In muscles that have been chronically stimulated for several weeks, mRNA levels generally reflect the concentration of their corresponding proteins (Pluskal and Sreter 1983; Hu et al. 1995; Brownson et al. 1988; Jaschinski et al. 1998; Fluck and Hoppeler 2003). This observation indicates the importance of pretranslational mechanisms and is demonstrated in figure 6.2 for the proteins SERCA1a and phospholamban (PLB; from the work of Hu et al. 1995) and in figure 6.3 for MHCs (from the work of Jaschinski et al. 1998). In chronically stimulated rat muscle, mRNA for MHC IIb is reduced within 1 day of initiation of stimulation and increases rapidly after stimulation cessation. At the same time, transcription of the IIa gene increases and decreases accordingly (Kirschbaum et al. 1990). Similarly, citrate synthase mRNA increases more than twofold after 3 days of stimulation of rabbit tibialis anterior (Annex et al. 1991). This shows that pretranslational processes are influenced early in adaptation. In stimulated rabbit muscles, the changes in mRNAs for MHCs reflect changes in the corresponding proteins after 3 weeks of stimulation and 12 days of recovery (Brownson et al. 1992).

This information is more readily forthcoming from stimulation studies than from exercise studies because in the former the stimulus is more continuous, whereas in the latter the stimulus is short lived and intermittent. The mRNA level in both types of study depends on, among other factors, the intensity of the stimulus, the time relative to the stimulus at which the mRNA is measured, and the half-life of the transcript. The stimulus is more constant in electrical stimulation studies, and so these studies allow us to compare mRNA with protein, as is done in figures 6.2 to 6.5. Such comparisons are more complex with exercise training, during which mRNA changes occur in pulses and with various time courses of increase and decrease (Williams and Neufer 1996).

The abundance of mRNA also depends on its half-life, which can be influenced by a number of factors, including nutrient levels, cytokines, hormones, temperature fluctuations, and even viral infections (Fluck and Hoppeler 2003). During the first few days of chronic electrical stimulation, there is some evidence of an increase in mRNA stability, at least for certain proteins (Freyssenet et al. 1999).

These events apply to all nuclei present throughout the length of the muscle fiber. There is evidence, however, that myonuclei are compartmentalized, in that nuclei located under the nerve terminal express some proteins involved in synaptic function (such as the alpha-subunit of the ACh receptor and alpha-dystrobrevin) that are

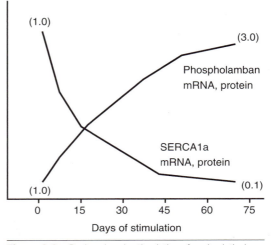

Figure 6.2 During chronic stimulation of canine latissimus dorsi at 1 Hz, the time courses for the change in protein and its corresponding mRNA are similar for SERCA1a, which decreases to 10% of original (i.e., from 1.0 to 0.1), and PLB, which increases threefold (from 1.0 to 3.0).

Data from Hu et al. 1995.

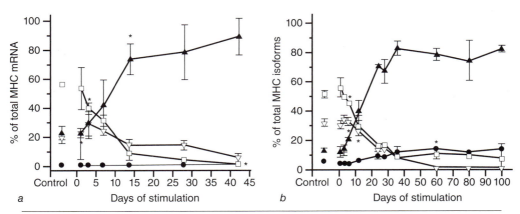

Figure 6.3 Changes in *(a)* MHC mRNAs and *(b)* corresponding proteins during chronic electrical stimulation (10 Hz for 10 h/d) of rat extensor digitorum longus. Symbols indicate IIb (open squares), IIx (open triangles), IIa (filled triangles), and I (filled circles) MHCs. Note the differences in timescale.

Reprinted by permission from Jaschinski et al. 1998.

not expressed by other nuclei in the same fiber (Cameron-Smith 2002). These special subsynaptic nuclei may therefore be under the influence of substances secreted by the motor nerve terminal, although the nature of this influence is currently unknown. There is also some (although not abundant) evidence that proximal and distal fiber ends can express proteins differently, producing fibers that are not homogeneous throughout their length.

An interesting case to consider involves the adaptations that occur in mitochondria in response to aerobic-type endurance training. While some mitochondrial proteins are coded on mitochondrial DNA, most are coded on nuclear DNA. With endurance training, mitochondria proliferate, and along with this, mitochondrial DNA increases in content. Thus, the increase in the synthesis of mitochondrial proteins that occurs as training progresses involves increased transcriptional activity for proteins coded by nuclear DNA. For proteins coded by mitochondrial DNA, however, the ratio of mRNA to DNA stays relatively constant, and the higher mRNA levels reflect higher amounts of mitochondrial DNA (Fluck and Hoppeler 2003). So, we have a situation in which increased mRNA can mean there is an increased rate of transcription or an increased amount of DNA.

HYPOXIC TRAINING
TO MAXIMIZE MUSCLE RESPONSES?

Many athletes use the training method of live low, train high (which refers to living at sea level and training at altitudes 2,500 meters and above) to maximize training responses and prepare for competition at altitude. Although it would seem obvious that the extra metabolic stress of training at altitude is beneficial for athletes, even athletes performing at sea level, just how beneficial is it? Vogt and Hoppeler (2010) at the University of Bern at Switzerland attempted to answer that question by reviewing the available literature on the subject. One critical message from their work is that the variability of experimental conditions in the available research literature, including variations in altitude, training duration, and beginning fitness level of the subjects, prevents us from arriving at definitive conclusions as to the benefits of this practice. The unique profile of altered gene expression in muscle resulting from exercise at hypoxia, and not normoxia, indicates that hypoxia alone has a unique stimulatory effect on the expression of several genes associated with improved metabolism and performance. This effect is probably mediated via the induction of hypoxia-inducible factor-1 (HIF-1). This factor in turn is involved in upregulating the expression of proteins such as those involved in glycolysis, pH regulation, and angiogenesis. With respect to aerobic endurance performance, there is evidence that performance at the training altitude is improved, although benefits for improvements in performance at sea level are equivocal. The authors suggested that the decision to live low and train high is an important one, and factors to consider are the initial health and fitness of the subjects, the application of appropriate nutrition (considering the carbohydrate and vitamin needs of the athlete), and the reduction of the normoxic training.

TRANSLATIONAL CONTROL

Changes in translational control of protein synthesis involve changes in the amount of protein synthesized per unit of mRNA (Booth et al. 1998). This control is exerted by influences on the processes of translation: initiation, elongation of the nascent peptide, and termination. Factors that may influence these processes include adenine and guanidine nucleotide levels, nicotinamide adenine dinucleotide phosphate ($NADP^+$ and NADPH) levels, or receptor-binding and mitogen-activated protein kinase (MAPK) signaling systems (Thomason 1998).

One example of changes in translational control concerns the increased translational capacity that occurs early in adaptation. In stimulated rabbit fast muscles, for example, total RNA, which is mostly ribosomal RNA, increases up to sevenfold during the first 2 weeks of stimulation (Takahashi et al. 1998; Neufer et al. 1996; Brownson et al. 1988). Such a change tends to favor increased protein production at the same mRNA level. During the initial days of electrical stimulation of rabbit tibialis anterior, citrate synthase increases faster than its mRNA increases, leading to the conclusion that an enhanced translation of the existing messenger occurred (Pette and Vrbova 1992).

Hu and colleagues (1995) demonstrated another example of translational control and its consequences on muscle phenotype (see figure 6.4). In chronically stimulated dog latissimus dorsi, the increased concentrations of SERCA2a and PLB follow time courses that are almost identical, while their mRNA concentrations do not. Specifically, the SERCA2a mRNA was almost finished increasing by 14 days, while the protein level increased continually between 7 and 42 days. This was in contrast to all other proteins examined in this study, the levels of which followed mRNA levels very closely (see figure 6.2). This finding suggests that, under certain circumstances, translation may be altered in order to coordinate the expression of two proteins whose functions

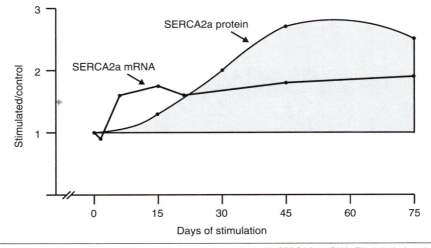

Figure 6.4 In chronically stimulated (1 Hz) canine latissimus dorsi, SERCA2a mRNA (filled circles) and protein (shaded area) show different time courses. Thus, SERCA2a protein levels are subject to posttranscriptional control.

Data from Hu et al. 1995.

are closely linked, as is the case with SERCA2a and PLB. The authors presented the possibility that this control may take the form of PLB regulating the translation of the SERCA2a transcript. This would be a means of ensuring the coordinated production of two gene products whose transcription rates differ.

POSTTRANSLATIONAL MODIFICATIONS

Posttranslational modifications are any modifications of the synthesized protein. They can include phosphorylation, subunit assembly, transport, or degradation (Booth et al. 1998).

A difference between the synthesis rate of a protein (measured using radioactively labeled precursors) and its incorporation into the fiber as a functional component indicates that posttranslational mechanisms are involved. For example, Termin and Pette (1992) noted that the rate of incorporation of labeled methionine into MHCs IIx and IIa, measured in vitro, increases faster than the increase in the amount of the corresponding muscle proteins during the initial 15 days of chronic stimulation of rat tibialis anterior (see figure 6.5). Their interpretation was that synthesis exceeded the rate at which these heavy chains replaced the IIb heavy chains that were being removed (which have a half-life of around 15 days) and thus these newly-synthesized heavy chains degraded at an increased rate, resulting in their increased turnover.

A similar situation was noted for the decrease in parvalbumin that occurs during the initial phase of stimulation-induced adaptation in rat tibialis anterior and extensor digitorum longus (Huber and Pette 1996). The rapid decline in parvalbumin mRNA (half-life of 1-2 days) and parvalbumin synthesis (half-life of 2-3 days) is not paralleled by a concomitant decrease in protein content (which does not decrease appreciably

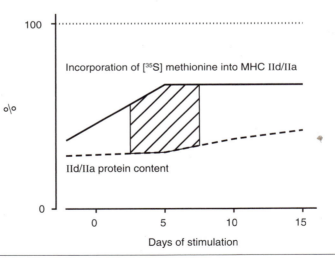

Figure 6.5   Chronic electrical stimulation (10 Hz for 10 h/d) of rat tibialis anterior. Incorporation of labeled methionine into MHC isoforms IIx and IIa (IId and IIa in figure; solid line) significantly increases after 2 d, whereas increased protein content (dashed line) is not apparent until 8 d following the initiation of stimulation. Diagonal lines show where methionine incorporation is significantly elevated but protein content is not.

Data from Termin and Pette 1992.

before 7 days; see figure 6.6). This suggests once again that protein degradation rate is altered (in this case decreased), perhaps to allow coordination of the parvalbumin decrease with the decrease in the content of MHC IIb.

One way to influence posttranslational modifications is the ubiquitin proteasome system. This system is the principal protein degradation mechanism in muscle fibers and can be activated to provide posttranslational modification of target protein levels. Ubiquitin serves as the marker for proteins destined for degradation, while degradation is performed by the activated proteasome complex. Ordway and colleagues (2000) showed that chronic electrical stimulation of rabbit tibialis anterior for up to 28 days results in a gradual increase in 20S proteasome as well as two of its regulatory proteins, PA700 and PA28. This increase is accompanied by a corresponding increase in the mRNA of a representative proteasome subunit.

Expression of the cytosolic heat shock protein Hsp70 is induced rapidly within the first day of the beginning of chronic stimulation at 6 to 10 hertz. Hsp70 remains elevated in the muscle for up to 21 days of stimulation (Ornatsky, Connor, and Hood 1995; Neufer et al. 1996). This increase in protein can occur without a corresponding increase in mRNA, thus implying increased protein stability as a posttranslational strategy. Hsp70 is one of several molecular chaperones that are involved in the binding of polypeptides on ribosomes and in the folding, stabilization, and trafficking of various proteins (Locke and Noble 1995; Ornatsky, Connor, and Hood 1995).

Along the same lines, precursor proteins are imported into the mitochondrial matrix of rat hind-limb muscles at higher rates after the first 2 weeks of chronic stimulation (Takahashi et al. 1998). This increase occurs as a result of modifications in the expression of components that are important parts of the import machinery.

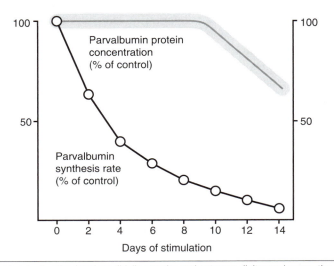

Figure 6.6　In chronically stimulated rat tibialis anterior and extensor digitorum longus, the decrease in parvalbumin concentration is slower than the decrease in the relative synthesis rate. This finding implicates posttranscriptional mechanisms.

Data from Termin and Pette 1992.

SIMULTANEOUS EXPRESSION OF ISOFORMS

The simultaneous expression of several MHC isoforms is unusual in control muscle fibers but not in adapting muscle fibers. Jacobs-El, Ashley, and Russell (1993) demonstrated the presence of mRNAs for IIx and I MHCs in 22% of rabbit tibialis anterior fibers 7 days after the beginning of stimulation, whereas there were only 3% of such fibers in controls. This observation was confirmed by Peuker, Conjard, and Pette (1998) for stimulated rabbit tibialis anterior, in which more than 50% of the fibers demonstrated mRNAs for 3 or 4 MHC isoforms simultaneously. This has also been demonstrated for expression of isoforms of the sarcoplasmic reticulum ATPase (Leberer et al. 1989).

Evidence for the coexistence of isoforms, as opposed to coexpression, is also more abundant for chronically stimulated fibers (Peuker, Conjard, and Pette 1998; Staron and Pette 1987; Williams et al. 1986) and fibers of humans undergoing aerobic endurance training (Klitgaard et al. 1990; Andersen and Schiaffino 1997). However, given the relatively long half-life of MHCs (about 2 weeks), we might expect a rather extended coexistence of several individual isoforms in the same fiber, in spite of the expression of only one.

ADAPTATIONS CAN OCCUR EX VIVO

Some of the adaptations to chronic electrical stimulation seen in normally innervated muscles in vivo can be reproduced quite closely in denervated muscles (Windisch et al. 1998; Schiaffino and Reggiani 1996; Kilgour, Gariepy, and Rehel 1991) and in culture (BartonDavis, LaFramboise, and Kushmerick 1996). Thus, the adaptive neuromuscular responses to aerobic-type endurance training do not necessarily require intact innervation (which is missing in denervated muscles and muscle cell cultures) or the in vivo environment with all the factors that are involved (body temperature, hormones, circulating factors). Since many of the adaptations to endurance training seem to be common to all models, it is also clear that muscle does not have to be voluntarily activated for training adaptations to occur, although simultaneous adaptations in the nervous system with endurance training via voluntary movements may be significant in enhancing performance.

ADAPTATIONS APPEAR IN A SPECIFIC SEQUENCE

The proteins that handle calcium are among the first to be altered during chronic electrical stimulation. Changes include rapid decreases in the calcium ATPase of sarcoplasmic reticulum (during the first day); decreases in the proteins of the junctional triad membrane (by 10 days); and decreases in parvalbumin, calsequestrin, and SERCA1a (by 10-15 days). These alterations and the corresponding effects on cellular calcium levels are accompanied by rapid changes in the muscle twitch time course, which becomes longer. The changes in cellular calcium that accompany these adaptations may provide one of the metabolic signals triggering subsequent adaptations, as is discussed later (Heilmann and Pette 1979; Huber and Pette 1996; Hicks et al. 1997; Carroll, Nicotera, and Pette 1999).

The membrane enzyme Na^+/K^+ ATPase also increases quickly in chronically stimulated muscle (by 5 days). Green, Ball-Burnett, and colleagues (1992) showed that this change precedes the increased concentration of the mitochondrial enzyme citrate synthase in the same muscles (which increases significantly only after 10 days).

In stimulated rabbit tibialis anterior, the appearance of Type IIx MHC mRNA precedes that of Type I MHC mRNA during the first 21 days. During this time, progress from expression of Type IIx to Type I appears to follow increases in fiber oxidative capacity, giving the impression that a certain oxidative potential is a necessary prerequisite for fiber adaptations to continue (JacobsEl, Ashley, and Russell 1993). The proposed transition of MHC isoforms in the sequence IIb to IIx to IIa to I (Pette and Düsterhöft 1992; Windisch et al. 1998; Skorjanc, Traub, and Pette 1998), which was based on combinations of isoforms in fibers of stimulated muscles, has been confirmed at the mRNA level (Jaschinski et al. 1998; Peuker, Conjard, and Pette 1998; see figure 6.3).

It appears that it is more difficult to evoke the Type II to I fiber transformation with chronic electrical stimulation in the rat than it is in the rabbit (Jaschinski et al. 1998; Peuker, Conjard, and Pette 1998). However, changes among the Type II fiber types are similar in these species, once again supporting the concept of a sequential transformation of MHC from IIb to I through IIx and IIa.

In fact, changes in oxidative enzyme activities may follow the changes in myosin isoforms and may be related to the energy demands imposed on the fiber because of these isoforms. During long-term stimulation, oxidative enzymes increase and then subsequently decline to levels that are still above those of control muscles (Henriksson et al. 1986; see figure 6.7). In rabbit muscles stimulated for 10 months at 2.5 hertz, oxidative enzyme (citrate synthase, succinate dehydrogenase, hydroxyacyl-CoA

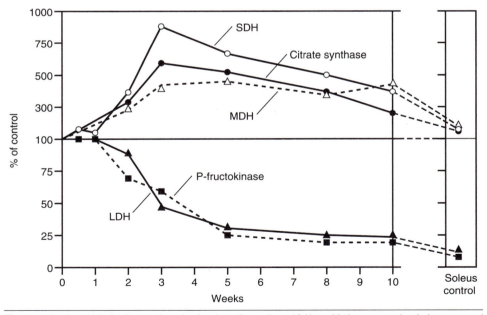

Figure 6.7   In rabbit tibialis anterior stimulated continuously at 10 Hz, oxidative enzyme levels increase and then gradually decrease to levels that are still higher than levels observed in controls. Glycolytic enzyme activities decrease to plateau levels similar to those seen in slow soleus.

Reprinted by permission from Henriksson et al. 1986.

dehydrogenase) activities are higher than they are in muscles stimulated at 10 hertz. In addition, muscles stimulated at 5 hertz have higher or lower activities of these enzymes depending on the degree of myosin isoform changes that have taken place (higher levels of oxidative enzymes are found in muscles with less transformation from Type II to Type I). These results have led us to think that ATP demands drive mitochondrial proliferation in this model and that a shift in myosin isoform to the less costly slow type results in a late decrease in oxidative potential that occurs after the initial increase (Sutherland et al. 1998; see figure 6.8).

THRESHOLDS OF ACTIVITY FOR ADAPTATION

As previously mentioned, the threshold for alterations in MHCs may involve an increase in oxidative capacity (JacobsEl, Ashley, and Russell 1993). Sutherland and colleagues (1998) stimulated rabbit tibialis anterior for 10 months at 2.5, 5, or 10 hertz in different groups of animals. Several of their findings showed quite clearly that thresholds of activity exist for the adaptation of many properties. For example, while the changes in myosin isoforms and contractile speed from fast to slow were extensive in the 10-hertz group and almost absent in the 2.5-hertz group, the 5-hertz group did not show intermediate values; some resembled animals in the 2.5-hertz group while

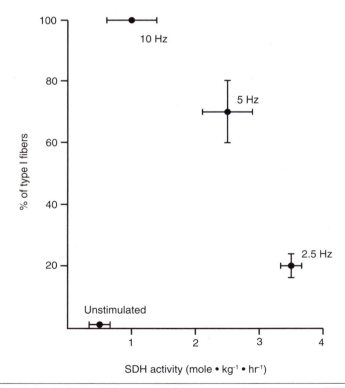

Figure 6.8 Responses of mitochondrial enzymes (in this case succinic dehydrogenase, SDH) in rabbit tibialis anterior to 10 months of chronic stimulation depend on the frequency of stimulation. The more transformation from Type II to I, the lower the oxidative enzyme level.

Data from Sutherland et al. 1998.

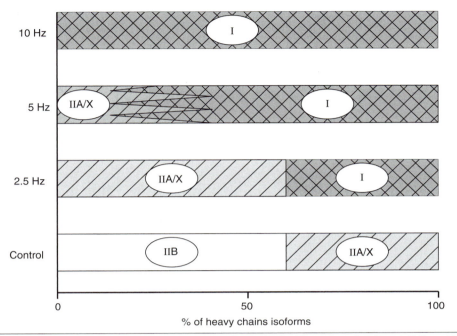

Figure 6.9 Effect of chronic stimulation of rabbit tibialis anterior for 10 months at 2.5, 5, and 10 Hz on MHC isoforms. Note the large variability in the 5 Hz group (indicated by the horizontal jagged lines). This variability suggests there is a threshold activation level for transformation to occur.

Data from Sutherland et al. 1998.

others resembled animals in the 10-hertz group (see figure 6.9). In addition, there were very few differences between the muscles stimulated at 2.5 hertz for 10 months and the muscles stimulated at 2.5 hertz for 10 weeks, signifying that the muscles had stabilized at a level of adaptation corresponding to the activity level imposed.

CHRONIC STIMULATION AND ATROPHY

One of the phenomena occurring with chronic stimulation that does not seem to occur with aerobic-type endurance training in humans is muscle atrophy. The atrophic effect of chronic stimulation on muscle is evidenced by the frequently reported finding of decreased weight, mean fiber area, and tetanic strength of the stimulated muscle (Kernell et al. 1987; Donselaar et al. 1987; Sutherland et al. 1998; see figure 6.10). Although part of this decrease may be attributable to pathological mechanisms resulting in a decrease in cell number (discussed later), it appears that a major part of this decrease may be due to missing periodic high-tension events. Even in a normally innervated muscle subjected to chronic stimulation, muscles would be maintained in a chronically fatigued state, to the point that even normal activation during non-stimulation periods would be expected to produce attenuated force responses. This interpretation is supported by the finding that the atrophic effect is attenuated when low-frequency stimulation is supplemented with short, high-frequency bursts that evoke high contractile forces, even when these high-frequency episodes are of very

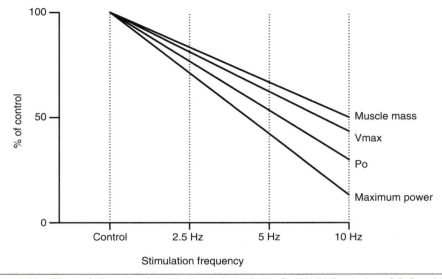

Figure 6.10 Effects of 10 months of chronic electrical stimulation of rabbit tibialis anterior at 2.5, 5, and 10 Hz on muscle mass, maximum shortening velocity (Vmax), maximum isometric tetanic force (Po), and maximum power.
Data from Sutherland et al. 1998.

limited total duration (0.5% of daily time; Kernell et al. 1987). Thus, the atrophic response is less when 40-hertz bursts of intermittent stimulation are used instead of 10-hertz continuous stimulation, even when the total number of impulses in the former case is four times the latter (Kernell et al. 1987). This would explain why endurance runners do not have atrophied leg muscles—the periodic high-tension contractions associated with the yield phase of running counter the atrophic effect of chronic stimulation provided by the nervous system.

As we shall see in chapter 8, signals associated with stress and strain on the extracellular and cytoskeletal matrix of the muscle fiber are most likely involved in stimulating protein synthesis and thus promoting muscle hypertrophy. These signals might be virtually absent in a chronically stimulated muscle, diminished to a level less than that occurring in a normal, nonexercising muscle with the capacity to generate periodic high-tension contractions.

In support of these ideas, it has been demonstrated that early during chronic stimulation (up to 21 days), actin mRNA decreases in tibialis anterior and extensor digitorum longus of rabbits (Brownson et al. 1988). In humans performing aerobic exercise at 40% of $\dot{V}O_2$max for 4 hours, muscle protein degradation increases by 37% during the exercise and by 85% during a 4-hour recovery period (Carraro et al. 1990). The authors of this study, however, proposed that muscle mass does not gradually decrease during a training program involving this type of exercise, since they found that the fractional protein synthesis rate actually increases during recovery. Muscle mass is lost in many chronic stimulation experiments perhaps partly because of the lack of recovery.

Another possible explanation for the antihypertrophic effect of aerobic-type endurance training on muscle may relate to the activation of AMPK that occurs during endurance exercise (refer to the discussion of the role of AMPK as an energy sensor in chapter 3. One of the targets of this kinase is the protein kinase B (PKB, also known as *Akt)* and TOR protein kinase pathway, which is stimulated by growth-enhancing factors and which appears vital for the production of muscle hypertrophy. AMPK activation reduces protein synthesis in muscle homogenates at the same time that activation of Akt, mTOR, ribosomal s6 kinase, and eukaryotic translation initiation factor 4E-binding protein (which are all involved in increased protein synthesis and hypertrophy) are also decreased (Freyssenet 2007). Since AMPK is activated by endurance exercise (but not resistance exercise, because it does not last long enough) and AMPK inhibits Akt and TOR signaling pathways, endurance exercise does not provide the metabolic environment favoring muscle hypertrophy.

METABOLIC SIGNALS AND THE ADAPTIVE RESPONSE

Adaptation of skeletal muscle requires a stress that is sustained or repeated frequently enough to result in changes in protein synthesis or degradation. There are a multitude of biochemical and mechanical events that occur when a muscle changes from the resting to the contracting state that could be candidates for these signals. But what exactly are the signals that evoke the cascade of metabolic events responsible for eventual changes in muscle phenotype? I have separated these possible signals into four families (discussed in the following four sections), with the understanding that all four may be involved to varying degrees in the adaptive response.

Metabolite Signaling

Repetitive muscle contractile activity of the type that is evoked during voluntary aerobic-type endurance activity and chronic electrical stimulation increases metabolic rate. This increase is accompanied by changes in the steady-state concentrations of metabolites that might serve as signals for alterations in gene expression.

During chronic low-frequency electrical stimulation, the phosphorylation potential of the adenylic acid system (which is the ratio of ATP to the combined amount of free ADP and free inorganic phosphate) is persistently depressed throughout stimulation extending from 15 minutes to 50 days (Green, Düsterhöft, et al. 1992; Hood and Parent 1991). This behavior contrasts with that of many other metabolites, which show a recovery after an initial change. The phosphorylation potential of the adenylic acid system thus qualifies as a candidate for a signal that mediates phenotypic changes throughout the chronic stimulation. At the single-fiber level, the ratio of ATP to free ADP increases in the order I < IIa < IIx < IIb in normal (unstimulated) fibers. This finding adds support to the hypothesis that this ratio may be involved in the determination of the MHC profile (Conjard, Peuker, and Pette 1998).

Yaspelkis and colleagues (1999) have shown the possible importance of ATP concentration in the training-induced increases in GLUT4 and citrate synthase in the rat. In their experiment, rats that endurance trained while being given clenbuterol showed an attenuated increase in these two proteins that was attributable to higher ATP levels at rest and a decreased drop in ATP during exercise. When this effect was countered with beta-guanidinopropionic acid (β-GPA), ATP levels were lower at rest and after exercise, and the adaptations in GLUT4 and citrate synthase were more pronounced. Thus, alterations in ATP levels may serve as an important signal.

AMPK is activated by 5'-AMP, an effect that is antagonized by high levels of ATP (Hardie and Sakamoto 2006). Since the ratio of AMP to ATP is a sensitive measure of the cell's energy state (see chapter 3), AMPK is activated when the cell's energy state is disrupted in the direction of increased ATP consumption or decreased ATP production (as is the case during exercise and hypoxia, which result in an increased ratio of AMP to ATP). AMPK may also be stimulated or the stimulatory effect of the altered ratio of AMP to ATP may be amplified by elevated levels of the cytokine interleukin-6 (IL-6), which is produced by exercising muscles (Hardie and Sakamoto 2006). AMPK is activated in muscles following electrical stimulation and voluntary exercise (Jorgensen et al. 2007). AMPK may also regulate the expression of specific genes. For example, AMPK has been implicated in the expression of genes for GLUT4 and hexokinase II and the expression of mitochondrial proteins (see figure 6.11).

PPARs are nuclear receptors that influence the transcription of a number of genes by binding to DNA. FFA have been shown to bind PPARs, and thus changes in FFA concentrations may influence gene expression of, for example, mitochondrial proteins and glucose transporters (see figure 6.11). There are currently three known isoforms of PPAR, and these show some specificities as to the targeted genes. In general, however, when activated by metabolic events associated with aerobic-type endurance exercise, they act to increase mitochondrial biogenesis, mitochondrial metabolism, and endurance performance (Freyssenet 2007).

Calcium Signaling

Changes in base calcium level or in calcium transients in response to stimulation have been proposed as signals that alter gene expression during increased contractile activity. This hypothesis is difficult to verify, however, in an in vivo model. In chronically stimulated extensor digitorum longus of rabbits, intracellular calcium (measured using a calcium-sensitive electrode) increases, reaching a peak that is about five times the normal level by 15 days of stimulation, and then decreases to a base level that is slightly elevated (Sreter et al. 1987). If the stimulation is intermittent (8 hours per day instead of 24), calcium levels do not show this pattern, and conversion to slow fibers does not occur. This observation suggests a possible link between these two events. Interestingly, when stimulation is terminated, calcium shows a similar magnitude but shorter time course of response during the time when a reversal of the histochemical profile to an increased proportion of fast fibers is occurring. Obviously, the calcium story is complicated.

Carroll, Nicotera, and Pette (1999) have demonstrated calcium changes in chronically stimulated rat muscles that are quite different from those previously reported in rabbit muscles. It may be that the calcium changes in rabbits are a partial reflection of the fiber injury that occurs in this species during chronic electrical stimulation. In rat extensor digitorum longus, which can be chronically stimulated with very little fiber damage, intracellular calcium concentration increases during the first 2 hours of stimulation to a level that is 2.5 times that seen in control extensor digitorum longus; the concentration increase in stimulated muscle remains for at least 10 days. This increased calcium concentration is due in part to decreased calcium ATPase activity in the sarcoplasmic reticulum and decreased parvalbumin content. Interestingly enough, this increased calcium concentration is similar to that found in control slow-twitch fibers of the soleus. The constancy of calcium concentration for up to 10 days during chronic stimulation reinforces the idea that calcium may indeed be involved in altering gene expression in this model.

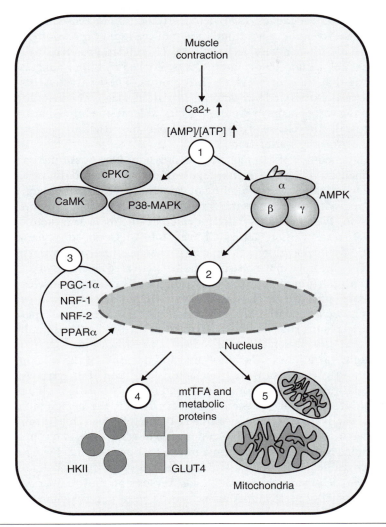

Figure 6.11   The possible central role of AMPK as a signal for adaptation in aerobic-type endurance training and chronic stimulation. 1. AMPK is activated by changes in the ratio of AMP to ATP and in calcium levels during exercise, as are PKC, calcium/calmodulin-dependent protein kinase (CaMK), and p38 MAPK. 2. These changes result in increased transcription of nuclear genes encoding proteins and other transcription modulators. 3. These changes lead to activation of nuclear genes encoding various mitochondrial and metabolic proteins as well as activation of mitochondrial genes. 4. and 5. Increased mitochondriogenesis and synthesis of metabolic enzymes occur.

Reprinted by permission from Jorgensen, Jensen, and Richter 2007.

In myotubes in culture, increasing base calcium levels using the calcium ionophore A23187 causes a reversible increase in expression of MHC I, cytochrome c, and citrate synthase and a decrease in expression of MHC II and the glycolytic enzyme glyceraldehyde 3-phosphate dehydrogenase (Freyssenet, Di Carlo, and Hood 1999; Meissner et al. 2000). The altered gene expression of MHC associated with increased calcium occurs within 24 hours. However, the levels of calcium in these studies may be high compared with those found in vivo.

Abu-Shakra and colleagues (1994) demonstrated that increased intracellular calcium consequent to cholinergic stimulation induces rapid expression of an immediate early gene, *zif268,* in myotubes in culture. This effect, produced by exposing the cells to carbachol, a cholinergic agonist, was blocked by alpha-bungarotoxin, thus demonstrating that it occurred via ACh receptors. These investigators also found that the effect was modulated by ryanodine and dantrolene, which modulate calcium release from the sarcoplasmic reticulum. Thus, at least one immediate early gene is responsive to calcium released from the sarcoplasmic reticulum during at least the initial phase of chronic stimulation.

How might calcium work to alter gene expression? One possibility is via involvement of calcineurin, which is a calcium-regulated serine/threonine phosphatase. When this enzyme is activated via the binding of calcium to a calmodulin and calcineurin complex, it activates in turn specific transcription factors that promote changes in gene expression. The major substrates for calcineurin are nuclear factor of activated T cells (NFAT) transcription factors (Chin et al. 1998). Dephosphorylation of NFAT results in NFAT translocation to the nucleus, where it stimulates expression of target genes. In transgenic mice in which expression of activated calcineurin was enhanced, an increased proportion of Type I fibers was found in hind-limb muscles (Naya et al. 2000). Similarly, administration of the calcineurin antagonist cyclosporin A to rats for 6 weeks caused a transition from slow to fast fibers in the soleus muscle (Chin et al. 1998). These results suggest that calcineurin is involved in the signaling required for maintenance as well as increase of MHC I gene expression.

Interestingly, variations in the amplitude and duration of the calcium signal can have different effects on transcription factor activation. For example, in B lymphocytes, the transcription factors NF-κB and c-Jun N-terminal kinase (JNK) are activated by large calcium transients, whereas NFAT, the principal substrate for calcineurin, is activated by a low, sustained calcium plateau (Dolmetsch et al. 1997). If such a scheme is also present in skeletal muscle fibers, it adds to our understanding of how differences in intensity and type of muscle contractions during the training process might influence the phenotypic adaptations that ultimately take place. Chin and colleagues (1998) have proposed that fast versus slow muscle fiber types might in fact be determined by the effects of fiber activation patterns on calcium levels. According to their hypothesis, regular sustained neuronal firing patterns in motoneurons innervating slow fibers provide the sustained elevated muscle fiber calcium levels necessary for NFAT activation through calcineurin. Elevated dephosphorylated NFAT, in collaboration with other transcription factors such as MEF2, increases transcription of genes coding for slow muscle fiber proteins (Wu et al. 2000; Chin et al. 1998).

Another possible pathway for calcium involvement is via mechanisms that involve MAPK activation. Thus, calcium may increase the phosphorylation and activation of CaMK II, which in turn phosphorylates and activates the p38 MAPK. Activated p38 could subsequently activate PGC-1 and MEF2, thus increasing mitochondrial biogenesis, as well as activate PGC-1 transcription factor binding, thus increasing GLUT4 expression (Wright 2007). Increased cytoplasmic calcium levels may also activate PKC, which may also activate the MAPK signaling system (Freyssenet, Di Carlo, and Hood 1999; discussed next).

Mechanical Signaling

Signals that transduce muscle activity into alterations in gene expression may involve the activation of MAPK signaling pathways, which are activated via a number of stresses that might include those associated with chronic muscle activity. An example of one such pathway is that of the stress-activated protein kinases, also known as the *JNKs.* These kinases, once activated by any number of stressors, translocate to the nucleus, where they increase transcription of several transcription factors, among them c-Jun (Aronson, Dufresne, and Goodyear 1997). In one study, stimulation of rat hind-limb muscle resulted in a sixfold increase in JNK activity within 15 minutes, and the activity remained at this level for the 60 minutes during which the measurements were taken. Interestingly, JNK was also quickly activated and remained elevated in rats during 60 minutes of treadmill exercise (Goodyear et al. 1996). In humans, the MAPK signaling cascade is activated in response to a 60-minute bout of ergometer exercise at 70% of $\dot{V}O_2$max (Aronson et al. 1998). The stress to which this system is responding is not known but could involve autocrine mediators, metabolic changes such as the phosphorylation potential, altered cell calcium levels, or mechanical shear stresses. If the latter is involved, this signal would most likely attenuate with time, as force production decreases in the chronically stimulated muscle. Thus, it would not constitute a good candidate for a signal promoting adaptations later during the chronic stimulation. It is possible that MAPK signaling cascades (such as the extracellular signal-regulated protein kinase and p38 kinase systems) are more important in muscle adaptations to short, high-intensity activities involving higher-tension contractions. This is discussed in more detail in chapter 8.

The involvement of the products of immediate early genes (IEGs), such as c-Fos, c-Jun, and EGR1 in response to aerobic-type endurance training has not been established unequivocally. The IEG products c-Fos and c-Jun form dimers to bind to activator protein 1 (AP-1) sites within specific promoters, while ERG1 is known to be activated during differentiation of bone and nerve cells (Michel, Ordway, et al. 1994). In chronically stimulated rabbit muscle, there is a transient increase in expression and concentration of these IEG products during the first 24 hours, with a secondary and sustained rise in the proteins but not the corresponding mRNAs for up to 21 days of stimulation (Michel, Ordway, et al. 1994). The relationships of these two different phases of activation of the IEGs in response to the muscle adaptations that occur during chronic stimulation have yet to be determined. Within a few hours postexercise in human vastus lateralis, IEGs are elevated up to 20-fold (Hoppeler and Flück 2002).

After a single 30-minute bout of aerobic endurance exercise, an increase in c-Fos and c-Jun proteins and in their corresponding mRNAs is evident for several hours. c-Fos shows a greater relative response than c-Jun shows (Puntschart et al. 1998).

Changes in gene expression may occur during recovery from activity. For example, transcription rates for GLUT4, hexokinase II, c-Fos, alpha-beta-crystallin, Hsp70, myoglobin, and citrate synthase are increased in the hours following cessation of chronic stimulation (Neufer et al. 1998). Increases in gene expression have been reported during recovery from exercise (Koval et al. 1998; Booth et al. 1998).

Hormones and Autocrine or Paracrine Factors

Factors such as insulin-like growth factor 1 (IGF-1) or mechanical growth factor that are released from muscle cells during exercise may play a role in altering gene expression. In addition, there are a host of other intracellular changes that might constitute signals for alterations in gene expression, such as hypoxia, increased hydrogen ion concentration, generation of ROS during contractile activity, and changes in concentration of a number of intermediary metabolites (Williams and Neufer 1996; Baar et al. 1999; Freyssenet 2007). For example, HIF-1 is a transcription factor that is activated in muscle fibers during levels of hypoxia that occur during voluntary exercise. This transcription factor increases expression of the gene for vascular endothelial growth factor (VEGF) in muscle fibers, thus promoting angiogenesis (Gustafsson et al. 1999), as well as increases expression of GLUT4 (Freyssenet 2007) and glycolytic enzymes (Koulmann and Bigard 2006). In addition, the production of ROS due to increased activity of the electron transport chain during exercise may serve to activate genes of antioxidants such as mitochondrial superoxide dismutase and inducible NO synthase via signaling pathways that include MAPK and NFκB (Ji, GomezCabrers, and Vina 2007). It is obvious that several signals are most likely in operation at any one time via several mechanisms and that the time course of involvement of the factors is variable, with some factors important early in exercise and others important in more prolonged exercise conditions.

DEGENERATIVE AND REGENERATIVE PROCESSES

Some evidence of degeneration and regeneration is present in chronically stimulated muscles, such as a transient appearance of embryonic and neonatal MHCs and the invasion of fibers by nonmuscle cells (Maier et al. 1988; Kirschbaum et al. 1990; Maier and Pette 1987). In stimulated rabbit tibialis anterior, 15% to 20% of fibers are myotubes or contain central nuclei after 30 and 60 days of stimulation, indicating an ongoing degeneration and regeneration process throughout the adaptation period (Schuler and Pette 1996). The degenerative and regenerative changes are more marked in some species than they are in others (Kirschbaum et al. 1990).

Sutherland and colleagues (1998) assumed that a major proportion of the decrease in specific tension of rabbit tibialis anterior after 10 months of stimulation at 10 hertz was due to degenerative loss of fibers. Their interpretation was that the loss of fibers was due to the hypoxia induced by the constant semitetanic contractions imposed by the stimulation; the loss of specific tension occurred mostly after 12 weeks of stimulation. Very little damage was seen in muscles stimulated at lower frequencies.

Lexell and colleagues (1992) performed an in-depth study of chronically stimulated rabbit tibialis anterior and concluded that the contribution of degeneration and regeneration to conversion of muscle fiber type is insignificant. More recently, however, studies of chronically stimulated rabbit extensor digitorum longus suggested that degeneration and regeneration may play a substantial role, depending on the muscle (Schuler and Pette 1996). Thus the significance of degeneration and regeneration to whole-body endurance training is not entirely clear, although training must by definition include a higher level of these processes, since protein synthesis and degradation rates are higher in trained muscles.

SUMMARY

Whole-body endurance training is very complex, due to the involved complexities of motor unit recruitment, and thus adaptations found using the biopsy technique in humans are primarily descriptive. More insight into mechanisms of adaptation is afforded by examining adaptive responses and their associated mechanisms following chronic muscle electrical stimulation. Chronic activity results in changes in protein synthesis at the pretranslational, translational, and posttranslational levels. Several possible biochemical signals that change in concentration during acute contractile activity, either alone or in combination, may promote changes in protein synthesis that characterize the adaptation process.

A summary of the changes that occur with chronic electrical stimulation is presented in table 6.1. The early, middle, and late phases of adaptation are relative and differ among the species that have been studied and among continuous stimulation, intermittent stimulation, and whole-body endurance training. As previously pointed out, the acute stimuli associated with the onset of chronic stimulation are not expected to be the same during chronic stimulation and endurance training. We might expect, for example, that some of the signals that exert their influence during the first minutes of electrical stimulation would be of special significance during endurance training, in which the repeated transitions between rest and exercise would reactivate these pathways at each training session. In addition, although the total tension-time integral would be expected to be larger for electrical stimulation (8-24 hours of maximal contractile activity per day) compared with endurance training (2 hours of submaximal activation per day), a higher integral per unit time would be characteristic of endurance training, which would thus deliver a more intense and most likely different stimulus than chronic stimulation delivers.

Not all of the muscle responses to endurance training are included in table 6.1. For example, endurance training appears to result in an increase in the antioxidant enzymes superoxide dismutase and glutathione peroxidase and an increase in the concentration of intracellular glutathione. The resulting increased capacity to reduce the persistence of free radicals during and following exercise is probably significant in reducing the cellular damage that these substances can produce (Powers, Ji, and Leeuwenburgh 1999). This most likely occurs during the middle phase of adaptation, although this response has not been studied in the electrical stimulation model. It is certainly an important adaptation and one that needs to be investigated in more detail.

In spite of the many differences between electrical stimulation and whole-body endurance training, it appears that the pattern and temporal order of adaptations are very similar between the two. Thus, the chronic stimulation model will most likely continue to provide us with information regarding the adaptive capacity of skeletal muscles to increased activity that we can apply to the infinitely more complex model of endurance training.

Neural Mechanisms in Aerobic Endurance Training

It is increasingly recognized that many of the adaptations caused by aerobic-type exercise training that result in better endurance performance may reside at physiological levels other than muscle. Our progress in determining adaptations in the nervous control of muscles has been limited primarily by the lack of tools with which to collect the necessary data (compared with the relatively easier access that we have to muscle tissue samples). In this chapter, my goal is to emphasize adaptations to aerobic endurance training, at the neuromuscular junction, at the level of the motoneuron, in the spinal cord, and in the central nervous system that promote the development of neuromuscular endurance. As we move from near the muscle through to the central nervous system, information becomes gradually less plentiful, and examples from models other than endurance training that might involve similar mechanisms become more numerous.

ADAPTATION OF THE NEUROMUSCULAR JUNCTION

Neuromuscular junctions are different in slow and fast muscles. This phenomenon may be due to differences in levels of chronic activity. Morphologically, junctions in the fast muscles are larger and demonstrate some systematic differences in the terminal axons (shorter, less numerous, and of larger diameter than those in fast fibers) and the synaptic grooves and junctional folds (less extensive than those in fast fibers; Ogata 1988). Nonetheless, there is not total agreement as to the universality of these differences across muscles, fiber types, and species. Physiologically, there is no doubt as to the difference between fast and slow neuromuscular junctions; EPPs are larger in fast fibers but run down more quickly in response to repetitive stimulation (Gertler and Robbins 1978). This difference gives the fast fiber a higher safety factor for transmission. The safety factor is defined as $EPP / (Eap - Em)$, where Eap is the threshold potential for generating an action potential and Em is the membrane potential (Ruff 1992). The safety factors for rat soleus and extensor digitorum longus neuromuscular junctions have been estimated as 3.5 and 5, respectively, from electrophysiological experiments (Wood and Slater 1997).

Several mechanisms can be evoked to explain this relatively higher safety factor in fast neuromuscular junctions compared with slow neuromuscular junctions. Quantal release is higher from fast terminals (Wood and Slater 1997). Fast postsynaptic membranes may also have higher densities of ACh receptors (Sterz, Pagala, and Peper 1983). The sodium current density at fast end plates is up to four times higher than that at slow end plates, due primarily to a correspondingly higher density in sodium channels (Ruff 1992). Sodium channels at end plates appear also to be qualitatively different between fast and slow fibers. The voltage dependences of activation and of fast and slow inactivation of the sodium current are different between these two fiber types; they are at more negative potentials in fast fibers (Ruff and Whittlesey 1993). This may indicate that fast fibers are more susceptible to a use-related reduction in membrane excitability, which would tend to limit the duration that these fibers could fire at high rates (Ruff 1996).

Neuromuscular junctions in fast muscles are also characterized by larger amounts of the asymmetric form of acetylcholinesterase (AChE) found attached to the post-synaptic membrane via a collagen tail and of the globular G4 form that is found in the

A COMBINATION OF LOCAL AND SYSTEMIC EFFECTS OF CHRONIC EXERCISE ON BRAIN FUNCTION?

Exercise has been demonstrated repeatedly to enhance learning and memory and executive function, to decrease depression, and to attenuate the age-associated decline in general brain function in humans. The effect of exercise on learning and memory has been demonstrated in rodents as well as in humans, via tasks designed to test these functions, thus giving researchers an experimental paradigm by which changes in tissue components can be correlated to behavioral outcomes. For example, in the dentate gyrus portion of the hippocampus, electrophysiological, cytoarchitectural, and biochemical changes accompany enhanced learning and memory behavior following an exercise regimen in rats. A significant component of this plasticity seems to involve systems that are in support of these localized adaptations, such as altered metabolism and blood flow to a variety of brain centers. The increase in growth factors such as IGF-1, VEGF, and brain-derived neurotrophic factor (BDNF) probably plays a major role in orchestrating these brain adaptations. In addition, regular exercise may exert an influence on brain plasticity by decreasing the deleterious effects of systemic inflammation associated with chronic disuse and being overweight. Together, these direct and indirect effects probably function in synchrony to enhance brain function and slow neurodegeneration. These thoughts have led many, including Cotman and associates at the University of California at Irvine, who have led this research, to believe that exercise may have a significant role to play in delaying the onset and attenuating the severity of many neurodegenerative conditions, such as Alzheimer's disease, Huntington's disease, and Parkinson's disease (Cotman, Berchtold, and Christie 2007).

perijunctional region (Gisiger, Bélisle, and Gardiner 1994; Sketelj et al. 1998). Whether fast and slow fibers within a muscle containing a mixture of fiber types exhibit these same differences is unknown.

Adaptations to Chronic Electrical Stimulation

Chronic electrical stimulation has been used to determine the adaptability of the neuromuscular junction much in the same way that it has been used to demonstrate muscle adaptability. Chronic stimulation changes the neuromuscular junctions of fast muscles so that they resemble, at least morphologically, junctions normally found in slow muscles. For example, Waerhaug and Lomo (1994) found that junctions of chronically stimulated extensor digitorum longus were reduced in area and had decreased density of terminal varicosities. Chronic stimulation of rat extensor digitorum longus resulted in an AChE profile resembling that found in soleus (lower A12 and G4 forms; Jasmin and Gisiger 1990). On the other hand, chronically stimulated (for 20 days) extensor digitorum longus exhibited an increased quantal content while maintaining the property of rapid decrease in EPP amplitude with increasing frequency of stimulation, suggesting that the change toward a slow junction (which would demonstrate a decrease in EPP amplitude and an increase in fatigue resistance of transmitter release) was not complete.

The research team of Harold Atwood at the University of Toronto has contributed valuable information to this issue. During the past 20 years, this team has studied synaptic adaptations in the neuromuscular junctions of the crayfish claw and abdominal muscles in response to chronic stimulation. The motoneurons innervating these muscles can be distinguished as phasic or tonic based on their morphological and physiological characteristics. With continued stimulation, phasic motoneurons show a high initial neurotransmitter release and a faster drop-off in transmitter release (lower fatigue resistance) than tonic motoneurons show. Thus this phasic–tonic distinction is similar functionally to the fast–slow distinction seen in mammals.

By chronically stimulating phasic motoneurons for as little as 3 days, for 2 hours per day, these researchers showed a phenomenon that they termed *long-term adaptation (LTA)* of the neuromuscular junction. This adaptation is characterized by an increased synaptic mitochondrial content, lower initial neurotransmitter release, and higher fatigue resistance in the phasic motoneuron, which thus becomes more like a tonic motoneuron (see figure 7.1).

The systematic search for the mechanisms involved in this adaptive response resulted in the following findings:

- Neurotransmitter release is not necessary for LTA. The adaptation was seen when axons were blocked with tetrodotoxin and stimulated central to the blockage so that the impulse did not even reach the terminal (Lnenicka and Atwood 1986).

- The production of action potentials by the stimulated motoneuron is not necessary for the expression of LTA. This conclusion was based on results of experiments in which LTA occurred after stimulation of sensory afferents that caused subthreshold depolarization in the phasic motoneuron (Lnenicka and Atwood 1988).

- LTA does not require a change in protein synthesis by the adapting motoneuron. LTA still occurred when the protein synthesis blocker cycloheximide was injected at the beginning or at the end of each stimulation session (Nguyen and Atwood 1990).

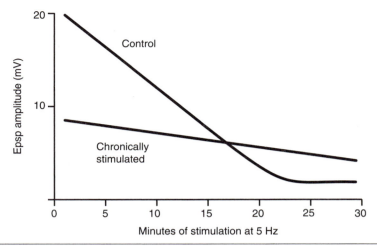

Figure 7.1   LTA of synaptic transmission in the crayfish claw closer muscle. Electrical stimulation of the innervating nerve was at 5 Hz for 2 h/d, for 3 d. There were amplitude changes in EPSP in stimulated and contralateral claw muscles during stimulation at 5 Hz, measured 2 d after cessation of the adaptation protocol.

Data from Nguyen and Atwood 1990.

- Examining different frequencies and total amounts of stimulation shows that the increase in neurotransmitter fatigue resistance has a lower threshold than the decrease in initial neurotransmitter release has. Thus, these two adaptive responses are distinct (Mercier, Bradacs, and Atwood 1992).

- The increase in synaptic fatigue resistance with LTA depends more on axonal transport than the decline in initial transmitter release depends on axonal transport. Thus, after establishment of LTA, axotomy resulted in a decrease in fatigue resistance that was more rapid than the change in initial transmitter release (Nguyen and Atwood 1992).

- The decrease in initial transmitter release, which occurred in vitro after 5 hours of depolarization of the cell body or axon, does not occur with simultaneous application of a calcium channel blocker. Thus, calcium influx during motoneuron depolarization is involved in this adaptive response (Hong and Lnenicka 1993).

Unfortunately, to date we have been unable to determine if these same mechanisms are involved in the endurance training model.

Adaptations to Aerobic Endurance Training

Research reports concerning morphological adaptations at the neuromuscular junction induced by whole-body endurance training are not numerous, and their conclusions are far from consistent (Deschenes et al. 1994). This is most likely a reflection of the difficulties involved in making quantitative morphological measurements, which is most likely the reason for a lack of consensus regarding the morphological differences between fast and slow neuromuscular junctions. The most consistent responses to endurance training to date appear to be an increased neuromuscular junction area and an increased complexity of nerve terminal branching (Deschenes et al. 1993, 1994).

Biochemically, endurance training induces increased AChE activity in fast muscles, which is evident in the end plate region of the muscle. This response has several notable and highly intriguing characteristics. First, it affects the G4 form of AChE selectively (Jasmin and Gisiger 1990; Fernandez and Donoso 1988). Correspondingly, in slow muscles such as the soleus, which contain very small amounts of G4, this response is absent. Second, the response is very rapid and reflects the level of activity within the preceding few days (Hubatsch and Jasmin 1997; Gisiger, Bélisle, and Gardiner 1994; Fernandez and Donoso 1988). Third, the patterns of adaptation in ankle flexors and extensors to treadmill running, compensatory overload, and swimming have led to the hypothesis that the adaptation may depend on the tonic (very little change in G4 AChE) versus phasic (increased G4) nature of the training stimulus (Gisiger, Bélisle, and Gardiner 1994; Jasmin, Gardiner, and Gisiger 1991; Gisiger, Sherker, and Gardiner 1991; Jasmin and Gisiger 1990). G4 appears to be located in the perijunctional sarcoplasmic reticulum, where it forms a conjunctional compartment that bathes the end plates in an AChE-rich environment (Gisiger and Stephens 1988). It has been proposed that G4 may play a significant role in removing ACh from the synaptic cleft during high-frequency, repetitive (i.e., phasic) activity and thus may help to avoid ACh receptor desensitization, which would continue to activate the end plate refractory phase. As previously discussed, end plate desensitization is considered a possible mechanism for fatigue of the neuromuscular system under certain conditions. The mechanism for the increased AChE in response to training is not known but may involve factors such as calcitonin gene-related peptide (CGRP) or ACh receptor–inducing activity (ARIA), which are released from nerve terminals during activation and which influence the composition of the postsynaptic membrane (Fernandez and HodgesSavola 1996; see figure 7.2).

Consistent with the proposed increase in the area of the neuromuscular junction associated with aerobic-type endurance training, the number of ACh receptors also increases in several muscles. These include the diaphragm (Desaulniers, Lavoie, and Gardiner 1998).

It appears that aerobic endurance training has consequences for the electrophysiology of the neuromuscular junction. Dorlöchter and colleagues (1991) examined single EPP amplitudes and responses to repetitive stimulation in in vitro extensor digitorum longus of mice that endurance trained in voluntary wheels for 2 to 8 months. Their results showed a doubling of EPP amplitude, which remained above control values during stimulation at 100 hertz for 1 second (see figure 3 in Dorlöchter et al. 1991). These authors also demonstrated that preparations from trained mice exhibited a higher resistance to both presynaptic (high magnesium, low calcium) and postsynaptic (curare) blockade, suggesting that the adaptation involved primarily increased neurotransmitter release. These results have been confirmed using the models of treadmill running (Desaulniers et al. 2001) and compensatory overload (Argaw et al. 2004) in rats.

Thus, a picture is slowly emerging in which the neuromuscular junction appears to adapt to aerobic endurance training by increasing both its presynaptic capacity (increased and more sustained transmitter release) and its postsynaptic efficiency (higher amount of G4 AChE and number of ACh receptors). The mechanisms for these adaptations are not yet known but may include neurotrophic and motoneuron-derived influences that are released at the neuromuscular junction and influence pre- and postsynaptic structures (see figure 7.2). The physiological consequences of these changes are also not yet known; technical constraints to date have prevented even the ability to determine whether the neuromuscular junction limits performance. However,

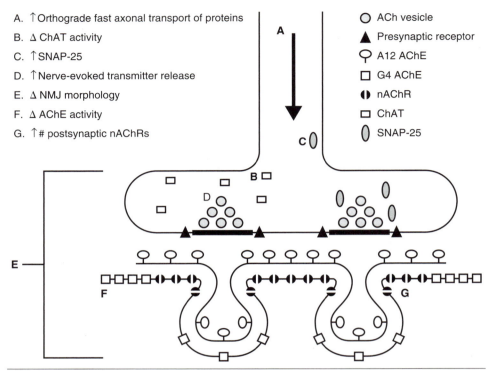

A. ↑ Orthograde fast axonal transport of proteins
B. Δ ChAT activity
C. ↑ SNAP-25
D. ↑ Nerve-evoked transmitter release
E. Δ NMJ morphology
F. Δ AChE activity
G. ↑ # postsynaptic nAChRs

○ ACh vesicle
▲ Presynaptic receptor
♀ A12 AChE
☐ G4 AChE
◖ nAChR
☐ ChAT
◖ SNAP-25

Figure 7.2 Summary of loci for demonstrated adaptations to increased chronic activity in the mammalian neuromuscular junction.

Reprinted from Panenic and Gardiner 1998.

teleologically speaking, adaptations found at this level suggest the strengthening of a potential weak link and thus shed some light on possible limiting components in the untrained state.

RESPONSES OF MOTONEURONS

Anterograde and retrograde axonal transport are functional properties of motoneurons by which materials are transported toward the nerve terminals and the soma, respectively. Anterograde axonal transport is classified as fast or slow based on the speed with which materials are transported. Fast and slow anterograde transports are also distinguishable by the materials that are moved and the energy requirements for transport. Fast transport (200-400 millimeters per day) involves transport of primarily membrane proteins and lipids, as well as neurotransmitters, while slow transport (0.2-10 millimeters per day) includes neurofilaments, microtubules, microfilaments, and metabolic enzymes (Siegel et al. 1989). We have some evidence that these fundamental nerve cell properties may be responsive to increased chronic activity. Adaptive changes have been found in fast axonal transport in motoneurons of endurance-trained rats (Kang, Lavoie, and Gardiner 1995; Jasmin, Lavoie, and Gardiner 1988). This experiment was performed by injecting radioactively labeled leucine into the ventral horn of the spinal cord of anesthetized rats and measuring the appearance of the radiolabeled proteins in the sciatic nerve 4 hours later. Peak and average transport velocities increased, as did the total amount of transported protein. This response may indicate elevated protein synthesis of the trained motoneuron, enhanced loading of protein onto the transport machinery, or both. We can postulate that motoneurons need more material in the nerve terminals of adapting neuromuscular junctions or that they need to enhance the delivery of trophic substances involved in presynaptic or postsynaptic adaptations.

It may be that certain proteins are preferentially targeted for rapid delivery to endurance training junctions because of their special roles in the adaptive process. Kang, Lavoie, and Gardiner (1995) found that a synaptosomal-associated protein, SNAP-25, was transported in higher amounts than other axonally transported proteins were transported in endurance-trained motoneurons (see figure 1 in Kang, Lavoie and Gardiner 1995). This protein plays an important role in the interaction of the synaptic vesicle with the presynaptic membrane of the nerve terminal and thus might be associated with the increased transmitter release found in trained nerve terminals. This protein would presumably also need to be upregulated to accommodate any neuromuscular junction remodeling that involves increased nerve terminal size or extent of branching.

Gharakhanlou, Chadan, and Gardiner (1999) demonstrated that motoneurons from endurance-trained rats have increased content and anterograde transport of CGRP (see figure 7.3). CGRP, produced by motoneurons and released at terminals, has been shown to stimulate ACh synthesis and may be involved in the control of muscle AChE.

The increase in fast axonal transport in general and in proteins such as SNAP-25 and CGRP in particular might explain why the sprouting response of motoneurons in the presence of denervated fibers is more robust in rats with increased daily activity levels (Seburn and Gardiner 1996; Gardiner, Michel, and Iadeluca 1984). It may also explain why the more active slow motoneurons appear to regenerate and reinnervate muscles faster than fast motoneurons do (Desypris and Parry 1990; Foehring, Sypert, and Munson 1986).

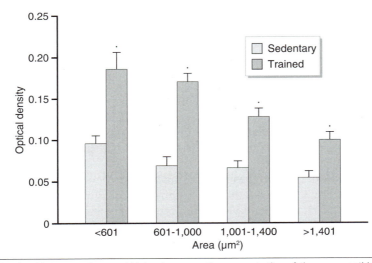

Figure 7.3 Aerobic-type endurance training increases the concentration of the neuropeptide CGRP in rat lumbar motoneurons.

Reprinted by permission from Gharakhanlou, Chadan, and Gardiner 1999.

There is some evidence that the soma of the motoneuron shows adaptations to aerobic endurance training. Motoneurons show signs of elevated protein synthesis (Edstrom 1957) and less stress in response to acute exhaustive exercise (Jasmin, Lavoie, and Gardiner 1988; Gerchman, Edgerton, and Carrow 1975). Whether metabolic enzyme systems in motoneurons are altered by aerobic endurance training is controversial. Since mitochondrial enzyme activity is inversely related to soma size (Ishihara, Roy, and Edgerton 1995; Suzuki et al. 1991) and slow fibers are innervated by smaller motoneurons, it seems reasonable to assume that endurance training might increase mitochondrial enzyme activity and perhaps even reduce motoneuron size. An early study (Gerchman, Edgerton, and Carrow 1975) showed that the activity of the enzyme malate dehydrogenase (MDH) was modestly elevated in the somata of hind-limb motoneurons of endurance-trained rats. More recent evidence that is based on more precise quantitative techniques and a different enzyme (SDH) seems to suggest that if motoneurons do become more oxidative in enzyme profile with aerobic endurance training, the effect is slight (Nakano et al. 1997; Seburn, Coicou, and Gardiner 1994; Suzuki et al. 1991). Motoneurons innervating muscles that are chronically stimulated to evoke classic muscle adaptations (such as slowing of contractile speed and increased oxidative enzyme activities) are not changed in size or soma SDH activity (Donselaar, Kernell, and Eerbeek 1986). Since stimulation of the muscle via the motor nerve evokes action potentials in the cell body via antidromic excitation, this result demonstrates clearly that the increase in the number of action potentials generated at the level of the soma does not constitute a stimulus for increased mitochondrial enzyme activity.

Does aerobic-type endurance training change the electrophysiology of motoneurons? Such changes might prove to be quite meaningful if they do occur. As we saw in a previous chapter, fundamental properties of motoneurons such as *Irh, Rin,* duration of afterpotentials, and susceptibility to phenomena such as late adaptation and bistability significantly influence how motor units are used during movement. Since these properties also influence the conscious effort necessary to drive motoneurons efficiently during effort and might affect fatigue, training-induced adaptations would prove meaningful. Unfortunately, there is currently very little information on this particular topic. Motoneurons of endurance-trained rats show changes (in the hyperpolarization, or more negative, direction) in the RMP as well as in the threshold potential at which the action potential is triggered (the voltage threshold), in the speed with which the action potential rises, and in the amplitude of the AHP (increased). While the functional consequences of these changes are not completely evident, they do indicate that motoneurons respond to increased activity (although these changes may represent changes in the opposite direction in motoneurons of control rats kept in conditions of severely limited activity). Our modeling studies have suggested that these changes indicate alterations in ion channels or their modulation in the motoneurons (Gardiner et al. 2006).

There is some evidence that the use of motoneurons differs between dominant and nondominant hands, which might be a reflection of motoneuronal changes. Adam, De Luca, and Erim (1998) examined motor unit recruitment and firing behavior in first dorsal interosseus of dominant and nondominant hands of human subjects (see table 7.1). Their finding that motor unit firing rates at a submaximal target force were lower in the dominant hands was attributed to the effect of prolonged use on the contractile properties of the muscle fibers. Thus, slower fibers would require a lower frequency of firing to obtain a given absolute force. They also found a lower mean force threshold, lower initial firing rates, and lower discharge variability in motor units of dominant hands. These findings suggest a change in fundamental motoneuronal properties that would tend to optimize performance. For example, a lower average force threshold suggests more motor units recruited to satisfy the same force. This would allow a lower firing frequency on the part of each unit and thus would avoid the problems associated with increasing firing rates (e.g., adaptation, neuromuscular junction and branch block failure, muscle fatigue). As seen previously, more closely spaced recruitment thresholds are characteristic of motoneurons that innervate fatigue-resistant, as opposed to fatigue-sensitive, muscle units (Bakels and Kernell 1994). Lower initial firing rates and reduced variability in firing may indicate fundamental changes in the components that control the AHP (see chapter 1). These interpretations from the work of Adam, De Luca, and Erim (1998) are based on the premise that handedness is a model of chronically increased neuromuscular usage.

We might think that, like muscle fibers, motoneurons become more like slow types in their electrophysiological properties in response to chronic stimulation and whole-body endurance training. The changes in motoneuron excitability and firing frequencies during activation under such conditions are unclear. The closest that we can come to resolving this issue is to examine the work of Munson and colleagues (1997), who investigated the electrophysiological properties of cat hind-limb motoneurons following chronic electrical stimulation to promote the classic adaptations in innervated muscles. After 2 to 3 months of chronic electrical stimulation of the medial gastrocnemius nerve, virtually all of the fibers of the medial gastrocnemius were converted to Type I fibers. Motoneurons innervating this muscle also showed tendencies to become more like slow motoneurons: AHP duration and Rin increased and Irh decreased (see figure 7.4). In addition, EPSPs in response to high-frequency stimulation changed from positive modulation (facilitation, characteristic of fast motoneurons) to negative modulation (depression, characteristic of slow motoneurons). The mechanisms involved in this response have not been determined but involve pre- and postsynaptic mechanisms.

Finally, the authors of this study found that these changes also occurred when the nerves that were stimulated innervated skin rather than muscle. Thus, it may be that stimulation of the peripheral nerve per se, involving afferent as well as efferent

Table 7.1   Dominant (D) Versus Nondominant (ND) Hand Differences in First Dorsal Interosseus Muscle and Motor Unit Properties

Whole-muscle performance	
MVC	D = ND
Force variability during sustained contraction at 30% of MVC	D < ND**
Motor unit properties	
Recruitment thresholds	D < ND*
Average firing rates	D < ND**
Initial firing rates	D < ND**
Discharge variability	D < ND**

*Significant differences at $p < 0.05$.

**Significant differences at $p < 0.01$.

Reprinted from Adam, De Luca, and Erim 1998.

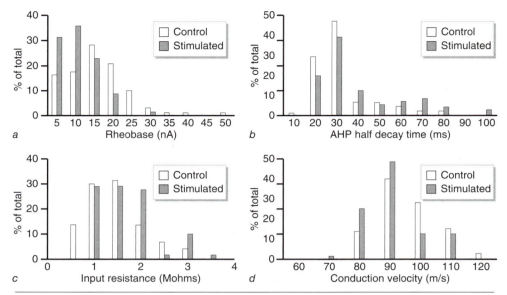

Figure 7.4 Electrophysiological properties of motoneurons of cat medial gastrocnemius tend to become slower after innervated muscle fibers are subjected to several months of chronic low-frequency stimulation. This effect is shown by a shift in the distribution of chronically stimulated motoneurons (solid bars) toward *(a)* lower *Irh, (b)* longer AHP, and *(c)* higher *Rin.* No clear trend was seen in *(d)* axon conduction velocities.

Reprinted by permission from Munson et al. 1997.

stimulation, evoked the recorded responses. However, it is exciting to think that the changes in muscle exerted a retrograde effect on the properties of the motoneurons, an interpretation that is supported in these results by the apparent lag between muscle adaptations and motoneuron adaptations. Such a retrograde effect could be exerted by neurotrophins such as neurotrophin-4 (NT4) produced by chronically active muscle fibers (Funakoshi et al. 1995). The results of experiments in which motoneurons innervating skin also demonstrate adaptations might be explained by a common or similar neurotrophic mediator that is produced by both slow or active muscle and skin and that has slowing effects on motoneuron properties.

Some evidence from the literature on spinal cord injury suggests that changes in alpha-motoneurons with step training of rats with spinal cord injury might be the basis for improved stepping that accompanies regular treadmill training. Petruska and colleagues (2007) found significant correlations between electrophysiological properties of motoneurons, such as the AHP amplitude and the heteronymous EPSP response, and success in treadmill training in rats that had received spinal cord transections as neonates. While successful training obviously involves more than adaptations in motoneuron properties, this type of experiment gives us a highly valuable window into adaptations that are possible in the nonlesioned system. A more detailed discussion of spinal cord adaptations to training in this injury model is included in the next section.

ADAPTATIONS OF SPINAL CORD CIRCUITS

Although aerobic endurance activities generally cannot be considered as highly skilled as, for example, activities demanding power (powerlifting) and perceptuomotor coordination (golf), there is a case for possible beneficial spinal cord adaptations to occur in endurance training. For example, the highly developed spinal circuitry governing the sequential activation of flexors and extensors must also be flexible enough to be capable of responding successfully to changes in afferent feedback, which are destined to occur during long-distance endurance activities due to changes in the muscle properties and in the functioning of the receptors themselves (see chapter 6). In addition, supraspinal inputs to the locomotion-generating spinal circuits change as overall effort increases during dropout of muscle fibers during fatigue. The outcome measure for successful spinal cord response to these changes during an endurance event is the capacity to maintain successful muscle activation patterns throughout the activity and to respond to unexpected perturbations in the locomotor system (such as tripping, stepping over a barrier, adjusting stride length, and the like). Chronic spinal cord adaptations (besides changes at the motoneuron level mentioned in the previous section) might include changes in the strength and proliferation of existing synaptic contacts with motoneuron pools or the uncovering of previously unused pathways. In the following sections, I summarize several sources of information that give evidence of the plasticity of spinal cord circuitry in response to repetitive activation (training).

USING ROBOTS TO EVOKE SPINAL CORD ADAPTATIONS

Following a complete or incomplete spinal cord injury, the information emanating from sensory receptors in the affected limbs takes on a more vital role in generating movements of the limbs below the lesion. Reggie Edgerton and colleagues at UCLA have for many years been experimenting with robotic systems that not only trace the trajectories of hind limbs during assisted treadmill locomotion in rats and mice with spinal cord injury but also produce trajectories of different patterns in order to promote locomotor adaptations. Their results have been seminal to our understanding of how the spinal cord adapts chronically to sensory signals coming from the limbs. For example, it appears that stereotypic movement patterns imposed on the limbs by the robotic device are less efficacious in promoting locomotor adaptations than are patterns that are presented in a more unpredictable manner (Edgerton and Roy 2009). The locomotor centers appear to thrive on step-to-step variations in sensory input in order to make network decisions on patterns of motor output. In addition, these experiments have shown that locomotion step duration changes systematically when different loads are placed in the hind limb via the robotics device; the spinal cord can sense load on the paw and can adapt to it even in the absence of supraspinal centers. The most recent versions of these robotic devices, which are used with rodent and human subjects, provide for assist-as-needed interventions. This type of intervention, in which assistance is provided when needed to maintain an effective stepping pattern, is similar to the type of intervention normally provided by a therapist. Tests of this type of device with rodents has shown that the assist-as-needed intervention is more efficient in promoting the development of locomotor patterns than the repetitive movement paradigms are. Although more research must be conducted to determine the best way to use this system, the use of robotics to understand the adapting spinal cord and to rehabilitate individuals with damaged spinal cords seems very promising.

Training of the Monosynaptic Reflex

Monkeys can be trained to either increase or decrease the amplitude of the monosynaptic stretch reflex response. This finding is the result of experimentation by Wolpaw and colleagues since the early 1980s (see figure 7.5). This series of experiments demonstrated hardwired changes in spinal cord circuitry that resulted from behavioral conditioning. For example, these investigators found that changes in the amplitude of H-reflexes were apparent when animals were anesthetized and for up to 3 days following removal of supraspinal influences by spinalization (Wolpaw and Lee 1989). They also ascertained that at least a portion of the adaptation may involve changes in intrinsic motoneuron properties. For example, in monkeys trained to decrease the amplitude of their triceps surae H-reflex, motoneurons had a more positive firing threshold and thus needed more depolarization to reach that threshold (Carp and Wolpaw 1994). The authors proposed that this change may be caused by a positive shift in the voltage for sodium channel activation in the involved motoneurons that is due to a change in activation of intraneuronal PKC (Halter, Carp, and Wolpaw 1995). The spinal changes produced by conditioning to upregulate and downregulate the amplitude of the H-reflex are not mirror images, since the latter but not the former is accompanied by changes detectable at the single-motoneuron level (Carp and Wolpaw 1995). Thus, these adaptive responses most likely include adaptations at several levels: motoneuron, afferent–motoneuron synapse, and interneuron.

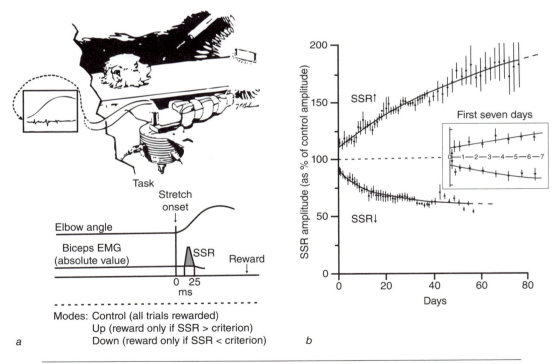

Figure 7.5 Monkeys can be trained to increase or decrease the amplitude of the monosynaptic reflex. *(a)* An extension torque is superimposed on a low-level tonic elbow flexion, evoking a spinal stretch reflex (SSR) at monosynaptic latency. *(b)* Time course of changes in SSR amplitude in animals trained to increase (upper line) or decrease (lower line) the SSR amplitude.

Reprinted by permission from Wolman and Carp 1990.

Locomotor Training Following Spinalization

During the past two decades, a general picture has been emerging that suggests that spinal cord plasticity is possible in the presence of repeated activation in the form of training. The preferred subject in this research has been the cat, which is capable of demonstrating coordinated activation of hind-limb locomotor muscles that is surprisingly similar to a normal pattern after spinal cord transection (Lovely et al. 1990; Barbeau and Rossignol 1987; Smith et al. 1982; Bélanger et al. 1996). A significant amount of literature has also been generated using the rat model (Edgerton et al. 2008). The quality of the locomotion evoked in the weeks and months following spinal cord transection (as measured by treadmill speed, stride length, amplitudes and temporal coordination of extensor and flexor bursts, incidence of failed steps) is greatly improved if daily training is imposed—that is, if the animal is supported on the treadmill and the hind-limb locomotor pattern is evoked by passive movements of the limbs or various means of sensory stimulation (Lovely et al. 1986; Chau, Barbeau, and Rossignol 1998; De Leon et al. 1998; see figure 7.6). Furthermore, the beneficial response to training is somewhat specific to the training modality; cats trained for weight support stand but do not walk better, while cats trained for walking improve in walking more than they improve in weight support (Hodgson et al. 1994; see their figure 2). Regular training after partial spinal cord injury, which includes many types of injury that humans sustain, that does not involve complete transection of the cord also improves the rate of motor recovery (Multon et al. 2003).

Similar findings, again in cats, describe the effect of axotomy of the nerves innervating ankle flexors on treadmill locomotor gait before and after spinalization (Carrier, Brustein, and Rossignol 1997). When these nerves, which are important in lifting the foot during the swing phase of locomotion, were cut in the otherwise intact normal cat, compensation for the deficit took the form of increased activation of knee flexors, hip flexors, or both. When these cats were spinalized 1 month later and then made to demonstrate hind-limb locomotor patterns on a treadmill, their locomotor patterns were disorganized and asymmetrical, with exaggerated knee flexion on the side of the axotomy. This locomotor pattern contrasts with the symmetrical locomotor pattern seen in spinalized cats that were not axotomized and in spinalized cats in which the axotomy procedure was performed following the spinal cord transection. The best interpretation of these results is that the cat made supraspinal adjustments (higher recruitment of knee or hip flexors) during locomotion to compensate for the deficit caused by axotomy, which gradually resulted in altered pathways in the spinal cord. The altered pathways were revealed when the supraspinal influences were removed via spinalization.

The mechanisms that foster these adaptive responses are not known completely. One factor that may play a significant role is the presence of neurotrophins and their effects on neural function. Gomez-Pinilla and colleagues have demonstrated, for example, that removal of afferents into the spinal cord results in a decrease in BDNF and neurotrophin-3 (NT-3) as well as a decrease in an important mediator of presynaptic function, synapsin I. Pertinent to the discussion of training effects, the levels of these factors and of GAP-43 were elevated in the cervical part of the spinal cord. This finding is consistent with an overuse of the forelimbs to compensate for loss of function of the hind limbs (Gómez-Pinilla et al. 2004). Supplementation of BDNF and NT-3 below the lesion in spinalized cats improves their locomotion even without training (Boyce et al. 2007).

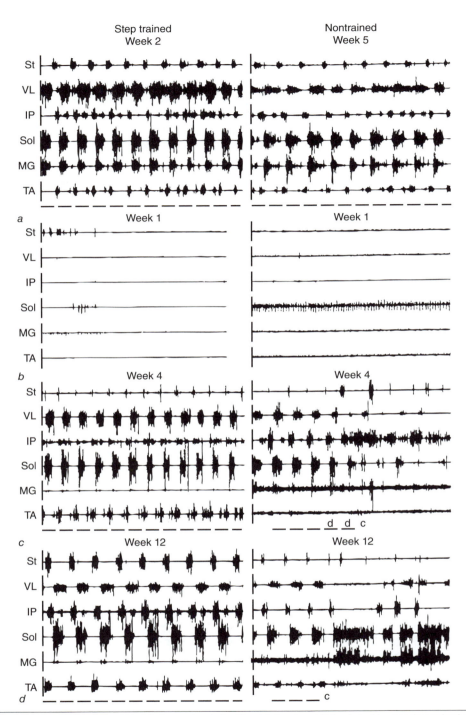

Figure 7.6 EMG activity of hind-limb muscles during bipedal stepping at 0.4 m/s in *(a)* a stepping-trained cat and *(b)* a nontrained cat. From top to bottom, records are from before, 1 wk after, 4 wk after, and 12 wk after spinal cord transection at T12 to T13. Training was 30 min/d, 5 d/wk, after spinal transection in the trained cat. Muscles are semitendinosus (St), vastus lateralis (VL), iliopsoas (IP), soleus (Sol), medial gastrocnemius (MG), and tibialis anterior (TA). Notice how the EMG pattern of the trained cat resembles the pretransection condition more closely than the EMG pattern of the nontrained cat resembles the pretransection condition.

Reprinted by permission from De Leon et al. 1998.

Adaptations of neurotrophins and their receptors and of synapsin I and GAP-43 to short durations (5-7 days) of daily treadmill exercise have been reported in normal nonlesioned rats (Gómez-Pinilla et al. 2001, 2002). BDNF has been implicated in synaptic plasticity and neural adaptations in the brain that underlie cognitive function; this may be comparable to the spinal cord learning that occurs with training following neural injuries (Vaynman and Gómez-Pinilla 2005).

One mechanism for better walking following training in animals with spinal cord transection may be the prevention, through regular activity, of the normal increase that occurs in the enzyme glutamic acid decarboxylase (GAD), which is responsible for the synthesis of the inhibitory neurotransmitter γ-aminobutyric acid (GABA). Less GABA would mean less inhibition, which could favor improved locomotor activity. This finding is consistent with other findings that blockers of inhibitory neurotransmitter receptors in cats with spinal cord transection improve the quality of their locomotion, but only if they are not previously trained (Tillakaratne et al. 2002). There is also evidence that training following spinal cord transection modifies Group I afferent influences on motoneurons (Coté and Gossard 2003).

These findings may have more significance for rehabilitative treatment than for endurance training; time will tell. They do tell us that adaptations at the spinal cord level can result in alterations in the way in which locomotor muscles are used. These demonstrated adaptations at the spinal cord level also suggest that more central adaptations may occur in response to increased activity. Central nervous system plasticity has been demonstrated following skill learning (Kleim, Barbay, and Nudo 1998; Bernardi et al. 1996; Pearce et al. 2000), voluntary endurance running (Van Praag et al. 1999; Cotman et al. 2007), peripheral nerve injury, spinal cord injury, and stroke (Liepert et al. 1998; Barbeau et al. 1998; Jones, Kleim, and Greenough 1996; Nudo et al. 1996; Jain, Florence, and Kaas 1998; Jones et al. 1999). Voluntary exercise for 7 days also increases axonal outgrowth from sensory (dorsal root ganglion) neurons in culture, which also demonstrate increased expression of BDNF, NT-3, synapsin I, and GAP-43 (Molteni et al. 2004). Exercise has beneficial effects on brain function that can be linked to the production of growth factors and neurotrophins and that promote synapse maintenance and formation and neurogenesis (Cotman et al. 2007). Globally, these types of plasticity may eventually prove to be highly functional and important adaptive responses to endurance training by which perception of fatigue is offset and rhythmic locomotor patterns kept more efficient during long-duration exercise.

SUMMARY

The adaptations within the nervous system (neuromuscular junction, alpha-motoneuron, spinal cord systems, and supraspinal mechanisms) are less well known than the changes that occur in skeletal muscle fibers, but they may prove important in explaining at least a portion of the increased performance that occurs as a result of training. Perhaps more importantly, these nervous system adaptations, once fully described, might be relevant for the use of endurance training in certain pathological states with nervous system involvement. At the level of the neuromuscular junction, morphological, biochemical, and physiological adaptations with aerobic endurance training serve to enhance endurance. These adaptations are both pre- and postsynaptic. Motoneurons show biochemical evidence of adaptations to increased activity, including increased content and axonal transport of factors that seem vital for optimal adaptation. However, changes resulting from increased activity in physiological properties that would determine how they are used during exercise are unknown at present. Spinal cord systems are responsive to increased usage. This is most evident in spinalized cats in which daily locomotor training with muscles distal to the lesion results in alterations in muscle usage patterns. This may indicate that locomotor actions can be improved in various ways through endurance training.

Muscle Molecular Mechanisms in Strength Training

The development of increased muscle mass and maximal voluntary muscle strength involves primary signals at the muscle level that differ from those involved during the development of neuromuscular endurance. The emphasis here is on primary signals, since, qualitatively speaking, many of the signals that evoke muscle adaptations to strength training and whole-body endurance training are probably similar. There are several factors that distinguish activities that promote increased neuromuscular strength and mass from those that promote neuromuscular endurance.

The main factor, of course, involves the forces to which structures are subjected during neuromuscular activation. Stress and strain on muscular structures can constitute a factor that, through mechanotransduction via several pathways, provides biochemical signals that can alter gene expression and ultimately phenotype, examples of which are included in this chapter. A second factor that distinguishes strength training from aerobic-type endurance training is the degree and pattern of recruitment of motor units during the training to induce hypertrophy or increased strength. This factor influences the outcome of the phenotype, not only by the virtue of the high forces and firing frequencies to which the individual muscle fibers are subjected but also by the apparent influence that recruitment patterns during this high degree of voluntary effort have on the associated nervous pathways (which we discuss in detail in chapter 10).

In this chapter, I summarize what I believe are the signals that are important in modulating protein synthesis to produce the phenotypic response to resistive overload. In doing so, I call upon several models of increased resistance exercise that promote hypertrophy, including chronic stretch of animal muscles in vivo, chronic and cyclic stretch of isolated cells (sometimes cells other than muscle cells, such as myocytes and epithelial cells) in vitro, chronic compensatory overload of muscles induced by ablation of synergistic muscles, electrical stimulation of both animal and human muscles, and actual strength training in both animals and humans. Clearly, some aspects of the models other than strength training per se are of limited value in furthering our knowledge of the latter. Consequently, I decided to leave out some information that others might think should be included in this chapter and to include other information for which the application to strength training in humans might be considered tenuous. As in the case of aerobic endurance training, a great deal can be gleaned concerning mechanisms by examining what happens in the minutes, hours, and days following the delivery of the first training stimulus.

Next, in chapter 9, I summarize the phenotypic changes that occur with strength training. I put some emphasis on issues such as hyperplasia as a mechanism and the specific effects of eccentric training, since these seem to be somewhat controversial subjects. Finally, in chapter 10, I discuss the neural influences that seem to be so important in the improved performance that occurs with resistance training.

ACUTE RESPONSES IN PROTEIN SYNTHESIS AND DEGRADATION

Compensatory overload of rat hind-limb muscles induced by tenotomy of synergists results in increases in both protein synthesis and degradation (Goldspink, Garlick, and McNurlan 1983). This response is also evident in resistance training in humans. Within 3 hours of a bout of resistance exercise involving concentric and eccentric contractions, fractional synthesis of mixed muscle proteins more than doubles and then remains elevated, but at a lower level, for at least 48 hours (MacDougall et al. 1992; Chesley et al. 1992; Biolo et al. 1995; see figure 8.1). The rapid increase is accompanied by an unchanged RNA content (thus increased RNA activity), signifying that alterations in posttranscriptional mechanisms are important contributors to the increased protein

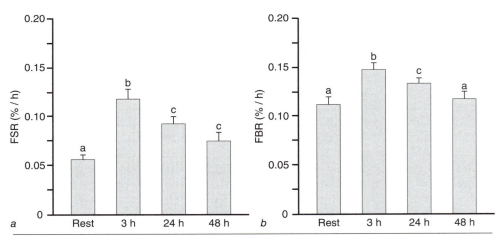

Figure 8.1   Effect of a single bout of resistance exercise (8 sets of 8 concentric or eccentric contractions at 80% of concentric 1-repetition max) on mixed muscle protein *(a)* fractional synthesis (FSR) and *(b)* fractional breakdown (FBR) rates. Means with different letters are statistically different.

Reprinted by permission from Phillips et al. 1997.

synthesis. Fractional degradation of proteins is also elevated in the hours following a bout of resistance exercise (Phillips et al. 1997, 1999; Biolo et al. 1995). The increased muscle protein synthesis is accompanied by increased passage of the amino acids leucine, lysine, and alanine from plasma into muscle, indicating that increased amino acid transport may constitute a significant promoter of this phenomenon.

Wong and Booth (1990a, 1990b) approached this problem of acute changes accompanying resistance training by electrically stimulating hind limbs of anesthetized rats so that muscles worked against a specified load. They then measured various components of protein synthesis at periods following the exercise. They reported several results: (1) Protein synthesis rates were higher than they were in control muscles 12 to 17 hours following activity involving 192 unloaded or loaded (concentric) contractions, (2) this elevated protein synthesis was still present 36 to 41 hours following the exercise, (3) protein synthesis rate increases were greater than the increases in RNA were, and (4) there was no relationship between the ability of the stimulation protocol to stimulate protein synthesis acutely and the ability of the stimulation protocol to increase muscle mass chronically.

The findings in general showed that protein synthesis and degradation are both elevated in muscles as a result of resistance overload and that these changes, induced by a relatively short exposure to the contractile stimulus, last for several days. Several mechanisms are involved in these changes; the major ones are discussed in the following sections.

Stretch as a Signal for Adaptation

One way in which resistance training differs from aerobic endurance training involves the stress and strain to which contractile and noncontractile tissues are exposed during an acute training bout as a consequence of the relatively high forces involved. Passive stretch constitutes a strong signal for changes in protein synthesis. When rabbit muscles are immobilized in a lengthened or stretched position, for example, muscle weight increases, primarily as a result of increased muscle fiber length (Dix and Eisenberg 1990). Under these conditions, much of the increased protein synthetic activity occurs in the distal end of the muscle, where sarcomeres are being added in series. During the first 6 days, increases in the concentration of polysomes are seen in the distal muscle region, attesting to increased protein synthesis. Molecules that are involved in anchoring the muscle fibers to the myotendinous junction, such as vinculin and talin, increase in expression. When electrical stimulation is added to the stretch, changes are more profound and are usually greater than the sum of electrical stimulation and stretch individually (Osbaldeston et al. 1995). This observation is indirect proof of the importance of stretch of noncontractile tissues in the adaptive response to resistance training.

The in vivo passive stretch model just described is somewhat limited in its value as a model for adaptations induced by resistance training, since it usually involves immobilization of the muscle to fix its length. In short-term experiments, however, before the atrophy of inactivity begins to express itself, the model does reveal that activation of components other than the contractile apparatus can have a mitogenic (i.e., growth-promoting) effect on muscle fiber protein synthesis. In this effect is found the fundamental difference in the signals involved in the adaptations to resistance training and aerobic endurance training. While the signals for endurance-related adaptations are probably linked to metabolic demand and supply, the signals for resistance-related adaptations are linked to forces exerted on structures that support the contractile machinery and perhaps on the contractile machinery itself.

There is converging evidence that stretch, independent of increased contractile activity, can stimulate protein synthesis via several pathways. The stretch of extracellular matrix generates intracellular signals that eventually lead to altered gene expression. Much of this evidence comes from experiments in which cell types other than skeletal muscle—such as cardiac myocytes, epithelial cells, and vascular smooth muscle cells—are exposed in vitro to controlled sustained or cyclic stretches. A general scheme of the intracellular signals relating to mechanical stimulation is shown in figure 8.2.

MAPKs

MAPKs constitute important communication links between events occurring at the cell surface membrane and transcriptional events in the nucleus. These kinases are activated by a variety of signals resulting from activation of G protein kinase–based or

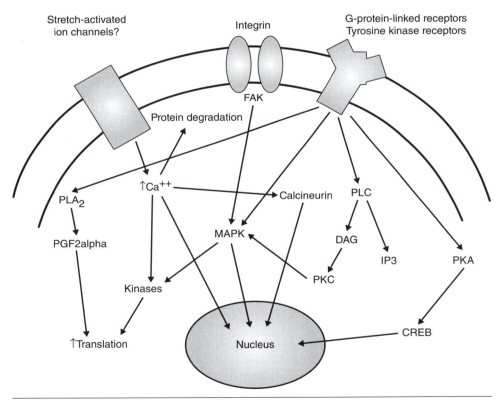

Figure 8.2 Stretch-related intracellular signals that may be involved in overload-induced muscle hypertrophy. FAK, focal adhesion kinase; PLC, phospholipase C; DAG, diacylglycerol; IP_3, inositol triphosphate; PKC, protein kinase C; PKA, protein kinase A; CREB, cyclic AMP response element binding protein; PLA2, phospholipase A2; PGF2α, prostaglandin F2α.

tyrosine kinase–based transmembrane receptors. These receptors in turn phosphory-late transcription factors and are thought to be involved in the induction of some of the IEGs, such as *FOS* and *JUN* mentioned further on (Seedorf 1995). An increasing number of MAPK-regulated transcription factors are being discovered (Sugden and Clerk 1998). MAPK may also play an important role in the posttranscriptional control of protein synthesis, allowing a more rapid response in protein synthesis than would be possible by relying on transcriptional mechanisms alone.

One hypothesis (Watson 1991) that might pertain to resistance training main-tains that passive muscle stretch, which causes a physical deformation of the three-dimensional configuration of the transmembrane receptors, results in an activation of that receptor, much the same as if by attachment of its ligand. In this way, passive muscle stretch would supply the intracellular compartment with mitogenic signals. This mechanism could help to explain how passive muscle stretch increases protein synthesis but has yet to be substantiated.

It has been suggested that one site where stretch is transduced into changes in protein synthesis is the integrin molecule (Carson and Wei 2000). Integrin is a trans-membrane protein that is attached on the extracellular surface to components of the extracellular matrix and on the intracellular surface to components of the cytoskeleton, including talin, vinculin, and alpha-actinin (Longhurst and Jennings 1998; Alberts et al. 1994). Integrins are clustered at sites termed *focal adhesions* (Burridge and ChrzanowskaWodnicka 1996). Intracellular signaling via integrins is accomplished by protein phosphorylation via a focal adhesion kinase (FAK), which can lead to MAPK activation, among other downstream events. Thus, the stretch signal that activates protein synthesis may be mediated via deformation of the integrin molecule during stretch of the membrane, which activates FAK and, as a result, MAPK and gene expression (Clark and Brugge 1995).

Very little information is available at present regarding MAPK activation in muscle; available information comes from in vitro studies with stretched myocytes (Sadoshima and Izumo 1993), epithelial cells (Papadaki and Eskin 1997), and osteoblasts (Schmidt et al. 1998). However, one recent report shows that MAPKs are activated robustly in intact muscles with passive stretch (Martineau and Gardiner 1999). One of the MAPKs, JNK, is activated by 60 minutes of bicycle ergometer exercise, with a corresponding increase in the expression of its downstream nuclear target, *JUN* (Aronson et al. 1998). More recently, it has been shown that JNK is particularly activated by eccentric con-tractions with knee extensors in humans (Boppart et al. 1999). In addition, electrical stimulation of rat muscles to perform strong, intermittent, isometric contractions (contractions with durations of 500 milliseconds at 100 hertz performed once per second) results in activation of both JNK and extracellular signal–regulated kinase (ERK), to different magnitudes and with different time courses. Thus, the signals resulting in activation of these MAPKs and their targets are distinct to some degree.

Although the idea of a signal or group of signals that activates the kinase system with stretch is appealing, we must consider the possibility that deactivation via a phosphatase system may also constitute a mechanism involved in increased protein synthesis and subsequent hypertrophy. For example, calcineurin, a calcium-regulated phosphatase, is necessary for the rapid gain in muscle mass and the severalfold increase in Type I fibers that occur during 4 weeks of compensatory overload in rat plantaris (Dunn, Burns, and Michel 1999).

IGF-1

IGF-1 is produced primarily by the liver (the splice variant IGF-1Ea), but it is also produced by muscle (splice variants IGF-1Ea and IGF-1Eb). IGF has stimulatory effects on gene transcription and translation, hyperplasia through satellite activation, and proliferation of myonuclei in vitro (DeVol et al. 1990). Mechano growth factor (MGF) is expressed in muscle subjected to activity and is derived from the *IGF1* gene by alternative splicing (Goldspink 2005). Most effects are mediated via the IGF-1 receptor, which, like the insulin receptor, is a tyrosine kinase. In cultures of avian myotubes, IGF-1 stimulates cell hyperplasia and cell hypertrophy (insulin does the same, but at a much higher, nonphysiological concentration, and IGF-2 stimulates hyperplasia but not hypertrophy under these conditions). One way in which IGF1 exerts its positive effect on protein synthesis is via enhancement of the formation of a complex of initiating factors (eIF4E–eIF4G; Vary, Jefferson, and Kimball 2000).

Many sources of evidence converge to suggest that the increased expression of IGF-1 by stretched or overloaded muscle fibers has significance for the proliferation of myonuclei. Yang and colleagues (1996) found that IGF-1 was upregulated within 2 hours after initiation of stretch in rabbit tibialis anterior and was located primarily in small fibers containing neonatal MHC. Thus, the ends of normal fibers are the regions where new longitudinal growth takes place, and IGF-1 takes part in this process.

However, the effect of IGF-1 is not isolated to myonuclei proliferation. Barton-Davis, Shoturma, and Sweeney (1999) showed that IGF-1 still exerts a significant hypertrophic effect (50% of the total effect) in mouse muscles in which myonuclei proliferation is prevented by previous irradiation.

The stretch component of strength training, and not merely the increased metabolic rate of exercise, is clearly the signal promoting IGF-1 production in muscle. Goldspink and colleagues (1995) found that the response of IGF-1 to stretch plus stimulation was different from the response to stimulation alone. Static stretch alone increased IGF-1 mRNA and stimulated protein synthesis. Low-frequency stimulation amplified this response considerably, whereas stimulation alone had no effect. In rat muscles stimulated to simulate 4 days of resistance training, IGF-1Ea and MGF mRNA increased, with higher increases noted when contractions were eccentric (Heinemeier, Olesen, Schjerling, et al. 2007).

Adams and Haddad (1996) showed that compensatory hypertrophy of rat plantaris resulted in increased IGF-1 mRNA and protein that peaked at 3 days at levels that were sixfold and fourfold that of controls, respectively. Increases in IGF-1 preceded the hypertrophic response. These authors also found an increase in DNA content that was proportional to the increased muscle mass at 3, 7, 14, and 28 days of overload and an increase in mitotically active nuclei, indicating activation of satellite cells. They concluded that IGF-1 upregulation by muscle fibers is involved in the increase in DNA content, perhaps by activation and incorporation of satellite cells. Their more recent work (Adams, Haddad, and Baldwin 1999) supports this conclusion by showing a close temporal relationship between increased IGF-1 expression and markers of satellite cell activation during the early period of overload. In related work, Adams and McCue (1998) found that localized infusion of IGF-1 directly onto a rat tibialis anterior muscle resulted in increased total muscle protein and DNA content.

Nonetheless, IGF-1 production by overloaded muscle is only part of the puzzle. Yan, Biggs, and Booth (1993) examined the effect of one acute bout of 192 electrically evoked eccentric contractions of the rat tibialis anterior on IGF-1 expression. IGF-1 protein became significant at levels three times higher than the control value by the fourth day after eccentric contractions. IGF-1 thus peaks after the increase in protein synthesis and therefore is not responsible for it. Levels of mRNA for IGF-1Ea, IGF-1Eb, and MGF are all increased following 5 weeks of resistance training, with MGF levels being the highest (Hameed et al. 2004).

Downstream signaling mechanisms for IGF-1 include the phosphatase calcineurin and the PI3K and Akt pathways. The calcineurin pathway was discussed in chapter 6—current thought is that this pathway is more important for endurance-related changes in muscle and may not be necessary for the development of muscle hypertrophy (Glass 2003).

The Akt and mTOR Cascade

The Akt and mTOR signaling pathway (figure 8.3) seems to be essential for muscle hypertrophy, since selective blockade of mTOR with rapamycin prevents muscle hypertrophy (Bodine et al. 2001; Glass 2003). Akt (a protein kinase that is also known as *protein kinase B,* or *PKB)* is activated by IGF-1, as well as by insulin, through the generation of PIP3, which results from activation of the enzyme PI3K. In essence, Akt is recruited to the plasma membrane by PIP3s, where it becomes phosphorylated and thereby activated (Sandri 2008). Activated Akt then activates mTOR through a

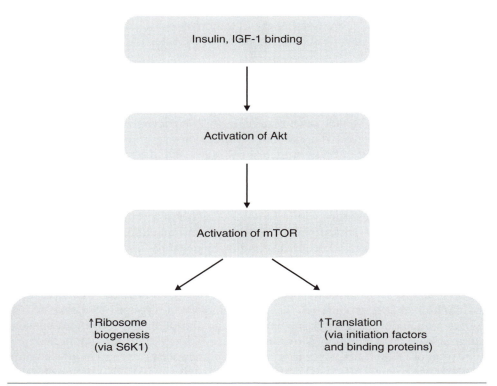

Figure 8.3 The role of mTOR in the response to resistance training.

series of intermediate reactions. Akt and mTOR are both phosphorylated during the minutes following a bout of resistance exercise. Downstream targets of mTOR include proteins involved in the control of mRNA translation, such as S6K1, and initiation factors eIF4G and the eIF4E binding protein 4EBP1 (Bolster et al. 2003). These targets are also activated in the minutes following cessation of resistance exercise (Bolster et al. 2003). All of these changes appear to demonstrate a bell-shaped increase, then decrease, in the recovery period following the exercise, suggesting that an important response to acute resistance training is a transient increase in translation initiation that, when repeated after each training session, results in muscle hypertrophy.

PKC

PKC is a widely distributed kinase with broad substrate specificity that is activated by calcium, phospholipids, and DAG and is therefore activated upon stimulation of phospholipase C (PLC; Haller, Lindschau, and Luft 1994). Increased calcium levels result in increased binding of PKC to membrane lipids. Among its functions is negative feedback, via phosphorylation of tyrosine receptors, that decreases signaling activity of these tyrosine receptors (Haller, Lindschau, and Luft 1994). PKC also increases calcium flux from the extracellular space and from intracellular stores and regulates several calcium-regulating enzymes, including calcium ATPase. PKC is also involved in mitogenic signaling during cell growth. Certain isoforms of PKC are translocated to the nucleus, where they may be involved in gene transcription (Haller, Lindschau, and Luft 1994).

Compensatory hypertrophy in the rat results in an increase in PKC that peaks at double the control value 4 days after the initiation of overload. PKC activity subsequently declines to control values by about 9 days (Richter and Nielsen 1991).

Prostaglandins

Vandenburgh and colleagues (1995) contributed significantly to our knowledge regarding prostaglandin synthesis during stretching of myotubes. The techniques used in their series of studies included cyclic stretching of avian myofiber cultures embedded in an extracellular matrix and maintained in vitro for several days (Vandenburgh, Swasdison, and Karlisch 1991). Stretched cultures respond by producing two forms of prostaglandins with different time courses. PDF2 stimulates protein synthesis, while PGE2 stimulates protein degradation. PDF2 production increases gradually from the initiation of cyclic stretching for at least 4 days, while PGE2 increases rapidly and then declines to baseline within 24 hours (Vandenburgh et al. 1990, 1995). These events are blocked by indomethacin, a blocker of prostaglandins synthesis. Thus, the increased production of these prostaglandins by muscle is not due to mobilization of presynthesized intramuscular stores. Prostaglandin production is also blocked by inhibiting phospholipases (Vandenburgh et al. 1993).

These changes are related to G proteins, since treatment of cultures with pertussis toxin inhibits PDF2 production, the activation of cyclooxygenase (the enzyme that converts arachidonic acid to prostaglandins), and phospholipase activation (Vandenburgh et al. 1995). In addition, activation of these pathways does not require electrical activation of membranes, since it occurs in the presence of tetrodotoxin, a sodium conductance channel blocker (Vandenburgh et al. 1993, 1995). Thus, it appears that mechanical stretch stimulates a G protein–linked cascade that results in prostaglandin synthesis. How this transduction takes place and what species of G proteins are involved are not known. PDF2 has been shown to stimulate the phosphorylation of the 40S ribosomal protein S6 via stimulation of the implicated kinases (p70S6K and p90RSK; Thompson and Palmer 1998). It may also stimulate phosphorylation of PHAS-1, a protein that, when not phosphorylated, binds to the mRNA cap binding protein eIF4E and prevents it from initiating translation (Thompson and Palmer 1998).

Vandenburgh and colleagues (1993) presented several possibilities as to how stretch might activate phospholipases. Stretch may increase the sensitivity of the muscle cells to growth factors that either (1) activate phospholipases, transducers such as G proteins that activate phospholipases, and calcium-sensitive lipases via alterations in intracellular calcium levels or (2) increase the accessibility of the phospholipase to its membrane substrate.

Stretch activates the phospholipases PLA2 and PLD and the phosphatidylinositol-specific PLC, which is activated by electrical rather than mechanical events (Vandenburgh et al. 1993). PLA2 results in the release of arachidonic acid and the production of PDF2 (Thompson and Palmer 1998).

We also know from these studies that indomethacin, which blocks prostaglandin synthesis, also reduces (but does not abolish) the initial decrease and subsequent increase in protein synthesis. Thus, prostaglandins play a role but are not the whole story concerning the increased protein synthesis that occurs with stretch.

Proto-Oncogenes *FOS, JUN,* and *MYC*

Evidence for a change in protein synthesis in stretched cells is given by a rapid increase after the initiation of the stimulus in the expression of several proto-oncogenes, the products of which are important transcription factors that influence the expression of several proteins involved in cell adaptation. Three of these have been investigated specifically with reference to muscle: *FOS, JUN,* and *MYC.* (Many others have been found in other tissues, but their responsiveness and importance have not been demonstrated in muscle.)

The proto-oncogene *MYC* codes for a phosphoprotein that can bind to DNA and that appears to play a role in mitosis (Hesketh and Whitelaw 1992). It is normally downregulated in fully differentiated muscle fibers, reflecting a reduced cell proliferation. The mRNA for *MYC* increases within 3 hours following the induction of compensatory overload and remains elevated for 24 hours before returning to baseline (Whitelaw and Hesketh 1992). The increase in *MYC* expression has been proposed to be involved in the activation of ribosome synthesis early during adaptation.

The *FOS* and *JUN* oncogenes are components of the AP-1 transcription complex that binds to promoters of several growth-associated genes. Both *FOS* and *JUN* are induced in response to mechanical stimulation. In stretched rabbit muscle, *FOS* and *JUN* expression increases transiently during the first 3 hours of stretch and then returns to baseline (Dawes et al. 1996; see figure 8.4). Interestingly, these changes in *FOS* and *JUN* expression are different in amplitude when chronically stretched muscles are also electrically stimulated (Goldspink et al. 1995; see figure 8.5). For example, rabbit muscles show a 20-fold and 15-fold increase in mRNAs for *FOS* and *JUN,* respectively, when passively stretched, but when electrical stimulation is added, their levels increase to 80-fold and 60-fold, respectively, even though no increase occurs

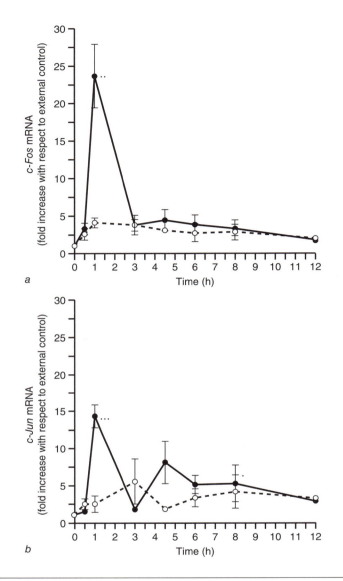

Figure 8.4   Effect of passive stretch of rabbit latissimus dorsi in vivo (to approximately 115% of optimal muscle length) on (a) c-Fos and (b) c-Jun mRNA. Filled circles are stretched; open circles are contralateral control muscles.

Reprinted by permission from Seedorf 1986.

with electrical stimulation alone (Goldspink et al. 1995). In addition, it is clear that *FOS* and *JUN* are not signals arising from the exact same stimulus. For example, when stretch is reapplied after a time without stretch, the responses of *FOS* and *JUN* are qualitatively as well as quantitatively different from their responses to the initial stretch. The responses of these two proto-oncogenes to stretch may therefore occur along two slightly different pathways. There is still much to learn about the importance of early changes in expression of *MYC*, *FOS*, and *JUN* with stretch and in resistance training.

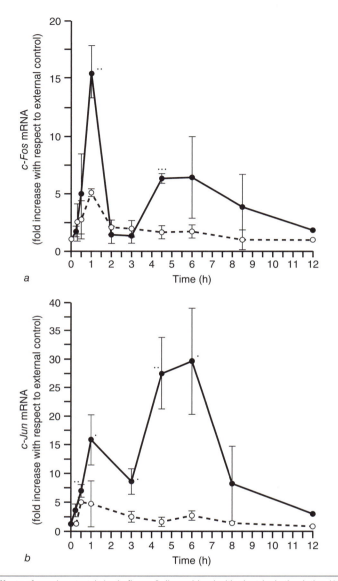

Figure 8.5   Effects of passive stretch (as in figure 8.4) combined with electrical stimulation (10 Hz) on *(a)* c-Fos and *(b)* c-Jun mRNA. Compare with figure 8.4.

Reprinted by permission from Osbaldston et al. 1995.

Muscle Regulatory Factor Genes

The products of the muscle regulatory factor (MRF) genes, which are active during muscle development and are present at negligible levels in normal adult muscles, include Myf-5, MyoD1, MRF4, and myogenin (JacobsEl, Zhou, and Russell 1995). In rat tibialis anterior subjected to stretch and electrical stimulation for 2 hours, mRNA for MRF4 and Myf5 increased significantly (11-fold and 6-fold, respectively). Since binding sites for these factors are found upstream of many muscle genes, this may signify another pathway in which stretch activates protein synthesis. In rat hind-limb muscles subjected to compensatory overload, and in human muscles subjected to an acute bout of electrical stimulation evoking strong contractions like those seen in resistance training, myogenin and *MyoD* expression increased within 12 hours of the stimulus, possibly as a result of increased IGF-1 production (Adams, Haddad, and Baldwin 1999; Bickel et al. 2005). The responses of the MRF genes peaked at 4 to 8 hours postexercise and then returned to control levels within 24 hours (Yang et al. 2005).

Myostatin Expression

Myostatin is a member of the transforming growth factor beta (TGF-β) superfamily, which exerts a negative influence on muscle growth. Myostatin mRNA is decreased following several days of resistance contractile activity and is decreased more for eccentric versus isometric and concentric contractions (Heinemeier, Olesen, Schjerling, et al. 2007).

Posttranscriptional Changes

We have already mentioned that increased RNA activity (in units of protein synthesis per unit time per unit RNA) occurs within hours of a bout of resistance exercise. Baar and Esser (1999) showed that 6 hours after electrically induced lengthening contractions of rat hind-limb muscles, there was an increase in the size of the polysome pool. This finding indicates increased initiation, elongation, or termination of protein synthesis. These investigators also found increased phosphorylation of the protein kinase p70S6K 3 and 6 hours after stimulation. Phosphorylation remained increased, but at lower levels, up to 36 hours postexercise. This kinase is involved in the phosphorylation of the small ribosomal subunit protein S6, which is involved in the regulation of protein synthesis and is mitogen stimulated (Sugden and Clerk 1998). This kinase also regulates the translation of a subset of mRNAs encoding translation factors and ribosomal proteins, with the result that protein production increases without an increase in steady-state mRNAs (Chen et al. 2002). In cardiac myocytes, the phosphorylation of this protein is required for stretch-induced increase in protein synthesis. In the Baar and Esser (1999) experiment, there was a relationship between the degree of phosphorylation of this protein at 6 hours and the increase in mass of the muscle after 6 weeks of the stimulation paradigm (see figure 8.6).

After 4 days of stimulation of rat muscle to simulate resistance training, the expression of large ribosomal protein is increased, as is the total RNA concentration. This finding signifies an increased ribosomal capacity for translation (Heinemeier, Olesen, Schjerling, et al. 2007). An increased RNA content is also found in human muscles after two sessions of resistance training (Bickel et al. 2005).

Control of protein synthesis at the translation level also depends to a large extent on the activation levels of numerous eukaryotic initiation factors (eIF) that facilitate peptide initiation at the ribosome. At 16 hours following an acute bout of strength training in rats, the activity of a member of this family, eIF2B, is elevated, along with protein synthesis (Farrell et al. 1999).

Esser and colleagues (Chen et al. 2002) have revealed the extent to which resistance exercise can activate posttranscriptional and translational mechanisms. Their studies have revealed that gene products activated by acute resistance exercise include elongation factor 1α (EF1α), a protein responsible for bringing tRNAs to the ribosome),

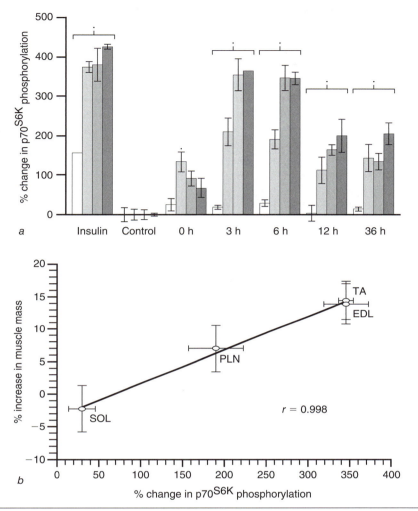

Figure 8.6   (a) Effect of a single bout of electrically evoked contractions (60 contractions at 100 Hz, each lasting 3 s) on phosphorylation of p70S6K in rat soleus (Sol; open bars), plantaris (Pln; lightly shaded bars), extensor digitorum longus (EDL; darkly shaded bars), and tibialis anterior (TA; filled bars). Dots indicate results different from controls. (b) Correlation between change in phosphorylation of p70S6K at 6 h after stimulation and muscle wet mass measured after stimulation 2 d/wk for 6 wk.

Reprinted by permission from Baar et al. 1999.

ribosomal protein L12 (a component of the large 60S ribosomal subunit), and c-Myc (which can induce transcription of many ribosomal protein mRNAs). Downregulated genes include elongation factor-2 kinase (which inactivates elongation factor 2), and cathepsin C (a lysosomal protease). These changes have an obvious net anabolic effect on muscle.

Intracellular Proteolytic Systems

Intracellular proteolytic systems include the calcium-activated neutral proteases (calpains). The lysosomal proteases and the ATP-ubiquitin-dependent pathway are also included in these systems.

Calpains I and II are proteases activated by micromolar and millimolar concentrations of calcium, respectively. There is some evidence that activation of calpain, which demonstrates a specificity toward myofibrillar and cytoskeletal proteins, is associated with the disruption of Z-lines in myofibrils from exhausted rat muscles (Belcastro, Albisser, and Littlejohn 1996; Belcastro 1993). Calpain activation may be secondary to the increased intracellular calcium that accompanies exercise-associated muscle damage. After 10 days of reduced load bearing, muscles exposed to 2 days of weight bearing demonstrate increased calpain II expression (Spencer and Tidball 1997). Interestingly, but not completely unexpectedly, the mRNAs for calpain I and II are elevated for at least 24 hours after acute resistance exercise, with the increase being more pronounced in Type I fibers (Yang et al. 2006).

Lysosomal proteases include cathepsins D, L, B, and H, as well as carboxypeptidases and aminopeptidases. During 3 days of muscle stretch, large increases (200%-400%) occur in activities of cathepsins B and L and dipeptidyl aminopeptidase, with or without stimulation (Goldspink et al. 1995). Perhaps these enzymes are instrumental in the muscle remodeling that occurs with stretch. Vandenburgh and colleagues (1990) found an increase in cathepsin H activity within 1 hour after the imposition of cyclic stretch in myotube cultures.

Apoptosis is programmed cell death. Following a bout of resistance exercise, the expression of the apoptotic genes *CASP3* and *BAX* and *BCL-2* are increased for up to 24 hours postexercise (Yang et al. 2006).

The ubiquitin proteasome system (see figure 8.7) may be the major source of nonlysosomal protein degradation. Ubiquitin is a protein that is known to form

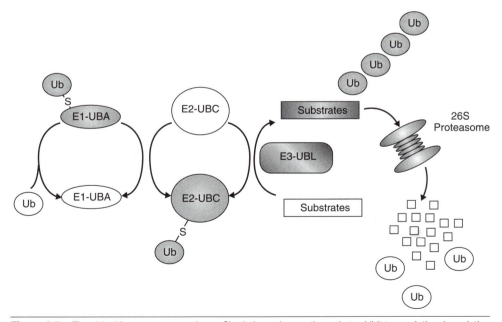

Figure 8.7 The ubiquitin proteasome pathway. Shaded proteins are those that exhibit transcriptional regulation in response to exercise. Ub, ubiquitin; E1-UBA, E1 protein or ubiquitin-activating enzyme; E2-UBC, E2 protein or ubiquitin-conjugating enzyme; E3-UBL, E3 protein or ubiquitin ligase.

Reprinted by permission from Reid 2005.

conjugates with abnormal proteins, thus tagging them for degradation by covalently binding and modifying the abnormal protein. The tagged proteins are then degraded by a large protease complex. Elevated ubiquitin levels (60%) have been found in the biceps brachii of subjects who had performed eccentric contractions 48 hours before (Thompson and Scordilis 1994). Following eccentric contractions, there is an increase in ubiquitin conjugates, ubiquitin protein, and ubiquitin mRNA and in the protein and mRNAs of 20S protease protein and E2 (ubiquitin carrier) protein (Reid 2005). Following a single bout of resistance exercise, the expression of muscle RING finger 1 (*MURF1*), an ubiquitin ligase in the ubiquitin proteasome pathway, is elevated for at least 24 hours (Yang et al. 2006). Much work remains to be done in the research of the ubiquitin proteasome system as an important component of muscle remodeling during training.

CONNECTIVE TISSUE RESPONSES

Four days of simulated resistance training via electrical stimulation of rat sciatic nerve increased the expression of TGF-β1 and collagens I and III in Achilles tendon, when measured 24 hours after the last exercise. In addition, expression of lysyl oxidase (*LOX*), an enzyme involved in cross-linking collagen, was elevated. Similar changes were evident in the muscle, although the changes, unlike in the tendon, were more pronounced with eccentric versus concentric or isometric types of contractions (Heine-meier, Olesen, Haddad, et al. 2007). In human muscle, the rate of collagen synthesis in patellar tendon increases by 6 hours after a resistance training bout using quadriceps and peaks at 24 hours postexercise (Miller et al. 2005). A similar time course occurs with muscle collagen synthesis (see figure 8.8).

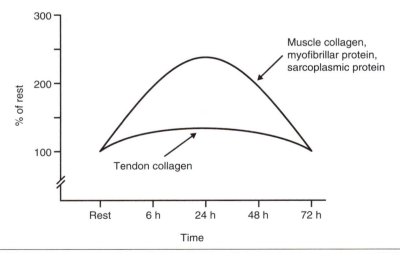

Figure 8.8 Fractional rates of synthesis of tendon collagen protein, muscle collagen protein, myofibrillar protein, and sarcoplasmic protein at rest and following 1 h of vigorous knee extensions at 67% of maximum workload.

Data from Miller et al. 2005.

ROLE OF MUSCLE DAMAGE

Some tissue disruption is bound to occur as a result of the high forces generated during resistance training. This is particularly evident during the eccentric phase of dynamic contractions, in which less muscle area is used to generate force than is used during the concentric phase (Gibala et al. 1995). Phenomena seen in muscles after eccentric contractile activity include cytoskeletal disruptions, Z-disc streaming, Aband disorganization, membrane damage, sarcoplasmic reticulum damage, hypercontracted regions, localized areas of increased intracellular calcium, increased incidence of desmin-negative fibers, loss of dystrophin organization, and invasion of cells (Saxton and Donnelly 1996; Komulainen et al. 1998; Fridén and Lieber 1992, 1996, 1998; Fridén, Seger, and Ekblom 1998; Warren III et al. 1995; Lowe et al. 1995; Yasuda et al. 1997; Moore et al. 2005; Crameri et al. 2007; Lovering and De Deyne 2004). It has been suggested that these signs of cell disruption may actually reflect adaptive remodeling of the cell, because ultramicroscopic evidence shows that Z-disc smearing and disruptions are actually greater 7 to 8 days following the exercise than they are 1 hour after the exercise (Wu et al. 2004). Effects are more severe in Type II fibers (Fridén and Lieber 1998). Muscle force capacity is reduced following eccentric damage and remains depressed for several days or up to several weeks (Lowe et al. 1995; Howell, Chleboun, and Conatser 1993; Brown et al. 1996). Interestingly, intrinsic proteolytic mechanisms possibly are not involved in the increased protein degradation that follows resistance exercise, since degradation does not increase during the 6 hours immediately following the insult (Lowe et al. 1995). Degradation of proteins

DELAYED ONSET MUSCLE SORENESS

The exact cause of delayed onset muscle soreness (DOMS) that follows strong lengthening contractions is not known. Comparisons of animal and human studies have proven problematic, since animal studies traditionally use electrical stimulation to evoke contractions in situ, while human studies involve voluntary contractions. Crameri and colleagues (2007) at the University of Copenhagen addressed this issue by studying the muscle responses to lengthening contractions evoked in the same subject by voluntary contraction in one leg and electrical stimulation in the other. Both conditions resulted in similar DOMS. However, significant disruption of cytoskeletal proteins, including desmin, a protein that has been implicated previously as a possible correlate to DOMS, occurred only in the electrically stimulated leg. The electrically stimulated leg also showed more intracellular disruption, destroyed Z-lines, and increased satellite markers when compared with the voluntarily recruited leg. One element that was similar in both legs, besides the extent of DOMS, was the increased staining for tenascin C, an intramuscular connective tissue associated with the extracellular matrix, and an increased number of inflammatory cells, CD68+ immunoreactive macrophages. These results strongly suggest that tears in the extracellular matrix and the resulting inflammatory response, and not necessarily other cytoskeletal protein disruptions that have previously been reported in animal and human studies, are closely associated with, and may cause, DOMS.

is temporally related to phagocytic infiltration into fibers during 24 to 120 hours following the injury (Lowe et al. 1995).

Following the damage induced by eccentric contraction, an inflammatory response ensues. This includes edema and the infiltration of the damaged fibers by neutrophils and macrophages. The attraction of inflammatory cells may occur via wound hormones released from the injured muscle (Tidball 1995). Activation of calpain via elevated calcium levels as a consequence of muscle damage may also generate a chemoattractive signal (Raj, Booker, and Belcastro 1998). Fibroblasts, macrophages, and neutrophils, as well as the damaged fiber itself, produce a number of substances that take part in the degeneration–regeneration process, including platelet-derived growth factor (PDGF), basic fibroblast growth factor (bFGF), IL, TGF, and TNF (Clarkson and Sayers 1999; Tidball 1995).

The importance of these events in the adaptive response of muscle to resistance training is not currently known. Phillips and colleagues (1999) have observed that the stimulatory effect of a single bout of eccentric knee extensions on muscle protein synthesis and breakdown is attenuated in trained subjects. The fact that the degree of muscle damage produced by acute eccentric exercise is also decreased in these subjects may suggest a link between muscle damage mechanisms and protein synthesis and degradation.

ROLE OF DIETARY SUPPLEMENTS

Of the more than 200 dietary supplements currently available in the marketplace, very few have been shown using scientifically rigorous techniques to increase the effects of resistance training by enhancing muscle mass and strength gains (Nissen and Sharp 2003). Creatine and β-hydroxy-β-methylbutyrate supplementation have been shown to increase lean body mass accretion with resistance training (Nissen and Sharp 2003). The literature regarding protein supplementation is less clear, primarily due to problems associated with placebo blinding, the precise definition and control of protein levels, and the possibility of inclusion of other nutrients in the protein mixture. Athletes who are resistance training appear to need more protein than their nontraining cohorts need (1.7-1.8 grams per kilogram per day) in order to ensure nitrogen balance, with any supplementation beyond this having little advantage for enhancing muscle mass increase. The indirect evidence that supplements taken before or after bouts of resistance training ultimately result in enhanced hypertrophy and strength gains is nevertheless quite strong. For example, increased amino acid ingestion following a bout of resistance training stimulates their net uptake and renders net muscle protein balance more positive (Elliot et al. 2006). The addition of supplements that stimulate insulin secretion has added benefit in enhancing protein synthesis (Manninen 2006). Thus, ingestion of carbohydrate and protein supplements before or after training appears to reduce catabolism and enhance glycogen resynthesis, but any interaction with the training stimulus to promote hypertrophy and strength gain has not been established unequivocally. Jeff Volek has proposed a pathway of adaptation (Volek 2004; see his figure 1) that includes many of the various issues that must be considered when examining the effects of nutrients on the responses to resistance training (Volek 2004). Readers are encouraged to consult several fine reviews of this subject (Kreider 1999; Clarkson and Rawson 1999; Nissen and Sharp 2003; Volek 2004; Manninen 2006).

CREATINE SUPPLEMENTATION AND RESISTANCE TRAINING

Creatine supplementation has been associated with improved muscle power during exercise and with improved responses to a resistance training program—responses such as increased satellite cell activation and myonuclei concentration. But how does creatine supplementation work? Does it influence the cell signaling that leads to the hypertrophic response? Louise Deldicque and colleagues (2008) at the Catholic University of Leuven decided to feed creatine supplements to subjects for 5 days before examining the acute response to a single standardized bout of resistance training (10 repetitions, 10 times, 80% of MVC, knee extensions). In biopsy material taken from the knee extensors, these authors found increases in the expression of several genes at rest before the resistance training, including *MHC I,* collagen I (á₁), and *GLUT4.* Following exercise, there were significant changes in the expression of many genes, as would be expected, but there were few differences in the mRNA responses between the creatine and control groups. These genes included markers for satellite cell proliferation, myogenic genes, and markers of MAPK and PKB pathways that are known to be involved in the cell signaling preceding muscle phenotypic responses. It appears that, although the positive effect of creatine supplementation on performance and training responses is clear, the principal mechanism of action of creatine supplementation may be to improve the muscle environment before exercise in order to provide the optimal substrate for increased muscle mass that resistance training evokes.

In one of the only studies of its kind, Andersen and colleagues (2005) looked at the effects of timed isoenergetic protein versus carbohydrate supplementation on changes in vastus lateralis hypertrophy and performance resulting from a 14-week resistance training program. The protein group but not the carbohydrate group showed fiber hypertrophy and improvement in squat jump height. Slow torque and countermovement jump heights increased similarly in both supplementation groups. Supplements were taken immediately before and after training sessions (three per week). In the elderly, it seems that protein supplements are particularly effective in enhancing the hypertrophic and strength-related responses to a 12-week resistance training program (Esmarck et al. 2001).

SUMMARY

The stimuli for altered muscle protein synthesis during strength training are related to the mechanical stress and strain to which the muscle is exposed during the activity. Protein synthesis and degradation both increase within hours of an acute bout of strength training. These changes involve primarily posttranscriptional mechanisms, including changes in initiation, elongation, or termination of protein synthesis. Protein degradation pathways that are activated include calpains, the ubiquitin proteasome system, and several lysosomal enzymes. While signals that promote the adaptation of muscles to aerobic-type endurance training appear to be related to the altered intracellular environment caused by prolonged contraction, the signals associated with muscle stretch and strain seem fundamental in the responses to strength training. Strength overload results in activation of several signals, including production of IGF-1, activation of the Akt and mTOR pathway, activation of phospholipases A and C, and activation of the kinase systems PKC, FAK, and MAPK. These signals promote changes in protein synthesis via activation of nuclear transcription factors, altered translation mechanisms, and activation of satellite cells. The production of chemical substances by invading cells consequent to the muscle damage produced during the exercise may also provide signals that promote an adaptive response to strength training. It is the combination of the effects of these stimuli on transcriptional and translational processes and on activation of degradative enzyme systems that determines the ultimate phenotypic response to strength training. Nutritional factors are also significant modulators of the response.

Muscle Property Changes in Strength Training

The results of strength training on muscle phenotype are presented in the sections that follow. For the most part, I have not made any attempt to separate the effects of dynamic and static types of training (or explosive and heavy resistance training). Although there are many similarities in the responses to these two types of resistance training, there are also some differences, which are highlighted when appropriate.

The phenotypic changes in response to resistance training are the result of changes in protein turnover. Resistance training for 8 weeks results in a chronic increase, in the resting state (i.e., 72 hours following the last training session), in mixed protein synthesis, as estimated using constant infusion of labeled phenylalanine. Interestingly, this increased protein synthesis appears to involve proteins other than myofibrillar proteins and may signify a more rapid turnover of nonmyofibrillar proteins (such as mitochondrial, soluble, and connective tissue proteins; Kim et al. 2005). An acute eccentric exercise bout stimulates myofibrillar protein synthesis to a greater extent than a concentric exercise bout matched to the same total work stimulates protein synthesis, while collagen synthesis increases to the same extent in both conditions (Moore et al. 2005).

INCREASED MUSCLE FIBER CROSS-SECTIONAL AREA

Although the maximum limit to which fibers can hypertrophy in response to resistance training is not known, increases in cross-sectional area of 30% to 70% are the most commonly reported. In bodybuilders, fibers can range up to 2.5 times larger than fibers in sedentary subjects (Alway et al. 1988). Generally speaking, Type II fibers show larger hypertrophic responses than Type I fibers show. As expected, training programs involving heavier resistances or eccentric actions demonstrate more pronounced cell hypertrophy than programs involving faster contractions or lighter loads without eccentric actions (Ewing et al. 1990; Hather et al. 1991; Hortobágyi, Hill, et al. 1996). The hypertrophic response can occur quite rapidly—magnetic resonance imaging (MRI) measurements show significant quadriceps hypertrophy (5.2%) after only 20 days of a 5-week resistance training program involving knee extensions (Seynnes et al. 2007). Protein supplementation immediately before and following training sessions results in larger training-induced increases in fiber size than carbohydrate supplementation results in, although muscle mechanical properties are not influenced differently by these two supplementation regimes (Russell et al. 2005).

FIBER TYPE COMPOSITION

The most common histochemical change in the proportions of fiber types, when there is one, is a decrease in IIx fibers and a corresponding increase in IIax/IIa or IIa proportions (Andersen and Aagaard 2000; Woolstenhulme et al. 2006). This change has been shown to occur most frequently in knee extensors, but it has also been shown in trained biceps and triceps brachii. This change has also been substantiated using electrophoresis to separate the isoforms in biopsy samples from vastus lateralis (see figure 9.1; Adams et al. 1993; Carroll et al. 1998; Woolstenhulme et al. 2006; Andersen and Aagaard 2000) and triceps brachii (Jürimäe et al. 1996). In women performing strength training of the trapezius, a decrease in MHC I and IIb with a corresponding increase in IIa has been reported (Kadi and Thornell 1999). More recently, Williamson

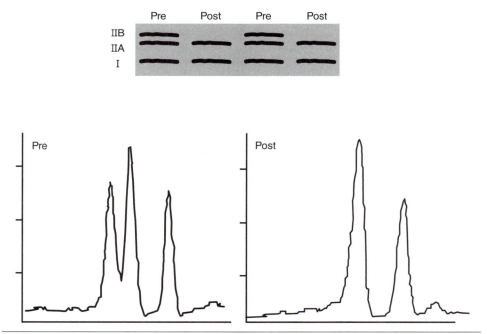

Figure 9.1   Heavy resistance training of the quadriceps for 19 wk alters MHC composition in vastus lateralis. Myofibrils were separated using polyacrylamide gel electrophoresis (PAGE), and the gel was subsequently subjected to densitometric scan. Posttraining biopsies contained negligible MHC IIb.

Reprinted by permission from Adams et al. 1993.

and colleagues (2001) reported a decrease in the proportions of hybrid fibers (fibers containing more than one MHC, such as I/IIa/IIx and IIa/IIx) with a concomitant increase in pure Type IIa fibers as a result of progressive resistance training of the knee extensors for 12 weeks.

MUSCLE FIBER NUMBER

Does hyperplasia (increased cell number) occur with resistance training? One way that this particular problem is investigated is to examine the muscles of bodybuilders who have produced extreme muscle hypertrophy through many years of resistance training. In groups of untrained subjects, intermediate-level bodybuilders, and elite bodybuilders, estimates of fiber numbers (based on average muscle area measured from tomographic scans divided by average fiber area taken from biopsy material) show that muscle (biceps brachii) cross-sectional area is more closely related to muscle fiber area $(r = 0.71\text{-}0.81)$ than it is to estimated fiber number $(r = 0.35\text{-}0.6$; Sale et al. 1987; MacDougall et al. 1984). Although the conclusion might be that hyperplasia does not play a role in extreme hypertrophic response, the wide variation in the numbers of muscle fibers among subjects, as well as the errors associated with the estimation techniques, leaves enough possibility that small degrees of hyperplasia might take place with training. This wide variation among individuals may signify genetic variation. The authors of these studies have pointed out that many of the subjects with the largest muscles also possess the largest number of fibers (MacDougall et al. 1984; see figure 9.2).

Other investigators (Larsson and Tesch 1986) reported that a larger muscle circumference of the biceps brachii in bodybuilders, with no difference in the average fiber area, signified that hyperplasia was involved. They also found evidence of a higher fiber density (using an intramuscular EMG technique by which the number of fibers belonging to the same motor unit within the recording range of the electrode was estimated), a finding that supports their claim. However, the number of subjects in their bodybuilder group was limited $(n = 4)$.

Hyperplasia is easier to investigate in cats than it is in humans for obvious technical reasons. In one experiment, cats were operantly conditioned to perform forearm flexion movements against a resistance for a food reward. Comparisons of fiber numbers between the left and right flexor carpi radialis, performed by counting fibers after maceration of the muscle via nitric acid digestion, revealed a difference of approximately 9% (Gonyea et al. 1986). Further experimentation with cats has confirmed satellite cell activation by uptake of tritiated thymidine in exercised muscles as well as the coexistence of larger and smaller fibers, which might indicate de novo fiber formation (Giddings, Neaves, and Gonyea 1985; Giddings and Gonyea 1992).

Hyperplasia occurs to a much greater extent in chickens and quail; this is shown using a technique of stretching the muscles that normally support the wing by application of a weight. When the stretch is intermittent (24 hours on, 48-72 hours off), hyperplasia is not present, even after the same total time of stretch application (Antonio and Gonyea 1993). The applicability of these findings to the mammalian system is not known.

A meta-analysis (Kelley 1996) regarding evidence for or against hyperplasia during muscle overload in animals came to the following conclusions: (1) Significant hyperplasia probably occurs (average increase in fiber number is 15%); (2) avian subjects

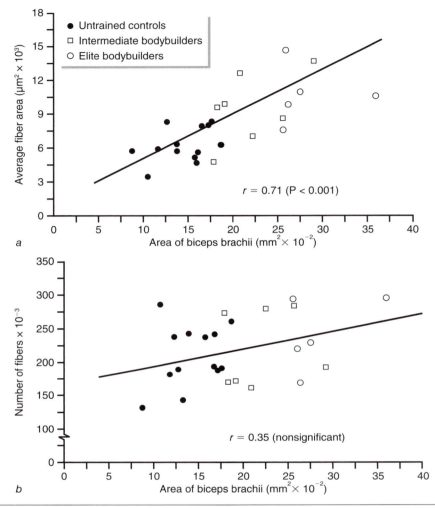

Reprinted by permission from MacDougall et al. 1984.

Figure 9.2 Correlations *(a)* between biceps brachii area and average muscle fiber area and *(b)* between biceps area and fiber number. Biceps area was determined using computerized tomographic scanning, and fiber areas were determined from biopsy material. The size of the biceps is better related to the size of the individual fibers than it is to the number of fibers.

show much more hyperplasia than mammalian subjects show (21% versus 8%); and (3) the degree of hyperplasia as a result of stretch exceeds that from compensatory overload, which in turn exceeds that from voluntary exercise (21% versus 12% versus 5%). We must consider, however, that these studies were published subsequent to the finding of a significant increase in fiber number. How many set out to find an increase but did not find one and therefore did not publish?

It has been shown that satellite cell activation appears to be necessary for the hypertrophic response and that nuclear proliferation occurs with resistance overload. Both of these phenomena are essential conditions for hyperplasia. Rosenblatt, Yong, and Parry (1994) showed that compensatory hypertrophy of the rat tibialis anterior in response to ablation of its synergists does not occur if the muscle is first exposed to radiation, which inhibits mitotic activity and thus satellite cell proliferation. Phelan and Gonyea (1997) substantiated this and showed that no small fibers were produced during the compensatory overload after muscle irradiation. Both compensatory overload (Allen et al. 1995) and resistance training (MacDougall et al. 1980; Kadi and Thornell 2000) result in an increase in myonuclei (figure 9.3). More recently, Crameri and colleagues provided evidence that muscle satellite cells are activated by a single bout of resistance exercise by finding an increase in the neural cell adhesion molecule (NCAM), which is a marker for these cells, as well as an increase in the membrane-bound

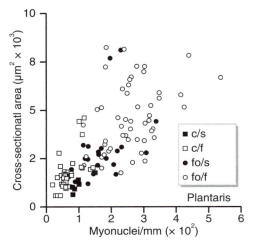

Figure 9.3 Relationship between muscle fiber area and number of myonuclei per millimeter in cat plantaris. Control slow (c/s) and fast (c/f) fibers and functionally overloaded (for 3 months) slow (fo/s) and fast (fo/f) fibers are shown.

Reprinted by permission from Allen et al. 1995.

protein fetal antigen 1 (FA1), which is normally activated in muscle-derived stem cells that are distinct from satellite cells (Crameri et al. 2004). Whether increased mitotic activity in response to resistance overload leads to the production of new muscle fibers and what is the ultimate fate of these fibers vis-a-vis their innervation and participation within existing motor units remain to be demonstrated.

MUSCLE COMPOSITION

We have already considered the adaptations in fiber type composition that, in combination with changes in fiber areas, have implications for muscle composition. There is some evidence that the percentage of collagenous and noncontractile proteins decreases in extremely hypertrophied fibers (Sale et al. 1987). Jones and Rutherford (1987) presented evidence from computerized tomographic scans that resistance-trained muscles were slightly more dense in myofibrillar packing. The activities of most energetic enzymes measured are unchanged after resistance training, including magnesium-stimulated ATPase, creatine kinase, phosphofructokinase, citrate synthase, lactate dehydrogenase, hexokinase, and succinate dehydrogenase (Thorstensson et al. 1976; Tesch, Thorsson, and Colliander 1990; Ploutz et al. 1994). Myokinase activity may increase slightly after resistance training (Thorstensson et al. 1976). The content of the muscle protein desmin increases more than 80% during a progressive resistance training program involving the knee extensors (see figure 9.4). Desmin is a protein of the cytoskeleton that connects myofibrils to the sarcolemma via structures known as *costameres.* It is responsible for the transmission of force across sarcomeres to the exterior of the muscle (Clark et al. 2002). Similar increases have been noted in trained knee extensors following 8 weeks of high-intensity cycle training (Woolstenhulme et al. 2005).

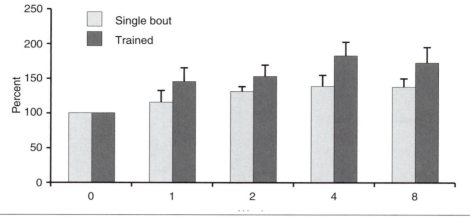

Figure 9.4 Percent change in desmin protein content in vastus lateralis during training performed for a single bout (open bars) or training performed 3 times per week (filled bars). Training consisted of lower-body resistance exercises involving knee extensions, leg presses, and hamstring curls at 60% to 80% of 1-repetition max (1RM).

Reprinted by permission from Woolstenhulme et al. 2006.

MUSCLE ARCHITECTURE

During resistance training involving bilateral knee extensions, fascicular lengths in vastus lateralis increased by 2.4 % after 10 days of training (3 times per week) and by 9.9% after 35 days of training (the end of the training). An increase in angle of pennation of 7.7% was also evident by the end of the training (Seynnes et al. 2007). Similar increases in fascicular length occur for both concentric and eccentric resistance training (Blazevich et al. 2007). Aagaard and colleagues (2001) used MRI and ultrasonography to examine the effects of 14 weeks of heavy resistance training of the lower limbs on muscle architectural indexes of the knee extensors. Besides the expected increases in muscle cross-sectional area and volume, they found that pennation angle of the vastus lateralis increased from 8° to 10.7°. They concluded that the increase in muscle fiber pennation allowed muscle fiber cross-sectional area and maximal contractile strength to increase more than was apparent from the anatomical muscle cross-sectional area and volume. In other words, due to these additional architectural adaptations, muscle became stronger than it appeared to be from muscle girth alone. Interestingly, high-speed plyometric training causes increases in fascicle length that are similar to the increases that have been reported for resistance training but also causes a decrease, as opposed to an increase, in fascicular angle (Blazevich 2006).

MUSCLE FIBER ULTRASTRUCTURE

Modest increases in the nuclei-to-fiber ratio have been found after resistance training that causes a 31% to 39% increase in fiber area (MacDougall et al. 1980). In compensatory overload in animals, in which hypertrophy is more extreme (a 2.8-fold increase in area, for example), a more than threefold increase in myonuclei per fiber can occur (Allen et al. 1995). This proliferation, which probably results from the activation and proliferation of satellite cells and their subsequent fusion with existing fibers, may be an adaptive attempt to maintain an adequate cytoplasmic volume per nucleus in the face of increased cytoplasmic volume (Allen et al. 1995).

MUSCLE GEOMETRY
——— AS A DETERMINANT OF PERFORMANCE ———

Muscle geometry in humans rarely includes simple structures such as fibers running in a straight line from origin to insertion. Most often, fibers are situated at an angle from the line of pull and are shorter than the distance between origin and insertion, being attached to the muscle aponeurosis. For a given muscle volume, having shorter fibers at higher angles to the line of pull favors force development, while having longer fibers with smaller angles favors shortening speed. Since the angle of pull determines force generation and the fiber length determines muscle shortening speed, variations in these parameters in, for example, a knee extensor can combine to exert significant effects on neuromuscular performance. A. Blazevich (2006) at Brunel University at Uxbridge (United Kingdom) has directed his attention to studying the geometry of human muscle fascicles via two-dimensional images obtained with ultrasound. He has observed that sprinters tend to have longer fascicles with smaller angles while endurance runners tend to have shorter fascicles with larger angles. These arrangements work to favor speed in the sprinters and economy in the endurance runners (since less shortening costs less ATP). In addition, there is evidence that fascicular lengths and angles may change with training. Several reports have shown that angles increase with resistance training and decrease with aerobic endurance training, although the reports to date are far from conclusive. It seems clear that the search for muscle properties that determine performance, such as fiber types and sizes and energetic enzyme levels, must also take muscle geometry into consideration. Fascicle geometry may even constitute another property with a genetic basis that can predispose individuals, even modestly, to success in specific types of athletic events.

Alway and colleagues (Alway et al. 1988; Alway, MacDougall, and Sale 1989) examined ultrastructural characteristics of muscle fibers before and after resistance training and in strength-trained athletes and controls. Their results suggest that significant hypertrophy can occur with very little change in the percentage of the fiber that is occupied by sarcoplasmic reticulum–transverse tubular network, cytoplasm, lipids, and myofibrils. However, gastrocnemius muscle fibers of strength-trained athletes had a mitochondrial fraction that was 30% less than that of controls. Recently, a possible decrease in mitochondrial enzyme (SDH) activity with intensive strength training was substantiated (Chilibeck, Syrotiuk, and Bell 1999). In addition to, or perhaps in spite of, fiber hypertrophy, capillary-to-fiber ratios increase after both isometric and dynamic resistance training (Hather et al. 1991; Rube and Secher 1990).

EVOKED ISOMETRIC CONTRACTILE PROPERTIES

Several reports have shown an increase in voluntary muscle strength after resistance training without a corresponding increase in the isometric force response evoked via electrical stimulation of the nerve (Sale, Martin, and Moroz 1992; Davies et al. 1985). Part of this discrepancy may have to do with neural effects that are called into play with voluntary contractions but are not seen when these contractions are evoked.

This issue is dealt with in more detail in chapter 10. Nonetheless, electrical stimulation can be used to investigate the basic muscle properties related to force and speed independently of changes in neural mechanisms.

The most systematic study to date regarding this issue is the classic work of Duchateau and Hainaut (1984), who trained adductor pollicis of subjects using either dynamic (fast adductions at a load corresponding to 33% MVC) or static (5-second MVC) overload for 3 months. These authors then measured evoked isometric contractile properties. Their findings showed unequivocally that the dynamic or isometric nature of the training stimulus affects intrinsic muscle properties (see figure 9.5). For example, isometric training results in larger increases in the properties associated with static strength (twitch and tetanic forces, maximal muscle power), whereas dynamic training results in preferential enhancement of speed-associated properties (faster time course of the isometric twitch, higher peak rate of tetanic force development and maximum shortening velocity).

The intramuscular signals that determine whether adaptations are dynamic-like or isometric-like are unknown. Behm and Sale (1993) showed that dynamic isokinetic training of the ankle flexors increased the maximal rate of twitch tension development and decreased the twitch time to peak tension of evoked contractions. Training also produced an increased isokinetic torque response that was most marked at the higher velocities. Interestingly, these adaptations, which would be expected according

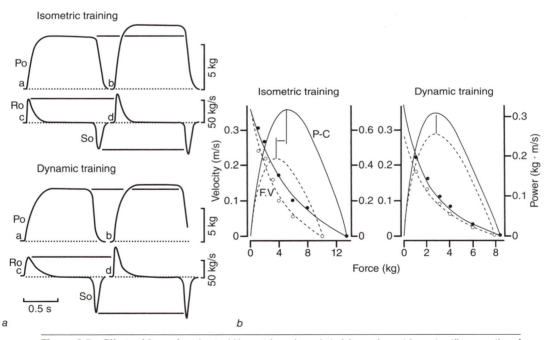

Figure 9.5 Effects of 3 mo of moderate (a) isometric or dynamic training on isometric contractile properties of adductor pollicis and effects of (b) isometric or dynamic training on force, velocity, and power characteristics of adductor pollicis. Isometric training resulted in a greater increase in isometric force, while maximal rate of tension development increased more with dynamic training. Po, maximum force; Ro, maximum rate of force development; So, rate of force relaxation.

Reprinted by permission from Duchateau and Hainaut 1984.

to the previous experiments of Duchateau and Hainaut (1984), occurred regardless of whether the contraction was isometric or isokinetic at a high speed (5.23 radians per second). In this experiment, however, the subjects were asked to generate force as quickly as possible, regardless of whether the ensuing contraction was dynamic or static. Thus, the intramuscular signal specifying dynamic-like or static-like muscle adaptations is not related to the degree or speed of muscle fiber shortening or to the load during the training. The signal may have to do with the rate of force development or the nature of the command from the nervous system.

Reports of the effects of resistance training on evoked contractile properties are mixed. Some show a more prolonged (Alway et al. 1988; Sale, Upton, et al. 1983) and others a faster (Alway, MacDougall, and Sale 1989; Sale et al. 1982) twitch response. Similarly, twitch force has been reported to decrease (Sale et al. 1982), increase (Sale, Upton, et al. 1983), or remain unchanged (Alway, MacDougall, and Sale 1989; Alway et al. 1988). These results vary because of differences in the subjects studied (some studies used trained subjects, while others compared sedentary subjects with strength-trained subjects), the nature of the training (dynamic or isometric), and the muscles examined.

The central message is that intrinsic muscle properties other than mass and strength change with resistance training. These changes are detectable using contractions evoked by electrical stimulation, and some of the properties that change (twitch tension and time course, rate of twitch tension development) have implications for muscle use during voluntary activation. These adaptations may be independent of the actual degree of shortening or lengthening performed by the muscle during the training but may depend on either the rate of activation of the contractile machinery or the nature of the signal reaching the muscle fiber from the nervous system.

CHANGES IN MUSCLE FORCE, VELOCITY, AND POWER

Evidence about changes in force, velocity, and power comes from studies of voluntary contractions and thus might reflect the partial influence of neural factors, which are discussed in detail in chapter 10. Nonetheless, certain general adaptive responses to resistance training are worth noting here, especially in light of our previous conclusion that some of the muscle changes are specific to the dynamic or static nature of the training.

Powerlifters and bodybuilders have higher torques (in plantar flexion, knee extension, and elbow extension) than controls have at both low and high velocities, with a fairly good correlation $(r = 0.84)$ between torque at high and torque at low velocities (Sale and MacDougall 1984). This suggests that a general training regimen involving a variety of loads and velocities increases strength throughout the entire range of muscle contractile speeds.

When speed of contraction during training is controlled, on the other hand, there is evidence that speed-specific adaptations occur. Caiozzo, Perrine, and Edgerton (1981) trained subjects using maximal contractions at either slow (1.68 radians per second) or fast (4.19 radians per second) isokinetic speeds for 4 weeks. Slow training involved improvements in torque at all velocities except the fastest ones and caused the torque–velocity curve to level off at a lower level. High-velocity training resulted in improvements at the higher velocities. The leveling-off phenomenon and its response to training suggested to the authors that motoneuron recruitment capacity was influenced by the training.

DOES PLYOMETRIC TRAINING EVOKE SPECIFIC MUSCLE ADAPTATIONS?

Plyometric training is used to enhance neuromuscular power and, for the legs, involves high-impact jumping exercises. The principle is to recruit the muscle ballistically in an eccentric-like fashion during the time when the muscle is lengthening as it yields to body weight, before the explosive jump phase (equivalent to jumping down from a height, but instead of landing softly, executing a maximum jump). Although increases in performance, particularly jump height and peak power, have been reported as a result of plyometric training, there has been very little literature on the specific morphological changes that occur, until recently. K. Vissing and colleagues (2008) at Aarhus University (Denmark) and Foure, Nordez, and Cornu (in press) at the University of Nantes (France) have shed some light on this issue. Their results show that, as expected, plyometric training increases muscle power and strength while causing minor changes in muscle fiber size, while conventional resistance training increases primarily strength and muscle fiber size. There were no differences between conventional resistance training and plyometric training in changes in fiber types or in muscle cross-sectional area. This suggests that some of the performance increases in ballistic force generation may be neural in nature. It also appears that the mechanical properties of tendons respond to plyometric training. After 14 weeks of training, the Achilles tendons of subjects showed a decreased dissipation coefficient (i.e., tendons dissipated less of the stored energy) and were stiffer, but there were no changes in the cross-sectional area of the tendon. Thus, plyometric training is a specific stimulus that results in the potential for more powerful contractions, most likely through changes in recruitment strategy rather than specific muscle adaptations, as well as through changes in the quality of the tendons such that they become more efficient in the transmission of muscle tension.

This velocity-specific training response has been confirmed by other investigators. Narici and colleagues (1989) had subjects train knee extensors for 60 days at a relatively slow isokinetic speed (2.9 radians per second). Torque increased at the training velocity and at velocities below it but not above it. Similarly, Ewing and colleagues (1990) found that the effects of a training program of 10 weeks on isokinetic torque were specific to the speed of training. Although the mechanism for this speed-specific response is not known, consider the findings of Behm and Sale (1993), who found a speed-specific adaptation of the torque–velocity relationship not only in subjects who performed the isokinetic movement during training but also in subjects who intended to make this movement but ended up making an isometric contraction instead.

Explosive training enhances the performance of powerful movements. Häkkinen, Komi, and Alén (1985) showed that explosive training increases power-related measurements, such as rates of isometric force development during maximal, isometric, ballistic, and voluntary contractions. Power measurements such as vertical jump height and drop jump performance (countermovement contractions) are also enhanced preferentially when the resistance training includes dynamic movement (Häkkinen and Komi 1983a; Colliander and Tesch 1990).

FATIGUE RESISTANCE

A commonly held belief is that hypertrophy induces increased strength at the expense of fatigue resistance. This is not borne out in the literature.

McDonagh, Hayward, and Davies (1983) measured fatigue of elbow flexors before and after 5 weeks of isometric training. Fatigue was measured by stimulating the muscle nerve using the Burke protocol. They found that fatigue resistance was significantly increased after training. Similarly, Rube and Secher (1990) found that a program of 5 weeks of static training using maximal isometric knee extensions increased fatigue resistance, as tested by maximal extensions every 5th second. Interestingly, when the test was performed bilaterally, improvement was seen only in the groups that had trained bilaterally, while improvement on the unilateral fatigue test was evident only in the group that had trained unilaterally with that leg.

Muscle fatigability was also measured by Hortobágyi, Barrier, and colleagues (1996) in women before and after either concentric or eccentric training of the quadriceps. The fatigue test consisted of 40 maximal isokinetic contractions. Training had no effect on fatigability, although it was demonstrated that concentric exercise was more fatiguing than eccentric exercise was.

Animal studies are rare in the strength-training literature, due to the difficulty of finding a model that mimics this type of training. Duncan, Williams, and Lynch (1998) studied the effects of weight training in rats; the training consisted of conditioning rats to climb a vertical grid repeatedly with a weight attached to their tails. The weight was made heavier as the animals adapted. The extensor digitorum longus muscles, which were hypertrophied as a result of the 26-week training program, were slightly more fatigue resistant in situ than those of controls were.

ROLE OF ECCENTRIC CONTRACTIONS

Not much has been said thus far regarding the role of eccentric (lengthening) contractions in the response to resistance training. Clearly, any resistance training using free weights or body weight and involving dynamic as opposed to isometric contractions includes an eccentric phase. Even if we rigorously control the speed of movement (which is not always the case in training studies), the simple act of lifting and lowering a weight is complicated by the fact that recruitment of motor units is different for these two phases of contraction against the same resistance (see the discussion of this point in chapter 2). Thus, to determine the importance of the concentric phase versus the eccentric phase of movement in the final training adaptations is no easy task. However, several investigators have conducted fairly controlled studies of the role of eccentric contractions as a training stimulus. Following is a summary of these findings.

Eccentric training results in bigger strength gains than concentric or isometric training with the same external load results in. Colliander and Tesch (1990) and Dudley and colleagues (1991) found this after comparing concentric and eccentric training with a program involving the same number of concentric-only contractions with the same load. Hortobágyi, Barrier, and colleagues (1996) demonstrated the superiority of eccentric to concentric training in optimizing strength development by showing that strengths gains with eccentric training were bigger even when the load was maximal for the concentric training groups but submaximal for the eccentric groups. The difference in the efficacy of eccentric and concentric programs for maximal strength development is less evident when strength testing is not specific to the type of training (i.e., when eccentrically trained subjects are tested with concentric contractions). This finding suggests that there are either muscle morphological or neural effects that are specific to the training mode. Neural effects are discussed in more detail in the next chapter.

Eccentric training may result in specific muscle morphological changes. Responses to eccentric training are more evident when testing in eccentric mode than when testing in concentric or isometric mode (Seger, Arvidsson, and Thorstensson 1998; Higbie et al. 1996; Hortobágyi, Barrier, et al. 1996; Hortobágyi, Hill, et al. 1996). While this result has been used as evidence for a neural component to resistance training, several investigators have also suggested that the mode-specific training response may be due at least in part to mode-specific changes in muscle morphology.

An acute bout of eccentric contractions, for example, produces a length-dependent deficit in force production that is most pronounced at shorter lengths and that lasts several days. This may indicate a stretching of myofibrillar and myotendinous structures that results in a physiological muscle lengthening (Saxton and Donnelly 1996). Whitehead and colleagues (1998) found a similar shift in the force–length curve toward longer lengths for optimal force development after an acute bout of eccentric exercise. They found, in addition, that this effect and the degree of muscle swelling were exacerbated in a group that had trained concentrically for 1 week before the acute experiment. Their conclusion was that the concentric training resulted in shorter sarcomeres, which would render the muscle more sensitive to damage induced by eccentric exercise.

Fridén (1984) reported that after 2 months of eccentric training involving the knee extensors, there were a myriad of changes in sarcomere ultrastructure, including a large variation in sarcomere length that was more pronounced in the Type II fibers. This author proposed that the described changes result in a better stretchability of the muscle fibers, thus reducing the risk of mechanical damage.

Hypertrophic patterns following concentric and eccentric training are different not only in extent but also in proximodistal pattern. Seger, Arvidsson, and Thorstensson (1998) used MRI to examine hypertrophy of the knee extensors and found that eccentric and concentric training resulted in different hypertrophic patterns. For the eccentrically trained group, muscle girth increased in the distal portion of the muscle only, while in the concentrically trained group, there was a tendency for increased girth only at the muscle belly (not in the distal portion).

The structural damage caused by eccentric exercise may help to explain some of the functional and structural effects that appear to be specific to eccentric exercise. Foley and colleagues (1999) used MRI to examine elbow flexors for up to 56 days following a single bout of eccentric exercise. They found that muscle compartment volume was decreased by 10% 2 weeks after the exercise and attributed this finding to a complete and irreversible necrosis of a population of fibers that were most susceptible. According to these investigators, such an occurrence would explain the repeated bout effect (discussed later in this section), in which a second eccentric exercise results in less evidence of muscle damage and pain. Fibers less susceptible to damage would be recruited during the second bout to replace the necrosed fibers, with reduced damage. This interesting hypothesis of the repeated bout effect remains to be substantiated.

Muscle hypertrophy may be more marked with eccentric versus concentric training (Dudley et al. 1991; Hather et al. 1991). In addition, the force–velocity curve changes after eccentric training of the elbow flexors in that its curvature becomes more pronounced and V_0 increases. The biceps become faster contracting. These changes should be considered in light of the morphological changes described earlier.

The performance of unfamiliar contractions such as eccentric contractions results in what is known as a *repeated bout effect*. In this phenomenon, a repetition of the same exercise some time after results in considerably less damage to the muscle. There are several hypotheses as to the explanation for the repeated bout effect, all of which have as a central strategy the reduction of muscle damage. The neural hypothesis is that motor unit recruitment patterns change because of learning from the first episode. The connective tissue or mechanical hypothesis maintains that cells respond by adapting connective tissue components in order to reduce damage during a similar contraction. The cellular hypothesis suggests that changes occur on sarcomere number or structure. None of these can fully explain the phenomenon (for example, the repeated bout phenomenon occurs with electrically stimulated muscles, thus questioning, but not necessarily alleviating, any neural involvement). McHugh (2003) has written a thorough review of the evidence for and against the neural, mechanical, and cellular hypotheses of the repeated bout effect. Readers are urged to consult this fine review, in which evidence is summarized in table format (McHugh 2003).

SUMMARY

Muscle fiber changes that occur with strength training include increases in fiber cross-sectional area and possible changes in muscle fiber composition toward more Type I and IIa MHCs. The protein composition of the fiber changes, with alterations occurring in some energetic enzymes and in desmin, which connects myofibrils to the sarcolemma via costameres. Myonuclei number per fiber increases. Architectural changes occur in the muscle, including changes in fascicular length and angle of pennation of the fibers, in directions that tend to increase muscle force capability per unit muscle girth and volume. In spite of increases in fiber area, capillary-to-fiber ratios stay unchanged or increase following resistance training. While isometric strength is influenced primarily by the static (isometric) component of the training, contractile speed is influenced by the dynamic component of the training. This specificity may not depend on the dynamics of the contractions but instead may depend on the rate of force (either dynamic or isometric) development generated voluntarily (either isometrically or anisometrically). When anisometric contractile speed during training is controlled, greatest strength gains occur at and near the training speed. Greatest strength gains occur when lengthening contractions are used during the training. Gains are greatest when measured using eccentric contractions. Eccentric training may involve muscle morphological changes not seen with other forms of strength training. The extent to which changes that are specific to training modes can be explained by muscle morphological factors as opposed to changes in neural mechanisms of recruitment is currently the subject of much debate. There is some evidence, nonetheless, that central nervous system changes that would be expected to improve performance occur with strength training. This is discussed in the next chapter.

CHAPTER 10

Neural Mechanisms in Strength Training

Much emphasis has been placed on the neural basis for increased performance following resistance training. Some of this evidence is somewhat indirect in that the neural component of the strength increase is arrived at by eliminating other possibilities (increased strength with no change in muscle girth, for example). Some of this evidence, however, is direct in that it is supported by neurophysiological data. In this chapter, I discuss the evidence for a neural component, progressing from what I consider to be the weakest to the strongest evidence.

GAINS IN STRENGTH VERSUS MUSCLE GIRTH

Strength increases faster than muscle girth increases during the initial stages of strength training, suggesting that factors other than morphological changes are involved. Even in untrained subjects, only about 50% of the variation in quadriceps MVC can be explained by quadriceps cross-sectional area (Jones, Rutherford, and Parker 1989), and it is difficult to believe that the remainder is due entirely to neural factors. Several investigators feel that even short-term resistance training might influence specific tension, angle of pennation, connective tissue content, and torque–length relationships (Howell, Chleboun, and Conatser 1993; Jones and Rutherford 1987; Jones, Rutherford, and Parker 1989). All these factors, singly or in combination, might produce this apparent discrepancy between increased maximal strength and increased muscle girth.

Jones, Rutherford, and Parker (1989) have proposed that connective tissue attachments might be altered by resistance training, with the result that force expression per fiber increases. Such an adaptation has yet to be demonstrated experimentally. In addition, Jones and Rutherford (1987) demonstrated evidence of increased quadriceps muscle density after resistance training and proposed that this might be due to decreased intramuscular fat content or to increased myofibrillar packing density, both of which would increase specific muscle tension. Changes in muscle architecture, as discussed in the previous chapter, might also explain at least some of the increase in strength that occurs out of proportion with increased muscle size.

Narici and colleagues (1989) measured cross-sectional area changes in the major knee extensors vastus lateralis, vastus intermedius, vastus medialis, and rectus femoris resulting from 60-day isokinetic resistance training of the knee extensors. Their nuclear magnetic resonance data, which they obtained by examining the cross-sectional area of these muscles at seven places along the femur, demonstrated that the increase in quadriceps cross-sectional area with resistance training is not uniform along its length and that the patterns of girth increase for the different muscles are not identical (more hypertrophic change was noted generally at the proximal and distal ends of the muscles, with less hypertrophy in the center). Thus, we must consider changes in the performance of this muscle group (which is by far the one most studied for resistance training) in the light of possible heterogeneous changes in the length–tension curves, cross-sectional areas, and angles of fiber pull on distal tendon that might occur among the different muscles of this complex group.

Ploutz and colleagues (1994), using contrast shift in MRI, estimated recruitment of fibers during contractions of quadriceps against various submaximal loads. Their results suggested that less absolute muscle cross-sectional area is used to accomplish the same submaximal load after resistance training (for 9 weeks, 2 times per week; see figure 10.1). Such a change might imply an increase in specific tension, as opposed to the decreased specific tension proposed by Kawakami and colleagues (1995). This change might also imply a difference in recruitment strategy, such that the same force is accomplished by recruiting fewer motor units at higher firing frequencies.

There are other, more compelling reasons to believe in a neural component to resistance training than comparison of increases in muscle girth and maximal strength. These are discussed next.

STRENGTH GAINS SHOW TASK SPECIFICITY

Increases in performance following resistance training are generally most evident when performing the task used during the training. This task specificity has been used as evidence for a neural component in resistance training. For example, 8 weeks of dynamic knee extensions resulted in a 67% increase in 1RM performance for squats, while isometric knee extension MVC increased only 13% (Thorstensson and Karlsson 1976). This specificity extends to the velocity at which the training is performed. Caiozzo, Perrine, and Edgerton (1981) and Ewing and colleagues (1990) found that training the knee extensors on an isokinetic device increased torque more at or near the training velocities. Similarly, Narici and colleagues (1989) found that knee extensor torque improved at isokinetic speeds that were at or below, but not above, the speed used for training.

Häkkinen, Komi, and Alén (1985) demonstrated task specificity of response by comparing the adaptations to heavy resistance training and to explosive training. Their results demonstrated that explosive training (jumping without a load or with a light load) increased not only maximal force and integrated EMG but also the maximal rates of force development and integrated EMG. While the value of absolute EMG measurements as an index of response to strength training is open to dispute (discussed later), changes in the rate of EMG development, corrected for the maximal measured value, seem intuitively more meaningful and valid.

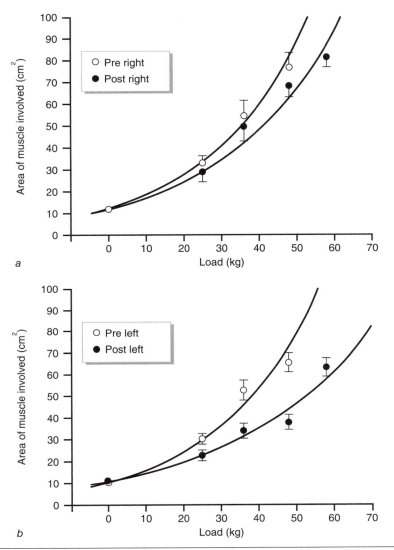

Figure 10.1 Average cross-sectional area of *(a)* right untrained and *(b)* left trained quadriceps recruited during concentric contractions with various loads. Activated muscle area was estimated from elevated spin-spin relaxation times in MRI following 5 sets of 10 repetitions with the load. Symbols indicate measurements before (open squares) and after (filled circles) 9 wk of training of the left quadriceps using knee extensions with heavy weights.

Reprinted by permission from Ploutz et al. 1994.

The task-specificity phenomenon extends to unilateral versus bilateral tasks. Bilateral training increases bilateral performance more than it increases unilateral performance, while unilateral training increases unilateral performance more than it affects bilateral performance (Häkkinen et al. 1996). This is another fairly convincing argument for changes in the nervous system induced by resistance training.

The model of eccentric training provides still another example of possible neural effects during resistance training. We have already seen that eccentric training is superior to concentric training in development of maximal strength, but only when strength is tested using the training mode. For example, eccentric training increases eccentric performance more than it increases concentric or isometric performance, while cross-mode effects (i.e., effects of eccentric training on concentric strength or of concentric training on eccentric strength) are not different between the two training modes (Hortobágyi, Hill, et al. 1996; see figure 10.2).

SURFACE EMG RESPONSE DURING MVC

Many reports have shown an increase in the integrated muscle EMG during an MVC after a resistance training program. The assumptions are that recruitment is less than maximal before training and that reproducible surface EMG measurements can be taken from the same subjects on different occasions separated by several months. The evidence seems quite strong that recruitment is maximal for most muscles tested in untrained subjects (see chapter 4), raising questions as to the source of the change in maximal EMG that occurs with resistance training. However, the frequency of reports of this response renders it difficult to ignore as a physiological adaptation. Several observations seem to support this increased EMG response as a real phenomenon. For instance, this change can occur without a concomitant change in the amplitude of the M-wave (Yue and Cole 1992; Van Cutsem, Duchateau, and Hainaut 1998; DelBalso and Cafarelli 2007). Also, integrated EMG (iEMG) increases in the contralateral untrained muscle performing an MVC (Narici et al. 1989; Moritani and De Vries 1979). In addition, in subjects who

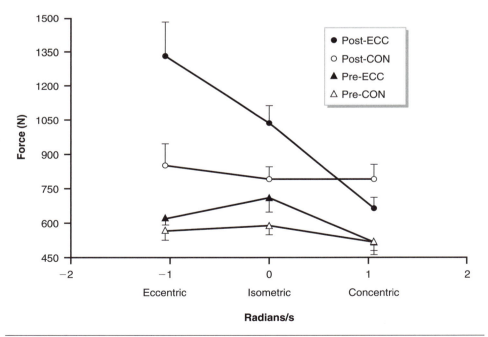

Figure 10.2 Changes in knee extensor force following eccentric (ECC) and concentric (CON) training. Training comprised 36 sessions performed over 12 wk using maximal concentric or eccentric contractions of one leg at a speed of 1.05 rad/s.

Adapted by permission from Hortobágyi et al. 1996.

train with both knee extensors, maximal iEMG is higher after training, but only when tested during the bilateral test (i.e., not when tested unilaterally). Similar specificity of response was found with subjects who trained with only one leg; the increase in iEMG was not seen when the test was with both legs (Häkkinen et al. 1996). The finding that the rate of EMG increase during development of MVC occurs in conjunction with the increased EMG amplitude at MVC (Aagaard et al. 2002a) lends further credence to these changes in surface EMG as real and important phenomena.

IMAGINARY STRENGTH TRAINING

In 1992, Yue and Cole published a somewhat controversial paper that demonstrated that strength increases could be produced when subjects imagined the training sessions. In this study of the abductor muscles of the fifth digit (abductor digiti minimi), subjects were asked during each session either to perform 15 isometric MVCs or to imagine performing 15 isometric MVCs while keeping the muscle totally relaxed. After 4 weeks (20 sessions), the group training with the MVCs increased force by 30%, while those who imagined performing the MVCs increased almost as much (22%). Other previously demonstrated effects of strength training were in evidence in this study: increased maximal EMG and an improvement in the contralateral muscle. The authors suggested an effect on the motor program that resulted in more complete recruitment of the muscle (see figure 10.3).

This result may be specific to muscles that are normally not used very much in everyday activities. For example, maximal isometric abduction of the little finger is not a normal action—nor is it easy to perform. Herbert, Dean, and Gandevia (1998) attempted to duplicate the result of Yue and Cole (1992) using the elbow flexors. After 8 weeks of imagined isometric training, effects such as those reported by Yue and Cole were not observed. Furthermore, training, either imaginary or real, had no influence on the extent of muscle activation during an MVC, which was always near 100% in all groups. These results demonstrate that if there is indeed an effect of imagined training, it does not occur with all muscle groups. This phenomenon will surely be revisited many more times before we can reach definite conclusions regarding its generality and mechanisms.

REFLEX ADAPTATIONS

In a previous chapter, we learned how a simple monosynaptic reflex can adapt to training (Wolpaw and Carp 1993). This paradigm has been applied successfully to humans (Zehr 2006). Resistance training is of course a different intervention than providing a reward for changing a reflex response, as Wolpaw and his colleagues did in their experiments. Still, there is some evidence from reflex studies that resistance training can change the nervous system.

Heavy resistance training with the legs for 14 weeks resulted in significant increases in evoked V-wave and H-reflex responses during maximal muscle contractions, with no change in H-reflex amplitude at rest. H-reflex amplitude is an estimate of net spinal excitability (DelBalso and Cafarelli 2007) and can vary due to the excitability of the motoneurons and to the degree of reflex inhibition of Ia afferent synapses. V-wave amplitude is thought to indicate neural drive in descending pathways. The changes noted suggest that motoneuron excitability was increased or Ia inhibition

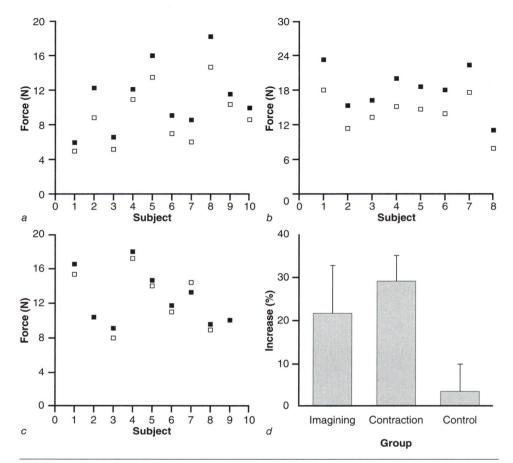

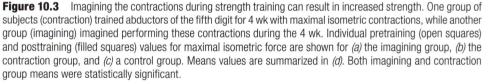

Figure 10.3 Imagining the contractions during strength training can result in increased strength. One group of subjects (contraction) trained abductors of the fifth digit for 4 wk with maximal isometric contractions, while another group (imagining) imagined performing these contractions during the 4 wk. Individual pretraining (open squares) and posttraining (filled squares) values for maximal isometric force are shown for *(a)* the imagining group, *(b)* the contraction group, and *(c)* a control group. Means values are summarized in *(d)*. Both imagining and contraction group means were statistically significant.

Reprinted by permission from Yue and Cole 1992.

was decreased and that neural drive from supraspinal centers was increased following resistance training (Aagaard et al. 2002b). The increased soleus V-wave has been substantiated more recently by DelBalso and Cafarelli (2007), who also showed that this adaptation occurs within 4 weeks following the initiation of training.

Lagerquist and colleagues (Lagerquist et al. 2006) have demonstrated that when one leg is resistance trained for 5 weeks, strength increases in both legs (see the following section on cross education), but H-reflex amplitude (measured at an M-wave amplitude of 5% of maximal) increases only on the trained side. Their conclusion was that the strength increase on the training side involves changes in spinal mechanisms, while the crossover effect involves supraspinal mechanisms (since H-reflex amplitude did not change but MVC did change on the untrained side). Clearly, the role of spinal mechanisms in strength increases resulting from resistance training and the way in which spinal mechanisms are reflected in reflex measurements have yet to be determined.

CROSS EDUCATION

Contralateral effects, also known as *cross education,* refer to the altered performance of the muscle contralateral to the trained one. A classic and often-quoted demonstration of this phenomenon is the work of Moritani and De Vries (1979). These researchers found increased MVC and maximal iEMG in the biceps brachii contralateral to one subjected to resistance (dynamic weight) training. This effect became evident by 2 weeks of training. In general, a survey of the literature reveals that the strength gain from cross education amounts to about 8% of the initial strength of the untrained side, which is equivalent to a strength gain on the trained side of 35% to 50% (Lee and Carroll 2007; Carroll et al. 2006; Munn et al. 2003). There are several other fascinating observations concerning this phenomenon:

1. Resistance training has a contralateral effect on force sensation. This was demonstrated by Cannon and Cafarelli (1987), who found that an isometric training program for one adductor pollicis affected the MVC of the contralateral muscle. MVC of the contralateral muscle increased 9.5% after voluntary training but not after training using electrical stimulation (MVC of both muscles increased by 15%). They also found a difference between voluntary training and electrical stimulation in the ability of the contralateral muscle to match trained muscle force, indicating a contralateral effect of voluntary resistance training on force sensation.

2. Häkkinen and colleagues (1996) examined the effects of unilateral and bilateral heavy resistance training of the knee extensors. They found that increases in MVC and maximal iEMG were higher when the test used was the same as that used during the training. Similarly, Rube and Secher (1990) examined fatigability after 5 weeks of static strength training with either one-leg or two-leg extension that resulted in a significant increase (36%-59%) in maximal extension torque; they found that fatigability was specific to the task used during the training. For example, if training was with both legs, fatigability was increased with a bilateral, but not a unilateral, test.

3. Contralateral effects extend to the coactivation of agonists and antagonists. Carolan and Cafarelli (1992) found a 32% increase in MVC with the trained leg and a 16% increase with the untrained leg after 8 weeks of unilateral isometric training involving 30 MVCs per session. They also found a significant decrease in coactivation of biceps femoris during knee extension in both trained and untrained legs.

4. Ploutz and colleagues (1994) used MRI to estimate the muscle cross-sectional area used during contractions of the knee extensors (see figure 10.1). Training decreased the absolute muscle area recruited per unit force in both the trained and the contralateral leg.

5. Seger, Arvidsson, and Thorstensson (1998) found a mode-specific effect of training on the contralateral limb. Significant increases in torque were found only at the isokinetic velocity at which training was performed and only when using the training mode (eccentric or concentric).

6. An imagined resistance training program increased MVC in the abductor muscles of the fifth digit but also increased the MVC of the contralateral muscle (Yue and Cole 1992).

7. Cross education of arm muscle strength is unidirectional in right-handed individuals. This finding attests to the motor learning–like nature of this phenomenon (Farthing et al. 2005).

Recent reviews of cross education (Carroll et al. 2006; Lee and Carroll 2007) suggest that this effect may be due to task-specific changes in the organization of motor pathways projecting to the contralateral muscles and that resistance training induces changes in cortical regions that are accessed during contralateral movements. Cortical, subcortical, and spinal levels are most likely all involved to various degrees in this phenomenon. Interestingly, when only one leg is trained, only that leg shows H-reflex changes, even though strength increases in both legs (Lagerquist et al. 2006). Demonstrations of this contralateral effect will surely continue to appear in the literature, and there is little doubt that it is a real phenomenon, although the underlying mechanisms remain elusive. The knowledge generated from experiments on cross education has obvious potential for application in the field of rehabilitation.

TRANSFER OF TRAINING EFFECTS TO THE OPPOSITE LIMB INVOLVES CORTICAL CHANGES

The transfer of strength (but not hypertrophic) effects of resistance training to the opposite untrained limb has been demonstrated repeatedly, with limited insight into the exact mechanisms involved. Clearly this phenomenon has implications for use in rehabilitation, and in fact has been used for rehabilitation, but knowledge of where the adaptations are occurring would benefit its specific application. Lee, Gandevia, and Carroll (2009) at the University of New South Wales and University of Queensland in Australia have demonstrated unequivocally for the first time the involvement of the motor cortex in cross education. Subjects performed unilateral wrist extension for 4 weeks using MVCs, after which strength and voluntary drive to the muscles were measured on both sides. Voluntary activation during contractions was assessed using the superimposed twitch technique, in which twitches evoked by stimulation of the cortex using TMS are superimposed on maximal contractions. The results demonstrated that the strength training slightly increased the strength of, and significantly increased the level of voluntary cortical drive to, the untrained wrist extensors. The finding of a strong cortical involvement in this phenomenon has implications for its clinical use not only in cases of training one side to affect an injured opposite side but also in cases where cortex might be involved, such as stroke and other neurological impairments.

DECREASED ACTIVATION OF ANTAGONISTS

Carolan and Cafarelli (1992) had subjects train for 8 weeks on a task requiring 30 MVCs of knee extensors per session. Effects of the training were evident as early as 1 week into the training program. Coactivation of biceps femoris decreased in both trained and untrained muscles during the first week of training. It was estimated that about 33% of the increase in MVC during the first week could be attributed to the decreased coactivation, and this contribution decreased to 10% by the end of the 8-week program. Thus, MVC continued to increase while coactivation did not change significantly after 1 week (see figure 10.4).

CHANGES IN MOTOR UNIT RECRUITMENT

Milner-Brown, Stein, and Lee (1975) were the first to demonstrate experimentally that strength training results in directly measurable changes in the way motor units are recruited during effort. In their experiment, they determined the degree of synchronization of motor unit recruitment during effort by comparing the firing pattern of single motor units, measured with intramuscular electrodes, with the pattern of the surface EMG. Generally, their experimental paradigm asked the question, When the single motor unit fires, does the surface EMG show that a lot of other units are firing at the same time? They compared weightlifters with untrained control subjects, and

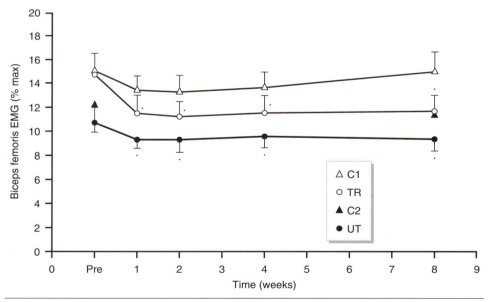

Figure 10.4   Strength training of the knee extensors decreases coactivation of the knee flexors. Decreased EMG of flexors (expressed as a percentage of maximal flexor EMG) during a maximal extension occurred 1 week after beginning the training program in the trained leg (TR). A smaller but still significant decreased coactivation was apparent in the nontrained contralateral leg (UT). The training consisted of 30 maximal isometric knee extensions per day, 3 times per week, for 8 weeks. C1 and C2 refer to the biceps femoris EMG responses in the control and contralateral legs, respectively.

Reprinted by permission from Carolan and Cafarelli 1992.

they also studied a group of subjects before and after 6 weeks of weight training. Their study revealed a higher degree of synchronization of motor units in the first dorsal interosseus muscle in trained subjects than in untrained subjects. They also found that weight training increased the amplitudes of V2 and V3 responses (reflexes evoked during voluntary activation) measured at the muscle in response to stimulation of the peripheral nerve. The V2 and V3 responses—with latencies of about 56 and 83 milliseconds, respectively, following nerve stimulation—represent transcortical reflexes. Thus, increased synchronization of motor units during effort in strength-trained individuals is accompanied by enhanced transcortical reflexes, suggesting that firing synchronization may be linked to a supraspinal mechanism.

A word on synchronization of motor units is warranted at this point. The time course of synchronization that I refer to here is of little consequence to the rapidity of force development: The relatively small difference in the degree of motor unit synchronization ratios between trained and untrained individuals does not translate into differences in power, as we might be inclined to conclude. For instance, a certain degree of asynchrony is expected in peak force generation among motor units, even if the appearance of their action potential in the EMG is perfectly synchronized, because of differences in contractile properties. Rather, synchronization is seen as an index of the strength of presynaptic influences from common sources on motoneurons that

NEURAL EFFECTS OF RESISTANCE TRAINING BENEFIT PATIENTS WITH MULTIPLE SCLEROSIS

Multiple sclerosis (MS) is a neuromuscular disease in which myelin is gradually destroyed, thus limiting the ability to recruit muscle fibers. Since resistance training has been demonstrated to have positive effects on the recruitment function of the nervous system, it might seem intuitive that this type of training would benefit patients with MS. Dr. Marius Fimland and colleagues at the Norwegian University of Science and Technology recruited patients with MS into a resistance training program that involved performing 4 repetitions, 4 times, of unilateral dynamic leg press and plantar flexion above 80% of MVC for 15 sessions spread over 3 weeks. Strength increases of about 20% in MVC were noted, as might be expected. More interesting, however, were the increases noted in the activation of the muscles by the nervous system. Specifically, maximal surface EMG amplitude, normalized to the amplitude resulting from nerve electrical stimulation, was increased by 40%, suggesting an increased capacity to recruit muscle fibers to generate a strong contraction. In addition, the amplitude of a reflex wave that estimates the strength of descending voluntary drive to motoneurons was enhanced by 55%, which brought this particular index to levels comparable with those in normal control subjects. Thus, it appears that the neural effects of resistance training are more than merely an experimental curiosity; these effects may be beneficial for subjects whose central recruitment of muscles by the nervous system is compromised. This group includes patients with MS but perhaps also includes other groups with neuromuscular deficits involving the central nervous system.

determine the degree of coincident generation of action potentials in motoneurons within the pool (Semmler and Nordstrom 1998).

Thus, the results of Milner-Brown, Stein, and Lee (1975) suggest an effect of strength training on supraspinal mechanisms that are responsible for the recruitment of motor units, possibly by an enhancement of the efficacy of synapses on motoneurons from supraspinal sources. These original findings have been supported by more recent studies.

Sale, MacDougall, and colleagues (1983) substantiated the findings of enhanced reflex responses after strength training. They found increased V1 and V2 responses in all muscles trained during the study (which included hypothenar muscles, extensor digitorum brevis, brachioradialis, and soleus) after 9 to 22 weeks of training. Since in this study the reflexes were evoked while the subjects were performing an MVC (unlike in the 1975 study of Milner-Brown, Stein, and Lee, in which the effort was submaximal), they suggested that the elevated reflex responses indicated increased excitation of motoneurons during MVC. This again suggests that activation is not maximal in untrained subjects. However we interpret the source of this training-induced effect, it does support the notion that the nervous system responds to strength training.

Motor unit synchronization appears to adapt to chronic activity. Motor unit synchronization is measured as the near-coincident discharge of pairs of motor units during voluntary movement and reflects the distribution of shared inputs to motor units from corticospinal pathways. Semmler and Nordstrom (1998) have confirmed an increased synchrony of motor unit activation in the first dorsal interosseus of strength-trained subjects and interpreted their findings as an increased corticospinal activity accompanying this task as a result of training. Increased motor unit synchronization would benefit the rate of force development during a voluntary contraction and the coordination of synergists during a complex movement (Semmler 2002). An interesting observation was that skilled individuals (experienced musicians who use their fingers extensively to play their instruments) demonstrated levels of motor unit synchronization, common drive to all motor units, and tremor amplitude during maintained submaximal contractions that were all less than those of controls and strength-trained individuals. This may signify that, in some cases, reduced motor unit synchronization might be a beneficial adaptation to allow for independent synergies among synergistic muscles that might be seen in, say, the hand muscles of a piano or flute player. For the skilled subjects, the task involved the trained muscles but not the task used in the daily training. For the strength-trained subjects, on the other hand, the task included neither trained muscles nor training task. While it is clear from this literature that the function of inputs to motor units adapts to chronic activity, the precise mechanisms remain to be elucidated. More recently, investigators have found adaptations in motor unit coherence, which is an estimate, once again using discharge properties of pairs of motor units, of the oscillatory input to motor units that originates in cortical and subcortical areas (Semmler et al. 2004). The finding that this measure also differs between dominant and nondominant hands and among individuals with differing training status suggests adaptations that involve higher nervous system levels.

Evidence from Van Cutsem, Duchateau, and Hainaut (1998) provides substantial evidence for a neural effect of training. In their study, they asked subjects to train their ankle dorsiflexors for 12 weeks (5 times per week) by moving a load representing 30%

to 40% of 1RM as quickly as possible. At the end of the study, recruitment of motor units during ballistic contractions was examined using intramuscular electrodes. The researchers found that ballistic contractions after the training program were faster, with a more rapid onset of EMG. They also found that maximal instantaneous firing rates of motor units during ballistic contractions were higher and showed less decrease in frequency after training. In addition, the percentage of motor units showing incidents of doublets (two spikes of the same motor unit separated by 5 milliseconds or less) increased from 5.2% of the control units to 32.7% of the trained units. The authors suggested that ballistic training causes increased motoneuron excitability that leads to the previously described changes during voluntary excitation.

More recent evidence from Carroll and colleagues (2002) suggests that resistance training alters the functional properties of the corticospinal pathway. In their experiment, they resistance trained the first dorsal interosseus muscles of 16 individuals for 16 weeks and measured the EMG responses to TMS and TES, at rest and at various levels of voluntary contraction, before and after the training. Since TMS-evoked responses are more influenced by the excitability state of the motor cortex than TES-evoked responses are, the investigators were able to estimate cortical and spinal mechanisms. Their conclusion was that resistance training changes the organization of synaptic circuitry in the spinal cord, which can lead to altered recruitment of motor units during the task, with no major adaptations occurring at the level of the cortex.

CHANGES IN MOTOR CORTEX

TMS is used to estimate the excitability of the motor cortex and upstream pathways. The idea is that when the cortex is stimulated at a submaximal intensity above the threshold for evoking a muscle EMG response, the peak-to-peak amplitude of the EMG response is indicative of the degree of excitation of cortical pathways. Thus, any change in this amplitude denotes change in excitation of these pathways. In a study by Griffin and Cafarelli (2007), subjects trained their right tibialis anterior muscle 3 times per week for 4 weeks. Training resulted in a significant increase in MVC of 18% as well as a significant increase in rate of maximal force development (subjects were asked to generate force as quickly as possible). Amplitudes of MEPs recorded from tibialis anterior increased up to 43% when measured during a sustained contraction of 10% of MVC. Interestingly, increases in MEP amplitudes were evident after 6 training sessions (2 weeks). Unfortunately, the study design did not allow the experimenters to rule out the possibility that the changes in excitability occurred in the spinal cord. In addition, the changes reported by Griffin and colleagues may not extend to other muscle groups—for example, Jensen and colleagues (2005) failed to demonstrate changes in cortical excitability following 12 weeks of resistance training of the biceps brachii. The effects of strength training on motor cortex thus remain elusive.

SUMMARY

It is believed that a proportion of the strength gain seen in resistance training is the result of changes in neural pathways and the way in which these changes alter recruitment of motor units. Evidence for neural effects of strength ranges from rather weak, indirect evidence (such as the discrepancy between strength gain and muscle mass increase, especially early in the training program) to more direct evidence such as electrophysiological changes in reflex pathways and motor unit recruitment patterns.

There is some argument that muscle compositional changes may explain at least a portion of these neural effects of strength training. Effects of strength training on cortex have been reported, but these changes, as well as the adaptation of neural pathways in general to strength training, have yet to be confirmed. If neural adaptations do occur with strength training, the implications for using exercise as a therapeutic modality in cases where the nervous system is compromised are obvious.

References

Aagaard, P., Andersen, J.L., Dyhre-Poulsen, P., Leffers, A., Wagner, A., Magnusson, S.P., Halkjaer-Kristensen, J., and Simonsen, E.B. 2001. A mechanism for increased contractile strength of human pennate muscle in response to strength training: Changes in muscle architecture. *J. Physiol. (Lond.)* 534:613–23.

Aagaard, P., Simonsen, E.B., Andersen, J.L., Magnusson, P., and Dyhre-Poulsen, P. 2002a. Increased rate of force development and neural drive of human skeletal muscle following resistance training. *J. Appl. Physiol.* 93:1318–26.

Aagaard, P., Simonsen, E.B., Andersen, J.L., Magnusson, P., and Dyhre-Poulsen, P. 2002b. Neural adaptation to resistance training: Changes in evoked V-wave and H-reflex responses. *J. Appl. Physiol.* 92:2309–18.

Abbruzzese, G., Morena, M., Spadavecchia, L., and Schieppati, M. 1994. Response of arm flexor muscles to magnetic and electrical brain stimulation during shortening and lengthening tasks in man. *J. Physiol.* 481:499–507.

AbuShakra, S.R., Cole, A.J., Adams, R.N., and Drachman, D.B. 1994. Cholinergic stimulation of skeletal muscle cells induces rapid immediate early gene expression: Role of intracellular calcium. *Mol. Brain. Res.* 26:55–60.

Adam, A., De Luca, C.J., and Erim, Z. 1998. Hand dominance and motor unit firing behavior. *J. Neurophysiol.* 80:1373–82.

Adams, G.R., and Haddad, F. 1996. The relationships among IGF1, DNA content, and protein accumulation during skeletal muscle hypertrophy. *J. Appl. Physiol.* 81:2509–16.

Adams, G.R., Haddad, F., and Baldwin, K.M. 1999. Time course of changes in markers of myogenesis in overloaded rat skeletal muscles. *J. Appl. Physiol.* 87:1705–12.

Adams, G.R., Hather, B.M., Baldwin, K.M., and Dudley, G.A. 1993. Skeletal muscle myosin heavy chain composition and resistance training. *J. Appl. Physiol.* 74:911–5.

Adams, G.R., and McCue, S.A. 1998. Localized infusion of IGFI results in skeletal muscle hypertrophy in rats. *J. Appl. Physiol.* 84:1716–22.

Alaimo, M.A., Smith, J.L., and Edgerton, V.R. 1984. EMG activity of slow and fast ankle extensors following spinal cord transection. *J. Appl. Physiol. Respir. Environ. Exerc. Physiol.* 56:1608–13.

Alberts, B., Bray, D., Lewis, J., Raff, M., Roberts, K., and Watson, J. 1994. *Molecular biology of the cell.* New York: Garland.

Aldrich, T.K. 1987. Transmission fatigue of the rabbit diaphragm. *Respir. Physiol.* 69:307–19.

Aldrich, T.K., Shander, A., Chaudhry, I., and Nagashima, H. 1986. Fatigue of isolated rat diaphragm: Role of impaired neuromuscular transmission. *J. Appl. Physiol.* 61:1077–83.

Alford, E.K., Roy, R.R., Hodgson, J.A., and Edgerton, V.R. 1987. Electromyography of rat soleus, medial gastrocnemius, and tibialis anterior during hind limb suspension. *Exp. Neurol.* 96:635–49.

Allen, D.G. 2004. Skeletal muscle function: Role of ionic changes in fatigue, damage and disease. *Clinical and Experimental Pharmacology and Physiology* 31:485–93.

Allen, D.G., Lamb, G.D., and Westerblad, H. 2008a. Impaired calcium release during fatigue. *J. Appl. Physiol.* 104(1): 296–305.

Allen, D.G., Lamb, G.D., and Westerblad, H. 2008b. Skeletal muscle fatigue: Cellular mechanisms. *Physiological Reviews* 88:287–332.

Allen, D.G., and Westerblad, H. 2001. Role of phosphate and calcium stores in muscle fatigue. *Journal of Physiology* 536:657–65.

Allen, D.L., Linderman, J.K., Roy, R.R., Bigbee, A.J., Grindeland, R.E., Mukku, V., and Edgerton, V.R. 1997. Apoptosis: A mechanism contributing to remodeling of skeletal muscle in response to hindlimb unweighting. *Am. J. Physiol. Cell Physiol.* 273:C579–87.

Allen, D.L., Linderman, J.K., Roy, R.R., Grindeland, R.E., Mukku, V., and Edgerton, V.R. 1997. Growth hormone IGFI and/or resistive exercise maintains myonuclear number in hindlimb unweighted muscles. *J. Appl. Physiol.* 83:1857–61.

Allen, D.L., Monke, S.R., Talmadge, R.J., Roy, R.R., and Edgerton, V.R. 1995. Plasticity of myonuclear number in hypertrophied and atrophied mammalian skeletal muscle fibers. *J. Appl. Physiol.* 78:1969–76.

Allen, G.M., Gandevia, S.C., and McKenzie, D.K. 1995. Reliability of measurements of muscle strength and voluntary activation using twitch interpolation. *Muscle and Nerve* 18:593–600.

Altenburg, T.M., Degens, H., vanMechelen, W., Sargeant, A.J., and deHaan, A. 2007. Recruitment of single muscle fibers during submaximal cycling exercise. *J. Appl. Physiol.* 103(5): 1752–6.

Alway, S.E., MacDougall, J.D., and Sale, D.G. 1989. Contractile adaptations in the human triceps surae after isometric exercise. *J. Appl. Physiol.* 66:2725–32.

Alway, S.E., MacDougall, J.D., Sale, D.G., Sutton, J.R., and McComas, A.J. 1988. Functional and structural adaptations in skeletal muscle of trained athletes. *J. Appl. Physiol.* 64:1114–20.

Andersen, J.L., and Aagaard, P. 2000. Myosin heavy chain IIX overshoot in human skeletal muscle. *Muscle and Nerve* 23:1095–104.

Andersen, J.L., and Schiaffino, S. 1997. Mismatch between myosin heavy chain mRNA and protein distribution in human skeletal muscle fibers. *Am. J. Physiol. Cell Physiol.* 272:C1881–9.

Andersen, L.L., Tufekovic, G., Zebis, M.K., Crameri, R.M., Verlaan, G., Kjær, M., Suetta, C., Magnusson, P., and Aagaard, P. 2005. The effect of resistance training combined with timed ingestion of protein on muscle fiber size and muscle strength. *Metabolism: Clinical and Experimental* 54:151–6.

Annex, B.H., Kraus, W.E., Dohm, G.L., and Williams, R.S. 1991. Mitochondrial biogenesis in striated muscles: Rapid induction of citrate synthase mRNA by nerve stimulation. *Am. J. Physiol. Cell Physiol.* 260:C266–70.

Antonio, J., and Gonyea, W.J. 1993. Role of muscle fiber hypertrophy and hyperplasia in intermittently stretched avian muscle. *J. Appl. Physiol.* 74:1893–8.

Appelberg, B., Hulliger, M., Johansson, H., and Sojka, P. 1983. Actions on gammamotoneurones elicited by electrical stimulation of group III muscle afferent fibres in the hind limb of the cat. *J. Physiol. (Lond.)* 335:275–92.

Argaw, A., Desaulniers, P., and Gardiner, P.F. 2004. Enhanced neuromuscular transmission efficacy in overloaded rat plantaris muscle. *Muscle and Nerve* 29:97–103.

Aronson, D., Boppart, M.D., Dufresne, S.D., Fielding, R.A., and Goodyear, L.J. 1998. Exercise stimulates cJun NH$_2$ kinase activity and cJun transcriptional activity in human skeletal muscle. *Biochem. Biophys. Res. Commun.* 251:106–10.

Aronson, D., Dufresne, S.D., and Goodyear, L.J. 1997. Contractile activity stimulates the cJun NH$_2$terminal kinase pathway in rat skeletal muscle. *J. Biol. Chem.* 272:25636–40.

Avela, J., and Komi, P.V. 1998. Reduced stretch reflex sensitivity and muscle stiffness after longlasting stretchshortening cycle exercise in humans. *Eur. J. Appl. Physiol. Occup. Physiol.* 78:403–10.

Aymard, C., Katz, R., Lafitte, C., Le Bozec, S., and Pénicaud, A. 1995. Changes in reciprocal and transjoint inhibition induced by muscle fatigue in man. *Exp. Brain Res.* 106:418–24.

Baar, K., Blough, E., Dineen, B., and Esser, K. 1999. Transcriptional regulation in response to exercise. *Exerc. Sport Sci. Rev.* 27:333–79.

Baar, K., and Esser, K. 1999. Phosphorylation of p70^{S6k} correlates with increased skeletal muscle mass following resistance exercise. *Am. J. Physiol. Cell Physiol.* 276:C120–7.

Babij, P., and Booth, F.W. 1988. Alphaactin and cytochrome c mRNAs in atrophied adult rat skeletal muscle. *Am. J. Physiol.* 254:C651–6.

Babault, N., Desbrosses, K., Fabre, M.S., Michaut, A., and Pousson, M. 2006. Neuromuscular fatigue development during maximal concentric and isometric knee extensions. *J. Appl. Physiol.* 100:780–5.

Bakels, R., and Kernell, D. 1993. Matching between motoneurone and muscle unit properties in rat medial gastrocnemius. *J. Physiol.* 463:307–24.

Bakels, R., and Kernell, D. 1994. Thresholdspacing in motoneurone pools of rat and cat: Possible relevance for manner of force gradation. *Exp. Brain Res.* 102:69–74.

Baker, L.L., and Chandler, S.H. 1987a. Characterization of hindlimb motoneuron membrane properties in acute and chronic spinal cats. *Brain Res.* 420:333–9.

Baker, L.L., and Chandler, S.H. 1987b. Characterization of postsynaptic potentials evoked by sural nerve stimulation in hindlimb motoneurons from acute and chronic spinal cats. *Brain Res.* 420:340–50.

Baldwin, K.M., Herrick, R.E., and McCue, S.A. 1993. Substrate oxidation capacity in rodent skeletal muscle: Effects of exposure to zero gravity. *J. Appl. Physiol.* 75:2466–70.

Balestra, C., Duchateau, J., and Hainaut, K. 1992. Effects of fatigue on the stretch reflex in a human muscle. *Electroencephalogr. Clin. Neurophysiol. Electromyogr. Motor Control* 85:46–52.

Bangart, J.J., Widrick, J.J., and Fitts, R.H. 1997. Effect of intermittent weight bearing on soleus fiber force–velocity–power and force–pCa relationships. *J. Appl. Physiol.* 82:1905–10.

Barany, M. 1967. ATPase activity of myosin correlated with speed of muscle shortening. *J. Gen. Physiol.* 50:197–218.

Barbeau, H., Norman, K., Fung, J., Visintin, M., and Ladouceur, M. 1998. Does neurorehabilitation play a role in the recovery of walking in neurological populations? *Ann. NY Acad. Sci.* 860:377–92.

Barbeau, H., and Rossignol, S. 1987. Recovery of locomotion after chronic spinalization in the adult cat. *Brain Res.* 412:84–95.

BarOr, O., Dotan, R., Inbar, O., Rothstein, A., Karlsson, J., and Tesch, P. 1980. Anaerobic capacity and muscle fiber type distribution in man. *Int. J. Sports Med.* 1:82–5.

Barstow, T.J., Jones, A.M., Nguyen, P.H., and Casaburi, R. 2000. Influence of muscle fibre type and fitness on the oxygen uptake/power output slope during incremental exercise in humans. *Exp. Physiol.* 85:109–16.

BartonDavis, E.R., LaFramboise, W.A., and Kushmerick, M.J. 1996. Activitydependent induction of slow myosin gene expression in isolated fasttwitch mouse muscle. *Am. J. Physiol. Cell Physiol.* 271:C1409–14.

Barton-Davis, E.R., Shoturma, D., and Sweeney, L. 1999. Contribution of satellite cells to IGF-1 induced hypertrophy of skeletal muscle. *Acta Physiol. Scand.* 167:301–5.

Bawa, P., and Lemon, R.N. 1993. Recruitment of motor units in response to transcranial magnetic stimulation in man. *J. Physiol.* 471:445–64.

Bawa, P., and Murnaghan, C. 2009. Motor unit rotation in a variety of human muscles, *J. Neurophysiol.* 102: 2265-2272.

Bawa, P., Pang, M., Olesen, K., and Calancie, B. 2006. Rotation of motoneurons during prolonged isometric contractions in humans. *Journal of Neurophysiology* 96:1135–40.

Bazzy, A.R., and Donnelly, D.F. 1993. Diaphragmatic failure during loaded breathing: Role of neuromuscular transmission. *J. Appl. Physiol.* 74:1679–83.

Behm, D.G., and Sale, D.G. 1993. Intended rather than actual movement velocity determines velocity-specific training response. *J. Appl. Physiol.* 74:359–68.

Behm, D.G., and St.-Pierre, D.M.M. 1997. Fatigue characteristics following ankle fractures. *Med. Sci. Sports Exerc.* 29:1115–23.

Bélanger, M., Drew, T., Provencher, J., and Rossignol, S. 1996. A comparison of treadmill locomotion in adult cats before and after spinal transection. *J. Neurophysiol.* 76:471–91.

Belcastro, A.N. 1993. Skeletal muscle calciumactivated neutral protease (calpain) with exercise. *J. Appl. Physiol.* 74:1381–6.

Belcastro, A.N., Albisser, T.A., and Littlejohn, B. 1996. Role of calciumactivated neutral protease (calpain) with diet and exercise. *Can. J. Appl. Physiol.* 21:328–46.

BelhajSaïf, A., Fourment, A., and Maton, B. 1996. Adaptation of the precentral cortical command to elbow muscle fatigue. *Exp. Brain Res.* 111:405–16.

Beltman, J.G.M., Sargeant, A.J., Haan, H., Van Mechelen, W., and De Haan, A. 2004. Changes in PCr/Cr ratio in single characterized muscle fibre fragments after only a few maximal voluntary contractions in humans. *Acta Physiol. Scand.* 180:187–93.

Beltman, J.G.M., Sargeant, A.J., Van Mechelen, W., and De Haan, A. 2004. Voluntary activation level and muscle fiber recruitment of human quadriceps during lengthening contractions. *J. Appl. Physiol.* 97:619–26.

Bellemare, F., Woods, J.J., Johansson, R., and BiglandRitchie, B. 1983. Motorunit discharge rates in maximal voluntary contractions of three human muscles. *J. Neurophysiol.* 50:1380–92.

Bennett, D.J., Hultborn, H., Fedirchuk, B., and Gorassini, M. 1998. Shortterm plasticity in hindlimb motoneurons of decerebrate cats. *J. Neurophysiol.* 80:2038–45.

Bentley, D., Smith, P., Davie, A., and Zhou, S. 2000. Muscle activation of the knee extensors following high intensity endurance exercise in cyclists. *Eur. J. Appl. Physiol.* 81:297–302.

Benton, C.R., Wright, D.C., and Bonen, A. 2008. PGC-1 alpha-mediated regulation of gene expression and metabolism: Implications for nutrition and exercise prescriptions. *Appl. Physiol. Nutr. Metab.* 33:843–62.

Berg, H.E., Dudley, G.A., Häggmark, T., Ohlsén, H., and Tesch, P.A. 1991. Effects of lower limb unloading on skeletal muscle mass and function in humans. *J. Appl. Physiol.* 70:1882–5.

Berg, H.E., Dudley, G.A., Hather, B., and Tesch, P.A. 1993. Work capacity and metabolic and morphologic characteristics of the human quadriceps muscle in response to unloading. *Clin. Physiol.* 13:337–47.

Berg, H.E., Larsson, L., and Tesch, P.A. 1997. Lower limb skeletal muscle function after 6 wk of bed rest. *J. Appl. Physiol.* 82:182–8.

Berg, H.E., and Tesch, P.A. 1996. Changes in muscle function in response to 10 days of lower limb unloading in humans. *Acta Physiol. Scand.* 157:63–70.

Bernardi, M., Solomonow, M., Nguyen, G., Smith, A., and Baratta, R. 1996. Motor unit recruitment strategy changes with skill acquisition. *Eur. J. Appl. Physiol. Occup. Physiol.* 74:52–9.

Bevan, L., Laouris, Y., Reinking, R.M., and Stuart, D.G. 1992. The effect of the stimulation pattern on the fatigue of single motor units in adult cats. *J. Physiol.* 449:85–108.

Bickel, C.S., Slade, J., Mahoney, E., Haddad, F., Dudley, G.A., and Adams, G.R. 2005. Time course of molecular responses of human skeletal muscle to acute bouts of resistance exercise. *J. Appl. Physiol.* 98:482–8.

BiglandRitchie, B., Dawson, N., Johansson, R., and Lippold, O.C.J. 1986. Reflex origin for the slowing of motoneurone firing rates in fatigue of human voluntary contractions. *J. Physiol. (Lond.)* 379:451–9.

Bigland-Ritchie, B., Fuglevand, A., and Thomas, C. 1998. Contractile properties of human motor units: Is man a cat? *Neuroscientist* 4:240–9.

BiglandRitchie, B., Furbush, F., and Woods, J.J. 1986. Fatigue of intermittent submaximal voluntary contractions: Central and peripheral factors. *J. Appl. Physiol.* 61:421–9.

BiglandRitchie, B., Johansson, R., Lippold, O.C.J., Smith, S., and Woods, J.J. 1983. Changes in motoneurone firing rates during sustained maximal voluntary contractions. *J. Physiol. (Lond.)* 340:335–46.

BiglandRitchie, B., Jones, D.A., and Woods, J.J. 1979. Excitation frequency and muscle fatigue: Electrical responses during human voluntary and stimulated contractions. *Exp. Neurol.* 64:414–27.

BiglandRitchie, B., and Woods, J.J. 1984. Changes in muscle contractile properties and neural control during human muscular contraction. *Muscle and Nerve* 7:691–9.

Billeter, R., Heizmann, C.W., Howald, H., and Jenny, E. 1981. Analysis of myosin light and heavy chain types in single human skeletal muscle fibers. *Eur. J. Biochem.* 116:389–95.

Binder, M.D., Bawa, P., Ruenzel, P., and Henneman, E. 1983. Does orderly recruitment of motoneurons depend on the existence of different types of motor units? *Neurosci. Lett.* 36:55–8.

Binder, M.D., Heckman, C.J., and Powers, R.K. 1996. The physiological control of motoneuron activity. In *Handbook of physiology: 12. Exercise, regulation and integration of multiple systems,* ed. L.B. Rowell and J.T. Shepherd, 3–53. New York: Oxford University Press.

Biolo, G., Maggi, S.P., Williams, B.D., Tipton, K.D., and Wolfe, R.R. 1995. Increased rates of muscle protein turnover and amino acid transport after resistance exercise in humans. *Am. J. Physiol. Endocrinol. Metab.* 268:E514–20.

Biro, A., Griffin, L., and Cafarelli, E. 2007. Reflex gain of muscle spindle pathways during fatigue. *Experimental Brain Research* 177(2): 157–66.

Blakemore, S.J., Rickhuss, P.K., Watt, P.W., Rennie, M.J., and Hundal, H.S. 1996. Effects of limb immobilization on cytochrome c oxidase activity and GLUT4 and GLUT5 protein expression in human skeletal muscle. *Clin. Sci.* 91:591–9.

Blazevich, A.J. 2006. Effects of physical training and detraining, immobilisation, growth and aging on human fascicle geometry. *Sport Med.* 36(12): 1003–17.

Blazevich, A.J., Cannavan, D., Coleman, D.R., and Horne, S. 2007. Influence of concentric and eccentric resistance training on architectural adaptation in human quadriceps muscles. *J. Appl. Physiol.* 103(5): 1565–75.

Blewett, C., and Elder, G.C.B. 1993. Quantitative EMG analysis in soleus and plantaris during hindlimb suspension and recovery. *J. Appl. Physiol.* 74:2057–66.

Bobbert, M.F., De Graaf, W.W., Jonk, J.N., and Casius, L.J.R. 2006. Explanation of the bilateral deficit in human vertical squat jumping. *J. Appl. Physiol.* 100:493–9.

Bodine, S.C. 2006. mTOR signaling and the molecular adaptation to resistance exercise. *Med. Sci. Sports Exerc.* 38(11): 1950–7.

Bodine, S.C., Stitt, T.N., Gonzalez, M., Kline, W.O., Stover, G.L., Bauerlein, R., Zlotchenko, E., Scrimgeour, A., Lawrence, J.C., Glass, D.J., and Yancopoulos, G.D. 2001. Akt/mTOR pathway is a crucial regulator of skeletal muscle hypertrophy and can prevent muscle atrophy *in vivo*. *Nature Cell Biology* 3:1014–9.

BodineFowler, S.C., Roy, R.R., Rudolph, W., Haque, N., Kozlovskaya, I.B., and Edgerton, V.R. 1992. Spaceflight and growth effects on muscle fibers in the rhesus monkey. *J. Appl. Physiol.* Suppl. no. 73:82S–9S.

Bolster, D.R., Kubica, N., Crozier, S., Williamson, D., Farrell, P.A., Kimball, S.R., and Jefferson, L.S. 2003. Immediate response of mammalian target of rapamycin (mTOR)-mediated signaling following acute resistance exercise in rat skeletal muscle. *J. Physiol. (Lond.)* 553:213–20.

Bonato, C., Zanette, G., Manganotti, P., Tinazzi, M., Bongiovanni, G., Polo, A., and Fiaschi, A. 1996. "Direct" and "crossed" modulation of human motor cortex excitability following exercise. *Neurosci. Lett.* 216:97–100.

Bonen, A., Han, X.X., Habets, J., Febbraio, M., Glatz, J., and Luiken, J.J.F.P. 2007. A null mutation in skeletal muscle FAT/CD36 reveals its essential role in insulin- and AICAR-stimulated fatty acid metabolism. *Am. J. Physiol. (Endocrinol. Metab.)* 292: E1740–9.

Bongiovanni, L.G., and Hagbarth, K.E. 1990. Tonic vibration reflexes elicited during fatigue from maximal voluntary contractions in man. *J. Physiol. (Lond.)* 423:1–14.

Bongiovanni, L.G., Hagbarth, K.E., and Stjernberg, L. 1990. Prolonged muscle vibration reducing motor output in maximal voluntary contractions in man. *J. Physiol. (Lond.)* 423:15–26.

Booth, F.W., and Kelso, J.R. 1973. Effect of hindlimb immobilization on contractile and histochemical properties of skeletal muscle. *Pflugers Arch.* 342:231–8.

Booth, F.W., and Kirby, C.R. 1992. Changes in skeletal muscle gene expression consequent to altered weight bearing. *Am. J. Physiol. Regul. Integr. Comp. Physiol.* 262:R329–32.

Booth, F.W., Lou, W., Hamilton, M.T., and Yan, Z. 1996. Cytochrome c mRNA in skeletal muscles of immobilized limbs. *J. Appl. Physiol.* 81:1941–5.

Booth, F.W., and Seider, M.J. 1979. Early change in skeletal muscle protein synthesis after limb immobilization of rats. *J. Appl. Physiol.* 47:974–7.

Booth, F.W., Tseng, B.S., Flück, M., and Carson, J.A. 1998. Molecular and cellular adaptation of muscle in response to physical training. *Acta Physiol. Scand.* 162:343–50.

Boppart, M., Aronson, D., Gibson, L., Roubenoff, R., Abad, L., Bean, J., Goodyear, L., and Fielding, R. 1999. Eccentric exercise markedly increases c-Jun NH_2-terminal kinase activity in human skeletal muscle. *J. Appl. Physiol.* 87:1668–73.

Bosco, C., Montanari, G., Ribacchi, R., Giovenali, P., Latteri, F., Iachelli, G., Faina, M., Colli, R., DalMonte, A., LaRosa, M., Cortili, G., and Saibene, F. 1987. Relationship between the efficiency of muscular work during jumping and the energetics of running. *Eur. J. Appl. Physiol.* 56:138–43.

Botterman, B.R., and Cope, T.C. 1988a. Maximum tension predicts relative endurance of fasttwitch motor units in the cat. *J. Neurophysiol.* 60:1215–26.

Botterman, B.R., and Cope, T.C. 1988b. Motorunit stimulation patterns during fatiguing contractions of constant tension. *J. Neurophysiol.* 60:1198–214.

Bottinelli, R., Betto, R., Schiaffino, S., and Reggiani, C. 1994a. Maximum shortening velocity and coexistence of myosin heavy chain isoforms in single skinned fast fibres of rat skeletal muscle. *J. Muscle Res. Cell Motil.* 15:413–9.

Bottinelli, R., Betto, R., Schiaffino, S., and Reggiani, C. 1994b. Unloaded shortening velocity and myosin heavy chain and alkali light chain isoform composition in rat skeletal muscle fibres. *J. Physiol.* 478:341–9.

Bottinelli, R., Canepari, M., Pellegrino, M.A., and Reggiani, C. 1996. Force–velocity properties of human skeletal muscle fibres: Myosin heavy chain isoform and temperature dependence. *J. Physiol.* 495:573–86.

Bottinelli, R., Canepari, M., Reggiani, C., and Stienen, G.J.M. 1994. Myofibrillar ATPase activity during isometric contraction and isomyosin composition in rat single skinned muscle fibres. *J. Physiol. (Lond.)* 481:663–76.

Bottinelli, R., and Reggiani, C. 1995a. Essential myosin light chain isoforms and energy transduction in skeletal muscle fibers. *Biophys. J.* 68:227S.

Bottinelli, R., and Reggiani, C. 1995b. Force–velocity properties and myosin light chain isoform composition of an identified type of skinned fibres from rat skeletal muscle. *Pflugers Arch.* 429:592–4.

Boyce, V.S., Tumolo, M., Fischer, I., Murray, M., and Lemay, M.A. 2007. Neurotrophic factors promote and enhance locomotor recovery in untrained spinalized cats. *Journal of Neurophysiology* 98(4): 1988–96.

BrasilNeto, J.P., Cohen, L.G., and Hallett, M. 1994. Central fatigue as revealed by postexercise decrement of motor evoked potentials. *Muscle and Nerve* 17:713–9.

Briggs, F.N., Poland, J.L., and Solaro, R.J. 1977. Relative capabilities of sarcoplasmic reticulum in fast and slow mammalian skeletal muscles. *J. Physiol.* 266:587–94.

Brooke, M.H., Williamson, E., and Kaiser, K.K. 1971. The behavior of four fiber types in developing and reinnervated muscle. *Arch. Neurol.* 25:360–6.

Brown, S.J., Child, R.B., Donnelly, A.E., Saxton, J.M., and Day, S.H. 1996. Changes in human skeletal muscle contractile function following stimulated eccentric exercise. *Eur. J. Appl. Physiol.* 72:515–21.

Brownson, C., Isenberg, H., Brown, W., Salmons, S., and Edwards, Y. 1988. Changes in skeletal muscle gene transcription induced by chronic stimulation. *Muscle and Nerve* 11:1183–9.

Brownson, C., Little, P., Jarvis, J.C., and Salmons, S. 1992. Reciprocal changes in myosin isoform mRNAs of rabbit skeletal muscle in response to the initiation and cessation of chronic electrical stimulation. *Muscle and Nerve* 15:694–700.

Brownstone, R.M., Jordan, L.M., Kriellaars, D.J., Noga, B.R., and Shefchyk, S.J. 1992. On the regulation of repetitive firing in lumbar motoneurones during fictive locomotion in the cat. *Exp. Brain Res.* 90:441–55.

Burke, R.E. 1967. Motor unit types of cat triceps surae muscle. *J. Physiol. (Lond.)* 193:141–60.

Burke, R.E. 1990. Motor unit types: Some history and unsettled issues. In *The segmental motor system,* ed. M. Binder and L. Mendell, 207–221. New York: Oxford University Press.

Burke, R.E., Dum, R., Fleshman, J., Glenn, L., LevTov, A., O'Donovan, M., and Pinter, M. 1982. An HRP study of the relation between cell size and motor unit type in cat ankle extensor motoneurons. *J. Comp. Neurol.* 209:17–28.

Burke, R.E., Levine, D.N., Tsairis, P., and Zajac, F.E. 1973. Physiological types and histochemical profiles in motor units of the cat gastrocnemius. *J. Physiol. (Lond.)* 234:723–48.

Burke, R.E., and Tsairis, P. 1974. The correlation of physiological properties with histochemical characteristics in single muscle units. *Ann. N Y Acad. Sci.* 228:145–59.

Burridge, K., and ChrzanowskaWodnicka, M. 1996. Focal adhesions, contractility, and signalling. *Ann. Rev. Cell Devel. Biol.* 12:463–519.

Butler, I., Drachman, D., and Goldberg, A. 1978. The effect of disuse on cholinergic enzymes. *J. Physiol. (Lond.)* 274:593–600.

Butler, J.E., Taylor, J.L., and Gandevia, S.C. 2003. Responses of human motoneurons to corticospinal stimulation during maximal voluntary contractions and ischemia. *J. Neurosci.* 23:10224–30.

Button, D., Gardiner, K., Marqueste, T., and Gardiner, P.F. 2006. Frequency-current relationships of rat hindlimb alpha-motoneurones. *J. Physiol. (Lond.)* 573:663–77.

Button, D.C., Kalmar, J.M., Gardiner, K., Cahill, F., and Gardiner, P.F. 2007. Spike frequency adaptation of rat hindlimb motoneurons. *J. Appl. Physiol.* 102(3): 1041–50.

Caiozzo, V.J., Baker, M.J., and Baldwin, K.M. 1998. Novel transitions in MHC isoforms: Separate and combined effects of thyroid hormone and mechanical unloading. *J. Appl. Physiol.* 85:2237–48.

Caiozzo, V.J., Haddad, F., Baker, M.J., Herrick, R.E., Prietto, N., and Baldwin, K.M. 1996. Microgravityinduced transformations of myosin isoforms and contractile properties of skeletal muscle. *J. Appl. Physiol.* 81:123–32.

Caiozzo, V.J., Perrine, J.J., and Edgerton, V.R. 1981. Traininginduced alterations of the *in vivo* force–velocity relationship of human muscle. *J.*

Appl. Physiol. Respir. Environ. Exerc. Physiol. 51:750–4.

Cairns, S.P. 2006. Lactic acid and exercise performance: Culprit or friend? *Sport Med.* 36:279–91.

Cairns, S.P., and Lindinger, M.I. 2008. Do multiple ionic interactions contribute to skeletal muscle fatigue? *Journal of Physiology* 586:4039–54.

Calbet, J.A.L., Gonzalez-Alonso, J., Helge, J.W., Sondergaard, H., Munch-Andersen, T., Boushel, R., and Saltin, B. 2007. Cardiac output and leg and arm blood flow during incremental exercise to exhaustion on the cycle ergometer. *J. Appl. Physiol.* 103:969–78.

Calbet, J.A.L., Jensen-Urstad, M., Van Hall, G., Holmberg, H., Rosdahl, H., and Saltin, B. 2004. Maximal muscular vascular conductances during whole body upright exercise in humans. *J. Physiol. (Lond.)* 558:319–31.

Callister, R.J., Sesodia, S., Enoka, R.M., Nemeth, P.M., Reinking, R.M., and Stuart, D.G. 2004. Fatigue of rat hindlimb motor units: Biochemical-physiological associations. *Muscle and Nerve* 30:714–26.

Calancie, B., and Bawa, P. 1985. Voluntary and reflexive recruitment of flexor carpi radialis motor units in humans. *J. Neurophysiol.* 53:1194–200.

Cameron-Smith, D. 2002. Exercise and skeletal muscle gene expression. *Clinical and Experimental Pharmacology and Physiology* 29:209–13.

Cannon, R.J., and Cafarelli, E. 1987. Neuromuscular adaptations to training. *J. Appl. Physiol.* 63:2396–402.

Canon, F., Goubel, F., and Guezennec, C.Y. 1998. Effects on contractile and elastic properties of hindlimb suspended rat soleus muscle. *Eur. J. Appl. Physiol. Occ. Physiol.* 77:118–24.

Capaday, C. 1997. Neurophysiological methods for studies of the motor system in freely moving human subjects. *J. Neurosci. Methods* 74:201–18.

Carolan, B., and Cafarelli, E. 1992. Adaptations in coactivation after isometric resistance training. *J. Appl. Physiol.* 73:911–7.

Carp, J.S., and Wolpaw, J.R. 1994. Motoneuron plasticity underlying operantly conditioned decrease in primate Hreflex. *J. Neurophysiol.* 72:431–42.

Carp, J.S., and Wolpaw, J.R. 1995. Motoneuron properties after operantly conditioned increase in primate Hreflex. *J. Neurophysiol.* 73:1365–73.

Carraro, F., Stuart, C.A., Hartl, W.H., Rosenblatt, J., and Wolfe, R.R. 1990. Effect of exercise and recovery on muscle protein synthesis in human subjects. *Am. J. Physiol. Endocrinol. Metab.* 259: E470–6.

Carrier, L., Brustein, E., and Rossignol, S. 1997. Locomotion of the hindlimbs after neurectomy of ankle flexors in intact and spinal cats: Model for the study of locomotor plasticity. *J. Neurophysiol.* 77:1979–93.

Carroll, S.L., Klein, M.G., and Schneider, M.F. 1997. Decay of calcium transients after electrical stimulation in rat fast and slowtwitch skeletal muscle fibres. *J. Physiol.* 501:573–88.

Carroll, S.L., Nicotera, P., and Pette, D. 1999. Calcium transients in single fibers of low-frequency stimulated fast-twitch muscle of rat. *Am. J. Physiol.* 277:C1122–9.

Carroll, T.J., Abernethy, P.J., Logan, P.A., Barber, M., and McEniery, M.T. 1998. Resistance training frequency: Strength and myosin heavy chain responses to two and three bouts per week. *European Journal of Applied Physiology and Occupational Physiology* 78:270–5.

Carroll, T.J., Herbert, R.D., Munn, J., Lee, M., and Gandevia, S.C. 2006. Contralateral effects of unilateral strength training: Evidence and possible mechanisms. *J. Appl. Physiol.* 101(5): 1514–22.

Carroll, T.J., Riek, S., and Carson, R.G. 2002. The sites of neural adaptation induced by resistance training in humans. *Journal of Physiology* 544:641–52.

Carson, J., and Wei, L. 2000. Integrin signaling's potential for mediating gene expression in hypertrophying skeletal muscle. *J. Appl. Physiol.* 88:337–43.

Castro, M., Apple, D., Staron, R.S., Campos, G.E.R., and Dudley, G.A. 1999. Influence of complete spinal cord injury on skeletal muscle within 6 mo of injury. *J. Appl. Physiol.* 86:350–8.

Chan, K.M., Andres, L.P., Polykovskaya, Y., and Brown, W.F. 1998. Dissociation of the electrical and contractile properties in single human motor units during fatigue. *Muscle and Nerve* 21:1786–9.

Chau, C., Barbeau, H., and Rossignol, S. 1998. Early locomotor training with clonidine in spinal cats. *J. Neurophysiol.* 79:392–409.

Chen, Y.W., Nader, G.A., Baar, K.R., Fedele, M.J., Hoffman, E.P., and Esser, K.A. 2002. Response of rat muscle to acute resistance exercise defined by transcriptional and translational profiling. *Journal of Physiology* 545:27–41.

Chesley, A., MacDougall, J.D., Tarnopolsky, M.A., Atkinson, S.A., and Smith, K. 1992. Changes in human muscle protein synthesis after resistance exercise. *J. Appl. Physiol.* 73:1383–8.

Chi, M.M.Y., Choksi, R., Nemeth, P.M., Krasnov, I., IlyinaKakueva, E., Manchester, J.K., and Lowry, O.H. 1992. Effects of microgravity and tail suspension on enzymes of individual soleus and tibialis anterior fibers. *J. Appl. Physiol.* Suppl. no. 73:66S–73S.

Chi, M.M.Y., Hintz, C.S., Coyle, E.F., Martin III, W.H., Ivy, J.L., Nemeth, P.M., Holloszy, J.O., and Lowry, O.H. 1983. Effects of detraining on enzymes of energy metabolism in individual human muscle fibers. *Am. J. Physiol. Cell Physiol.* 244(13): C276–87.

Chi, M.M.Y., Hintz, C.S., McKee, D., Felder, S., Grant, N., Kaiser, K.K., and Lowry, O.H. 1987. Effect of Duchenne muscular dystrophy on enzymes of energy metabolism in individual muscle fibers. *Metabolism* 36:761–7.

Chilibeck, P.D., Syrotiuk, D.G., and Bell, G.J. 1999. The effect of strength training on estimates of mitochondrial density and distribution throughout muscle fibers. *Eur. J. Appl. Physiol.* 80:604–9.

Chin, E., Olson, E., Richardson, J., Yang, A., Humphries, C., Shelton, J., Wu, H., Zhu, W., Bassel-Duby, R., and Williams, R. 1998. A calcineurin-dependent transcriptional pathway controls skeletal muscle fiber type. *Genes Development* 12:2499–509.

Christakos, C.N., and Windhorst, U. 1986. Spindle gain increase during muscle unit fatigue. *Brain Res.* 365:388–92.

Christova, P., and Kossev, A. 1998. Motor unit activity during longlasting intermittent muscle contractions in humans. *Eur. J. Appl. Physiol. Occup. Physiol.* 77:379–87.

Clamann, H.P., and Robinson, A.J. 1985. A comparison of electromyographic and mechanical fatigue properties in motor units of the cat hindlimb. *Brain Res.* 327:203–19.

Clark, B.D., Dacko, S.M., and Cope, T.C. 1993. Cutaneous stimulation fails to alter motor unit recruitment in the decerebrate cat. *J. Neurophysiol.* 70:1433–9.

Clark, E., and Brugge, J. 1995. Integrins and signal transduction pathways: The road taken. *Science* 268:233–9.

Clark, K., McElhinny, A., Beckerle, M., and Gregorio, C. 2002. Striated muscle cytoarchitecture: An intricate web of form and function. *Ann. Rev. Cell Devel. Biol.* 18:637–706.

Clarkson, P.M., and Rawson, E. 1999. Nutritional supplements to increase muscle mass. *Critical Reviews in Food Science and Nutrition* 39:317–28.

Clarkson, P.M., and Sayers, S.P. 1999. Etiology of exerciseinduced muscle damage. *Can. J. Appl. Physiol.* 24:234–48.

Clifford, P.S. 2007. Skeletal muscle vasodilatation at the onset of exercise. *J. Physiol. (Lond.)* 583:825–33.

Clifford, P.S., and Hellsten, Y. 2004. Vasodilatory mechanisms in contracting skeletal muscle. *J. Appl. Physiol.* 97:393–403.

Clifford PS, Kluess HA, Hamann JJ, Buckwalter JB, and Jasperse JL. 2006. Mechanical compression elicits vasodilatation in rat skeletal muscle feed arteries. *J Physiol.* 572: 561-567.

Coffey, V.G., Zhong, Z., Shield, A., Canny, B.J., Chibalin, A.V., Zierath, J.R., and Hawley, J.A. 2006. Early signaling responses to divergent exercise stimuli in skeletal muscle from welltrained humans. *FASEB J.* 20:190–2.

Colliander, E.B., and Tesch, P.A. 1990. Effects of eccentric and concentric muscle actions in resistance training. *Acta Physiol. Scand.* 140:31–9.

Conjard, A., Peuker, H., and Pette, D. 1998. Energy state and myosin heavy chain isoforms in single fibres of normal and transforming rabbit muscles. *Pflugers Arch.* 436:962–9.

Cooke, R. 2007. Modulation of the actomyosin interaction during fatigue of skeletal muscle. *Muscle and Nerve* 36(6): 756–77.

Cope, T.C., Bodine, S.C., Fournier, M., and Edgerton, V.R. 1986. Soleus motor units in chronic spinal transected cats: Physiological and morphological alterations. *J. Neurophysiol.* 55:1202–20.

Cope, T.C., Sokoloff, A.J., Dacko, S.M., Huot, R., and Feingold, E. 1997. Stability of motorunit force thresholds in the decerebrate cat. *J. Neurophysiol.* 78:3077–82.

Cope, T.C., Webb, C.B., Yee, A.K., and Botterman, B.R. 1991. Nonuniform fatigue characteristics of slowtwitch motor units activated at a fixed percentage of their maximum tetanic tension. *J. Neurophysiol.* 66:1483–92.

Cormery, B., Pons, F., Marini, J.F., and Gardiner, P.F. 1999. Myosin heavy chains in fibers of TTX-paralyzed rat soleus and medial gastrocnemius muscles. *J. Appl. Physiol.* 88:66–76.

Coté, M.-P., and Gossard, J.-P. 2003. Spinal cats on the treadmill: Changes in load pathways. *J. Neurosci.* 23:2789–96.

Cotman, C.W., Berchtold, N.C., and Christie, L.A. 2007. Exercise builds brain health: Key roles of growth factor cascades and inflammation. *Trends in Neurosciences* 30(9): 464–72.

Coyle, E.F., Coggan, A.R., Hopper, M.K., and Walters, T.J. 1988. Determinants of endurance in welltrained cyclists. *J. Appl. Physiol.* 64:2622–30.

Coyle, E.F., Costill, D.L., and Lesmes, G.R. 1979. Leg extension power and muscle fiber composition. *Med. Sci. Sports Exerc.* 11:12–5.

Coyle, E.F., Sidossis, L.S., Horowitz, J.F., and Beltz, J.D. 1992. Cycling efficiency is related to the percentage of type I muscle fibers. *Med. Sci. Sports Exerc.* 24:782–8.

Crameri, R.M., Aagaard, P., QvortrUp, K., Langberg, H., Olesen, J., and Kjaer, M. 2007. Myofibre damage in human skeletal muscle: Effects of electrical stimulation versus voluntary contraction. *J. Physiol. (Lond.)* 583(1): 365–80.

Crameri, R.M., Langberg, H., Magnusson, P., Jensen, C.H., Schroder, H.D., Olesen, J.L., Suetta, C., Teisner, B., and Kjaer, M. 2004. Changes in satellite cells in human skeletal muscle after a single bout of high intensity exercise. *Journal of Physiology* 558:333–40.

Czeh, G., Gallego, R., Kudo, N., and Kuno, M. 1978. Evidence for the maintenance of motoneurone properties by muscle activity. *J. Physiol. (Lond.)* 281:239–52.

Damiani, E., and Margreth, A. 1994. Characterization study of the ryanodine receptor and of calsequestrin isoforms of mammalian skeletal muscles in relation to fibre types. *J. Muscle Res. Cell Motil.* 15:86–101.

D'Aunno, D.S., Robinson, R.R., Smith, G.S., Thomason, D.B., and Booth, F.W. 1992. Intermittent acceleration as a countermeasure to soleus muscle atrophy. *J. Appl. Physiol.* 72:428–33.

Davies, C.T.M., Dooley, P., McDonagh, M.J.N., and White, M.J. 1985. Adaptation of mechanical properties of muscle to high force training in man. *J. Physiol.* 365:277–84.

Davies, C.T.M., Rutherford, L.C., and Thomas, D.O. 1987. Electrically evoked contractions of the triceps surae during and following 21 days of voluntary leg immobilization. *Eur. J. Appl. Physiol.* 56:306–12.

Davis, C.J.F., and Montgomery, A. 1977. The effect of prolonged inactivity upon the contraction characteristics of fast and slow mammalian twitch muscle. *J. Physiol. (Lond.)* 270:581–94.

Davis, H.L. 1985. Myotrophic effects on denervation atrophy of hindlimb muscles of mice with systemic administration of nerve extract. *Brain Res.* 343:176–9.

Davis, H.L., Bressler, B.H., and Jasch, L.G. 1988. Myotrophic effects on denervated fasttwitch muscles of mice: Correlation of physiologic, biochemical, and morphologic findings. *Exp. Neurol.* 99:474–89.

Davis, H.L., and Kiernan, J.A. 1981. Effect of nerve extract on atrophy of denervated or immobilized muscles. *Exp. Neurol.* 72:582–91.

Dawes, N.J., Cox, V.M., Park, K.S., Nga, H., and Goldspink, D.F. 1996. The induction of *cfos* and *cjun* in the stretched latissimus dorsi muscle of the rabbit: Responses to duration, degree and reapplication of the stretch stimulus. *Exp. Physiol.* 81:329–39.

DelBalso, C., and Cafarelli, E. 2007. Adaptations in the activation of human skeletal muscle induced by short-term isometric resistance training. *J. Appl. Physiol.* 103(1): 402–11.

Deldicque, L., Atherton, P., Patel, R., Theisen, D., Nielens, H., Rennie, M., and Francaux, M. 2008. Effects of resistance exercise with and without creatine supplementation on gene expression and cell signalling in human skeletal muscle, *J. Appl. Physiol.* 104: 371-378.

De Leon, R.D., Hodgson, J.A., Roy, R.R., and Edgerton, V.R. 1998. Locomotor capacity attributable to step training versus spontaneous recovery after spinalization in adult cats. *J. Neurophysiol.* 79:1329–40.

De Luca, C.J., and Erim, Z. 1994. Common drive of motor units in regulation of muscle force. *TINS* 17:299–305.

De Luca, C.J., Foley, P.J., and Erim, Z. 1996. Motor unit control properties in constantforce isometric contractions. *J. Neurophysiol.* 76:1503–16.

Del Valle, A., and Thomas, C.K. 2005. Firing rates of motor units during strong dynamic contractions. *Muscle and Nerve* 32:316–25.

Desaulniers, P., Lavoie, P., and Gardiner, P.F. 1998. Endurance training increases acetylcholine receptor quantity at neuromuscular junctions of adult rat skeletal muscle. *Neuroreport* 9:3549–52.

Desaulniers, P., Lavoie, P.A., and Gardiner, P.F. 2001. Habitual exercise enhances neuromuscular transmission efficacy of rat soleus muscle in situ. *J. Appl. Physiol.* 90:1041–8.

Deschenes, M.R., Covault, J., Kraemer, W.J., and Maresh, C.M. 1994. The neuromuscular junction: Muscle fibre type differences, plasticity and adaptability to increased and decreased activity. *Sports Med.* 17:358–72.

Deschenes, M.R., Maresh, C.M., Crivello, J.F., Armstrong, L.E., Kraemer, W.J., and Covault, J. 1993. The effects of exercise training of different intensities on neuromuscular junction morphology. *J. Neurocytol.* 22:603–15.

Desmedt, J.E., and Godaux, E. 1977. Ballistic contractions in man: Characteristic recruitment pattern of single motor units of the tibialis anterior muscle. *J. Physiol.* 264:673–93.

Desmedt, J.E., and Godaux, E. 1978. Ballistic contractions in fast or slow human muscles: Discharge patterns of single motor units. *J. Physiol.* 285:185–96.

Desypris, G., and Parry, D.J. 1990. Relative efficacy of slow and fast motoneurons to reinnervate mouse soleus muscle. *Am. J. Physiol.* 258: C62–70.

Devasahayam, S.R., and Sandercock, T.G. 1992. Velocity of shortening of single motor units from rat soleus. *J. Neurophysiol.* 67:1133–45.

DeVol, D.L., Rotwein, P., Sadow, J.L., Novakofski, J., and Bechtel, P.J. 1990. Activation of insulin-like growth factor gene expression during workinduced skeletal muscle growth. *Am. J. Physiol. Endocrinol. Metab.* 259: E89–95.

Diffee, G.M., Caiozzo, V.J., Herrick, R.E., and Baldwin, K.M. 1991. Contractile and biochemical properties of rat soleus and plantaris after hindlimb suspension. *Am. J. Physiol. Cell Physiol.* 260:C528–34.

Diffee, G.M., Caiozzo, V.J., McCue, S.A., Herrick, R.E., and Baldwin, K.M. 1993. Activityinduced regulation of myosin isoform distribution: Comparison of two contractile activity programs. *J. Appl. Physiol.* 74:2509–16.

Dinenno, F.A., and Joyner, M.J. 2003. Blunted sympathetic vasoconstriction in contracting skeletal muscle of healthy humans: Is nitric oxide obligatory? *J. Physiol. (Lond.)* 553:281–92.

Dix, D.J., and Eisenberg, B.R. 1990. Myosin mRNA accumulation and myofibrillogenesis at the myotendinous junction of stretched muscle fibers. *J. Cell Biol.* 111:1885–94.

Djupsjöbacka, M., Johansson, H., and Bergenheim, M. 1994. Influences on the gammamusclespindle system from muscle afferents stimulated by increased intramuscular concentrations of arachadonic acid. *Brain Res.* 663:293–302.

Djupsjöbacka, M., Johansson, H., Bergenheim, M., and Sjölander, P. 1995. Influences on the gammamusclespindle system from contralateral muscle afferents stimulated by KCl and lactic acid. *Neurosci. Res.* 21:301–9.

Djupsjöbacka, M., Johansson, H., Bergenheim, M., and Wenngren, B. 1995. Influences on the gammamuscle spindle system from muscle afferents stimulated by increased intramuscular concentrations of bradykinin and 5HT. *Neurosci. Res.* 22:325–33.

Dolmage, T., and Cafarelli, E. 1991. Rate of fatigue during repeated submaximal contractions of human quadriceps muscle. *Can. J. Physiol. Pharmacol.* 69:1410–5.

Dolmetsch, R.E., Lewis, R., Goodnow, C., and Healy, J. 1997. Differential activation of transcription factors induced by Ca^{2+} response amplitude and duration. *Nature* 386:855–8.

Donselaar, Y., Eerbeek, D., Kernell, D., and Verhey, B.A. 1987. Fibre sizes and histochemical staining characteristics in normal and chronically stimulated fast muscle of cat. *J. Physiol.* 382:237–54.

Donselaar, Y., Kernell, D., and Eerbeek, O. 1986. Soma size and oxidative enzyme activity in normal and chronically stimulated motoneurones of the cat's spinal cord. *Brain Res.* 385:22–9.

Dorlöchter, M., Irintchev, A., Brinkers, M., and Wernig, A. 1991. Effects of enhanced activity on synaptic transmission in mouse extensor digitorum longus muscle. *J. Physiol. (Lond.)* 436:283–92.

Dowling, J.J., Konert, E., Ljucovic, P., and Andrews, D.M. 1994. Are humans able to voluntarily elicit maximum muscle force? *Neurosci. Lett.* 179:25–8.

Duchateau, J. 1995. Bed rest induces neural and contractile adaptations in triceps surae. *Med. Sci. Sports Exerc.* 27:1581–9.

Duchateau, J., and Hainaut, K. 1984. Isometric or dynamic training: Differential effects on mechanical properties of a human muscle. *J. Appl. Physiol.* 56:296–301.

Duchateau, J., and Hainaut, K. 1987. Electrical and mechanical changes in immobilized human muscle. *J. Appl. Physiol.* 62(6): 2168–73.

Duchateau, J., and Hainaut, K. 1990. Effects of immobilization on contractile properties, recruitment and firing rates of human motor units. *J. Physiol. (Lond.)* 422:55–65.

Duchateau, J., and Hainaut, K. 1993. Behaviour of short and long latency reflexes in fatigued human muscles. *J. Physiol. (Lond.)* 471:787–99.

Duclay, J., and Martin, A. 2005. Evoked H-reflex and V-wave responses during maximal isometric, concentric, and eccentric muscle contraction. *Journal of Neurophysiology* 94:3555–62.

Dudley, G.A., Duvoisin, M.R., Adams, G.R., Meyer, R.A., Belew, A.H., and Buchanan, P. 1992. Adaptation to unilateral lower limb suspension in humans. *Aviat. Space Environ. Med.* 63:678–83.

Dudley, G.A., Tesch, P., Miller, B., and Buchanan, P. 1991. Importance of eccentric actions in performance adaptations to resistance training. *Aviat. Space Environ. Med.* June: 543–50.

Duncan, N.D., Williams, D.A., and Lynch, G.S. 1998. Adaptations in rat skeletal muscle following longterm resistance exercise training. *Eur. J. Appl. Physiol. Occup. Physiol.* 77:372–8.

Dunn, S.E., Burns, J.L., and Michel, R.N. 1999. Calcineurin is required for skeletal muscle hypertrophy. *J. Biol. Chem.* 274:21908–12.

Dunn, S.E., and Michel, R.N. 1997. Coordinated expression of myosin heavy chain isoforms and metabolic enzymes within overloaded rat muscle fibers. *Am. J. Physiol. Cell Physiol.* 273: C371–83.

DupontVersteegden, E.E., Houlé, J.D., Gurley, C.M., and Peterson, C.A. 1998. Early changes in muscle fiber size and gene expression in response to spinal cord transection and exercise. *Am. J. Physiol. Cell Physiol.* 275: C1124–33.

Dux, L. 1993. Muscle relaxation and sarcoplasmic reticulum function in different muscle types. *Rev. Physiol. Biochem. Pharmacol.* 122:69–147.

Edes, I., Dosa, E., Sohar, I., and Guba, F. 1982. Effect of plaster cast immobilization on the turnover rates of soluble proteins and lactate dehydrogenase isoenzymes of rabbit m. soleus. *Acta Biochim. Biophys. Acad. Sci. Hung.* 17:211–6.

Edes, I., Sohar, I., Mazarean, H., Takacs, O., and Guba, F. 1980. Changes in the aerobic and anaerobic metabolism of skeletal muscle subjected to plaster cast immobilization. *Acta Biochim. Biophys. Acad. Sci. Hung.* 15:305–11.

Edgerton, V.R., Barnard, R.J., Peter, J.B., Maier, A., and Simpson, D.R. 1975. Properties of immobilized hindlimb muscles of the *Galago senegalensis*. *Exp. Neurol.* 46:115–31.

Edgerton, V.R., Courtine, G., Gerasimenko, Y.P., Lavrov, I., Ichiyama, R.M., Fong, A.J., Cai, L.L., Otoshi, C.K., Tillakaratne, N.J.K., Burdick, J.W., and Roy, R.R. 2008. Training locomotor networks. *Brain Research Reviews* 57(1): 241–54.

Edgerton, V., and Roy, R. 2009. Robotic training and spinal cord plasticity, *Brain Research Bulletin* 78 ; 4-12.

Edgerton, V.R., Zhou, M.Y., Ohira, Y., Klitgaard, H., Jiang, B., Bell, G., Harris, B., Saltin, B., Gollnick, P.D., Roy, R.R., Day, M.K., and Greenisen, M. 1995. Human fiber size and enzymatic properties after 5 and 11 days of spaceflight. *J. Appl. Physiol.* 78:1733–9.

Edstrom, J.E. 1957. Effects of increased motor activity on the dimensions and the staining properties of the neuron soma. *J. Comp. Neurol.* 107:295–304.

Eken, T., and Kiehn, O. 1989. Bistable firing properties of soleus motor units in unrestrained rats. *Acta Physiol. Scand.* 136:383–94.

Elliot, T.A., Cree, M.G., Sanford, A.P., Wolfe, R.R., and Tipton, K.D. 2006. Milk ingestion stimulates net muscle protein synthesis following resistance exercise. *Med. Sci. Sports Exerc.* 38:667–74.

Ennion, S., Sant'ana Pereira, J., Sargeant, A.J., Young, A., and Goldspink, G. 1995. Characterization of human skeletal muscle fibres according to the myosin heavy chains they express. *J. Muscle Res. Cell Motil.* 16:35–43.

Esmarck, B., Andersen, J.L., Olsen, S., Richter, E.A., Mizuno, M., and Kjaer, M. 2001. Timing of postexercise protein intake is important for muscle hypertrophy with resistance training in elderly humans. *J. Physiol. (Lond.)* 535:301–11.

Essen, B., Jansson, E., Henriksson, J., Taylor, A.W., and Saltin, B. 1975. Metabolic characteristics of fiber types in human skeletal muscle. *Acta Physiol. Scand.* 95:153–65.

Evertsen, F., Medbo, J., Jebens, E., and Gjovaag, T. 1999. Effect of training on the activity of five muscle enzymes studies on elite cross-country skiers. *Acta Physiol. Scand.* 167:247–57.

Ewing, J.L., Wolfe, D.R., Rogers, M.A., Amundson, M.L., and Stull, G.A. 1990. Effects of velocity of isokinetic training on strength, power, and quadriceps muscle fibre characteristics. *Eur. J. Appl. Physiol.* 61:159–62.

Fahim, M.A. 1989. Rapid neuromuscular remodeling following limb immobilization. *Anat. Rec.* 224:102–9.

Fahim, M.A., and Robbins, N. 1986. Remodeling of the neuromuscular junction after subtotal disuse. *Brain Res.* 383:353–6.

Fallentin, N., Jorgensen, K., and Simonsen, E. 1993. Motor unit recruitment during prolonged isometric contractions. *Eur. J. Appl. Physiol.* 67:335–41.

Fang, Y., Siemionow, V., Sahgal, V., Xiong, F.Q., and Yue, G.H. 2004. Distinct brain activation patterns for human maximal voluntary eccentric and concentric muscle actions. *Brain Research* 1023:200–12.

Farrell, P., Fedele, M., Vary, T., Kimball, S., Lang, C., and Jefferson, L. 1999. Regulation of protein synthesis after acute resistance in diabetic rats. *Am. J. Physiol.* 276: E721–7.

Farthing, J.P., Chilibeck, P.D., and Binsted, G. 2005. Cross-education of arm muscular strength is unidirectional in right-handed individuals. *Med. Sci. Sports Exerc.* 37:1594–1600.

Fell, R.D., Steffen, J.M., and Musacchia, X.J. 1985. Effect of hypokinesiahypodynamia on rat muscle oxidative capacity and glucose uptake. *Am. J. Physiol. Regul. Integr. Comp. Physiol.* 249(18): R308–12.

Fernandez, H.L., and Donoso, J.A. 1988. Exercise selectively increases G_4 AChE activity in fasttwitch muscle. *J. Appl. Physiol.* 65:2245–52.

Fernandez, H.L., and HodgesSavola, C.A. 1996. Physiological regulation of G_4 AChE in fasttwitch muscle: Effects of exercise and CGRP. *J. Appl. Physiol.* 80:357–62.

Ferrando, A.A., Lane, H.W., Stuart, C.A., Davis-Street, J., and Wolfe, R.R. 1996. Prolonged bed rest decreases skeletal muscle and whole body protein synthesis. *Am. J. Physiol. Endocrinol. Metab.* 270: E627–33.

Fimland, M., Helgerud, J., Gruber, M., Leivseth, G., and Hoff, J. 2010. Enhanced neural drive after maximal strength training in multiple sclerosis patients, *Eur. J. Appl. Physiol.* 110 (2); 435-443.

Fitts, R.H. 1994. Cellular mechanisms of muscle fatigue. *Physiol. Rev.* 74:49–94.

Fitts, R.H. 2008. The cross-bridge cycle and skeletal muscle fatigue. *J. Appl. Physiol.* 104(2): 551–8.

Fitts, R.H., Bodine, S.C., Romatowski, J.G., and Widrick, J.J. 1998. Velocity, force, power, and Ca^{2+} sensitivity of fast and slow monkey skeletal muscle fibers. *J. Appl. Physiol.* 84:1776–87.

Fitts, R.H., Brimmer, C., HeywoodCooksey, A., and Timmerman, R. 1989. Single muscle fiber enzyme shifts with hindlimb suspension and immobilization. *Am. J. Physiol. Cell Physiol.* 25: C1082–91.

Fitts, R.H., Costill, D.L., and Gardetto, P.R. 1989. Effect of swim exercise training on human muscle fiber function. *J. Appl. Physiol.* 66:465–75.

Fluck, M., and Hoppeler, H. 2003. Molecular basis of skeletal muscle plasticity—from gene to form and function. *Reviews of Physiology, Biochemistry and Pharmacology* 146:159–216.

Foehring, R.C., and Munson, J.B. 1990. Motoneuron and muscleunit properties after longterm direct innervation of soleus muscle by medial gastrocnemius nerve in cat. *J. Neurophysiol.* 64:847–61.

Foehring, R.C., Sypert, G., and Munson, J.B. 1986. Properties of selfreinnervated motor units of medial gastrocnemius of cat: I. Longterm reinnervation. *J. Neurophysiol.* 55:931–46.

Foure, A., Nordez, A., and Cornu, C. 2010. Plyometric training effects on Achilles tendon stiffness and dissipation properties. *J. Appl. Physiol.*109: 849-854

Foley, J., Jayaraman, R., Prior, B., Pivarnik, J., and Meter, R. 1999. MR measurements of muscle damage and adaptation after eccentric exercise. *J. Appl. Physiol.* 87:2311–8.

Foster, C., Costill, D.L., Daniels, J.T., and Fink, W.J. 1978. Skeletal muscle enzyme activity, fiber composition and $\dot{V}O_2$max in relation to distance running performance. *Eur. J. Appl. Physiol.* 39:73–80.

Fournier, M., Roy, R., Perham, H., Simard, C., and Edgerton, V. 1983. Is limb immobilization a model of muscle disuse? *Exp. Neurol.* 80:147–56.

Freyssenet, D. 2007. Energy sensing and regulation of gene expression in skeletal muscle. *J. Appl. Physiol.* 102(2): 529–40.

Freyssenet, D., Connor, M., Takahashi, M., and Hood, D. 1999. Cytochrome c transcriptional activation and mRNA stability during contractile activity in skeletal muscle. *Am. J. Physiol.* 277: E26–32.

Freyssenet, D., Di Carlo, M., and Hood, D. 1999. Calcium-dependent regulation of cytochrome c gene expression in skeletal muscle cells: Identification of a protein kinase C–dependent pathway. *J. Biol. Chem.* 274:9305–11.

Fridén, J. 1984. Changes in human skeletal muscle induced by longterm eccentric exercise. *Cell Tissue Res.* 236:365–72.

Fridén, J., and Lieber, R.L. 1992. Structural and mechanical basis of exerciseinduced muscle injury. *Med. Sci. Sports Exerc.* 24:521–30.

Fridén, J., and Lieber, R.L. 1996. Ultrastructural evidence for loss of calcium homeostasis in exercised skeletal muscle. *Acta Physiol. Scand.* 158:381–2.

Fridén, J., and Lieber, R.L. 1998. Segmental muscle fiber lesions after repetitive eccentric contractions. *Cell Tissue Res.* 293:165–71.

Fridén, J., Seger, J., and Ekblom, B. 1998. Sublethal muscle fibre injuries after hightension anaerobic exercise. *Eur. J. Appl. Physiol.* 57:360–8.

Froese, E.A., and Houston, M.E. 1987. Performance during the Wingate anaerobic test and muscle morphology in males and females. *Int. J. Sports Med.* 8:35–9.

Fuglevand, A.J., Zackowski, K.M., Huey, K.A., and Enoka, R.M. 1993. Impairment of neuromuscular propagation during human fatiguing contractions at submaximal forces. *J. Physiol. (Lond.)* 460:549–72.

Funakoshi, H., Belluardo, N., Arenas, E., Yamamoto, Y., Casabona, A., Persson, H., and Ibáñez, C.F. 1995. Musclederived neurotrophin4 as an activitydependent trophic signal for adult motor neurons. *Science* 268:1495–9.

Gallego, R., Kuno, M., Nunez, R., and Snider, W.D. 1979a. Dependence of motoneurone properties on the length of immobilized muscle. *J. Physiol. (Lond.)* 291:179–89.

Gallego, R., Kuno, M., Nunez, R., and Snider, W.D. 1979b. Disuse enhances synaptic efficacy in spinal motoneurones. *J. Physiol.* 291:191–205.

Galler, S., Hilber, K., Gohlsch, B., and Pette, D. 1997. Two functionally distinct myosin heavy chain isoforms in slow skeletal muscle fibres. *FEBS Lett.* 410:150–2.

Galler, S., Schmitt, T.L., Hilber, K., and Pette, D. 1997. Stretch activation and isoforms of myosin heavy chain and troponinT of rat skeletal muscle fibres. *J. Muscle Res. Cell Motil.* 18:555–61.

Galler, S., Schmitt, T.L., and Pette, D. 1994. Stretch activation, unloaded shortening velocity, and myosin heavy chain isoforms of rat skeletal muscle fibres. *J. Physiol. (Lond.)* 478:513–21.

Gamrin, L., Berg, H.E., Essén, P., Tesch, P.A., Hultman, E., Garlick, P.J., McNurlan, M.A., and Wernerman, J. 1998. The effect of unloading on protein synthesis in human skeletal muscle. *Acta Physiol. Scand.* 163:369–77.

Gandevia, S.C. 1998. Neural control in human muscle fatigue: Changes in muscle afferents, motoneurones and motocortical drive. *Acta Physiol. Scand.* 162:275–83.

Gandevia, S.C., Allen, G.M., Butler, J.E., and Taylor, J.L. 1996. Supraspinal factors in human muscle fatigue: Evidence for suboptimal output from the motor cortex. *J. Physiol.* 490:529–36.

Gandevia, S.C., Petersen, N., Butler, J.E., and Taylor, J.L. 1999. Impaired response of human motoneurones to corticospinal stimulation after voluntary exercise. *J. Physiol. (Lond.)* 521:749–59.

Gandevia, S.C., and Rothwell, J.C. 1987. Knowledge of motor commands and the recruitment of human motoneurons. *Brain* 110:1117–30.

Gardetto, P.R., Schluter, J.M., and Fitts, R.H. 1989. Contractile function of single muscle fibers after hindlimb suspension. *J. Appl. Physiol.* 66:2739–49.

Gardiner, P.F. 1993. Physiological properties of motoneurons innervating different muscle unit types in rat gastrocnemius. *J. Neurophysiol.* 69:1160–70.

Gardiner, P., Dai, Y., and Heckman, C.J. 2006. Effects of exercise training on alpha-motoneurones. *J. Appl. Physiol.* 101:1228–36.

Gardiner, P.F., Favron, M., and Corriveau, P. 1992. Histochemical and contractile responses of rat medial gastrocnemius to 2 weeks of complete disuse. *Can. J. Physiol. Pharmacol.* 70:1075–81.

Gardiner, P.F., and Lapointe, M. 1982. Daily in vivo neuromuscular stimulation effects on immobilized rat hindlimb muscles. *J. Appl. Physiol.* 53:960–6.

Gardiner, P.F., Michel, R., and Iadeluca, G. 1984. Previous exercise training influences functional sprouting of rat hindlimb motoneurons in response to partial denervation. *Neurosci. Lett.* 45:123–7.

Gardiner, P.F., and Olha, A.E. 1987. Contractile and electromyographic characteristics of rat plantaris motor unit types during fatigue *in situ*. *J. Physiol.* 385:13–34.

Gardiner, P.F., and Seburn, K.L. 1997. The effects of tetrodotoxininduced muscle paralysis on the physiological properties of muscle units and their innervating motoneurons in rat. *J. Physiol. (Lond.)* 499:207–16.

Garland, S.J., Enoka, R., Serrano, L., and Robinson, G. 1994. Behavior of motor units in human biceps brachii during a submaximal fatiguing contraction. *J. Appl. Physiol.* 76:2411–9.

Garland, S.J., Garner, S.H., and McComas, A.J. 1988. Reduced voluntary electromyographic activity after fatiguing stimulation of human muscle. *J. Physiol.* 401:547–56.

Garland, S.J., Griffin, L., and Ivanova, T. 1997. Motor unit discharge rate is not associated with muscle relaxation time in sustained submaximal contractions in humans. *Neurosci. Lett.* 239:25–8.

Garland, S.J., and McComas, A.J. 1990. Reflex inhibition of human soleus muscle during fatigue. *J. Physiol. (Lond.)* 429:17–27.

Garnett, R.A.F., O'Donovan, M.J., Stephens, J.A., and Taylor, A. 1979. Motor unit organization of human medial gastrocnemius. *J. Physiol.* 287:33–43.

Gerchman, L.B., Edgerton, V.R., and Carrow, R.E. 1975. Effects of physical training on the histochemistry and morphology of ventral motor neurons. *Exp. Neurol.* 49:790–801.

Gerrits, H., De Haan, A., Hopman, M., Van Der Woude, L., Jones, D., and Sargeant, A. 1999. Contractile properties of the quadriceps muscle in individuals with spinal cord injury. *Muscle and Nerve* 22:1249–56.

Gertler, R.A., and Robbins, N. 1978. Differences in neuromuscular transmission in red and white muscles. *Brain Res.* 142:160–4.

Gharakhanlou, R., Chadan, S., and Gardiner, P.F. 1999. Increased activity in the form of endurance training increases calcitonin generelated peptide content in lumbar motoneuron cell bodies and in sciatic nerve in the rat. *Neuroscience* 89:1229–39.

Gibala, M.J., MacDougall, J.D., Tarnopolsky, M.A., Stauber, W.T., and Elorriaga, A. 1995. Changes in human skeletal muscle ultrastructure and force production after acute resistance exercise. *J. Appl. Physiol.* 78:702–8.

Gibson, J., Halliday, D., Morrison, W., Stoward, P.J., Hornsby, G.A., Watt, P., Murdoch, G., and Rennie, M. 1987. Decrease in human quadriceps muscle protein turnover consequent upon leg immobilization. *Clin. Sci.* 72:503–9.

Giddings, C.J., and Gonyea, W.J. 1992. Morphological observations supporting muscle fiber hyperplasia following weightlifting exercise in cats. *Anat. Rec.* 233:178–95.

Giddings, C.J., Neaves, W.B., and Gonyea, W.J. 1985. Muscle fiber necrosis and regeneration induced by prolonged weightlifting exercise in the cat. *Anat. Rec.* 211:133–41.

Giniatullin, R.A., Bal'tser, S.K., Nikol'skii, E.E., and Magazanik, L.G. 1986. Postsynaptic potentiation and desensitization at the frog neuromuscular junction produced by repeated stimulation of the motor nerve. *Neirofiziologiya* 18:645–54.

Gisiger, V., Bélisle, M., and Gardiner, P.F. 1994. Acetylcholinesterase adaptation to voluntary wheel running is proportional to the volume of activity in fast, but not slow, rat hindlimb muscles. *Eur. J. Neurosci.* 6:673–80.

Gisiger, V., Sherker, S., and Gardiner, P.F. 1991. Swimming training increases the G_4 acetylcholinesterase content of both fast ankle extensors and flexors. *FEBS Lett.* 278:271–3.

Gisiger, V., and Stephens, H.R. 1988. Localization of the pool of G_4 acetylcholinesterase characterizing fast muscles and its alteration in murine muscular dystrophy. *J. Neurosci. Res.* 19:62–78.

Glass, D.J. 2003. Signalling pathways that mediate skeletal muscle hypertrophy and atrophy. *Nature Cell Biology* 5:87–90.

Goldspink, D.F. 1977a. The influence of activity on muscle size and protein turnover. *J. Physiol.* 264:283–96.

Goldspink, D.F. 1977b. The influence of immobilization and stretch on protein turnover of rat skeletal muscle. *J. Physiol.* 264:267–82.

Goldspink, D.F., Cox, V.M., Smith, S.K., Eaves, L.A., Osbaldeston, N.J., Lee, D.M., and Mantle, D. 1995. Muscle growth in response to mechanical stimuli. *Am. J. Physiol. Endocrinol. Metab.* 268: E288–97.

Goldspink, D.F., Garlick, P.J., and McNurlan, M.A. 1983. Protein turnover measured *in vivo* and *in vitro* in muscles undergoing compensatory growth and subsequent denervation atrophy. *Biochem. J.* 210:89–98.

Goldspink, G. 2005. Mechanical signals, IGF-1 gene splicing, and muscle adaptation. *Physiology* 20:232–8.

Gómez-Pinilla, F., Ying, Z., Opazo, P., Roy, R.R., and Edgerton, V.R. 2001. Differential regulation by exercise of BDNF and NT-3 in rat spinal cord and skeletal muscle. *Eur. J. Neurosci.* 13:1078–84.

Gómez-Pinilla, F., Ying, Z., Roy, R.R., Hodgson, J., and Edgerton, V.R. 2004. Afferent input modulates neurotrophins and synaptic plasticity in the spinal cord. *Journal of Neurophysiology* 92:3423–32.

Gomez-Pinilla, F., Ying, Z., Roy, R.R., Molteni, R., and Edgerton, V.R. 2002. Voluntary exercise induces a BDNF-mediated mechanism that promotes neuroplasticity. *Journal of Neurophysiology* 88:2187–95.

Gonyea, W.J., Sale, D.G., Gonyea, F.B., and Mikesky, A. 1986. Exercise induced increases in muscle fiber number. *Eur. J. Appl. Physiol.* 55:137–41.

Gonzalez-Alonso, J., Mortensen, S.P., Jeppesen, T.D., Ali, L., Barker, H., Damsgaard, R., Secher, N., Dawson, E., and Dufour, S.P. 2008. Haemodynamic responses to exercise, ATP infusion and thigh compression in humans: Insight into the role of muscle mechanisms on cardiovascular function. *J. Physiol. (Lond.)* 586:2405–17.

Gonzalez-Alonso, J., Richardson, R.S., and Saltin, B. 2001. Exercising skeletal muscle blood flow in humans responds to reduction in arterial oxyhaemoglobin, but not to altered free oxygen. *J. Physiol. (Lond.)* 530:331–41.

Goodyear, L.J., Chang, P.Y., Sherwood, D.J., Dufresne, S.D., and Moller, D.E. 1996. Effects of exercise and insulin on mitogenactivated protein kinase signaling pathways in rat skeletal muscle. *Am. J. Physiol. Endocrinol. Metab.* 34: E403–8.

Gorassini, M.A., Bennett, D.J., and Yang, J.F. 1998. Selfsustained firing of human motor units.

Neurosci. Lett. 247:13–6.

Gorassini, M., Yang, J.F., Siu, M., and Bennett, D.J. 2002. Intrinsic activation of human motoneurons: Possible contribution to motor unit excitation. *Journal of Neurophysiology* 87:1850–8.

Goslow Jr., G.E., Cameron, W.E., and Stuart, D.G. 1977. The fast twitch motor units of cat ankle flexor: 1. Tripartite classification on basis of fatigability. *Brain Res.* 134:35–46.

Graham, S.C., Roy, R.R., Navarro, C., Jiang, B., Pierotti, D., BodineFowler, S., and Edgerton, V.R. 1992. Enzyme and size profiles in chronically inactive cat soleus muscle fibers. *Muscle and Nerve* 15:27–36.

Graham, S.C., Roy, R.R., West, S.P., Thomason, D., and Baldwin, K.M. 1989. Exercise effects on the size and metabolic properties of soleus fibers in hindlimbsuspended rats. *Aviat. Space Environ. Med.* 60:226–34.

Grana, E.A., ChiouTan, F., and Jaweed, M.M. 1996. Endplate dysfunction in healthy muscle following a period of disuse. *Muscle and Nerve* 19:989–93.

Granit, R., Kernell, D., and Shortess, G.K. 1963a. The behaviour of mammalian motoneurones during longlasting orthodromic, antidromic and transmembrane stimulation. *J. Physiol.* 169:743–54.

Granit, R., Kernell, D., and Shortess, G.K. 1963b. Quantitative aspects of repetitive firing of mammalian motoneurones, caused by injected currents. *J. Physiol.* 168:911–31.

Greaser, M., Moss, R., and Reiser, P. 1988. Variations in contractile properties of rabbit single muscle fibres in relation to troponin T isoforms and myosin light chains. *J. Physiol. (Lond.)* 406:85–98.

Green, H.J., Ball-Burnett, M., Chin, E., and Pette, D. 1992. Time-dependent increases in Na^+,K^+-ATPase content of low-frequency-stimulated rabbit muscle. *FEBS Lett.* 310:129–31.

Green, H.J., Duhamel, T.A., Holloway, G.P., Moule, J.W., Ranney, D.W., Tupling, A.R., and Ouyang, J. 2008. Rapid upregulation of GLUT-4 and MCT-4 expression during 16 h of heavy intermittent cycle exercise. *Amer. J. Physiol. Regul. Integr. C* 294(2): R594–600.

Green, H.J., Düsterhöft, S., Dux, L., and Pette, D. 1992. Metabolite patterns related to exhaustion, recovery and transformation of chronically stimulated rabbit fasttwitch muscle. *Pflugers Arch.* 420:359–66.

Gregor, R.J., Edgerton, V.R., Perrine, J.J., Campion, D.S., and DeBus, C. 1979. Torque–velocity relationships and muscle fiber composition in elite female athletes. *J. Appl. Physiol. Respir. Environ. Exerc. Physiol.* 47:388–92.

Griffin, L., and Cafarelli, E. 2007. Transcranial magnetic stimulation during resistance training of the tibialis anterior muscle. *J. Electromyogr. Kinesiol.* 17:446–52.

Griffin, L., Ivanova, T., and Garland, S. 2000. Role of limb movement in the modulation of motor unit discharge rate during fatiguing contractions. *Exp. Brain Res.* 130:392–400.

Grimby, G., Broberg, C., Krotkiewska, I., and Krotkiewski, M. 1976. Muscle fiber composition in patients with traumatic cord lesion. *Scand. J. Rehab. Med.* 8:37–42.

Grimby, L., Hannerz, J., and Hedman, B. 1981. The fatigue and voluntary discharge properties of single motor units in man. *J. Physiol. (Lond.)* 316:545–54.

Grossman, E.J., Roy, R.R., Talmadge, R.J., Zhong, H., and Edgerton, V.R. 1998. Effects of inactivity on myosin heavy chain composition and size of rat soleus fibers. *Muscle and Nerve* 21:375–89.

Guertin, P.A., and Hounsgaard, J. 1999. Nonvolatile general anaesthetics reduce spinal activity by suppressing plateau potentials. *Neuroscience* 88:353–8.

Gur, H., Gransberg, L., VanDyke, D., Knutsson, E., and Larsson, L. 2003. Relationship between in vivo muscle force and different speeds of isokinetic movements and myosin isoform expression in men and women. *Eur J Appl Physiol* 88, 487–496.

Gustafsson, B., Katz, R., and Malmsten, J. 1982. Effects of chronic partial deafferentation on the electrical properties of lumbar alphamotoneurones in the cat. *Brain Res.* 246:23–33.

Gustafsson, B., and Pinter, M.J. 1984a. An investigation of threshold properties among cat spinal alphamotoneurons. *J. Physiol. (Lond.)* 357:453–83.

Gustafsson, B., and Pinter, M.J. 1984b. Relations among passive electrical properties of lumbar alphamotoneurones of the cat. *J. Physiol.* 356:401–31.

Gustafsson, B., and Pinter, M.J. 1985. Factors determining the variation of the afterhyperpolarization duration in cat lumbar alphamotoneurons. *Brain Res.* 326:392–5.

Gustafsson, T., Puntschart, A., Kaijser, L., Jansson, E., and Sundberg, C.J. 1999. Exercise-induced expression of angiogenesis-related transcription and growth factors in human skeletal muscle. *American Journal of Physiology: Heart and Circulatory Physiology* 276: H679–85.

Gydikov, A., Dimitrov, G., Kosarov, D., and Dimitrova, N. 1976. Functional differentiation of motor units in human opponens pollicis muscle. *Exp. Neurol.* 50:36–47.

Haddad, F., Herrick, R.E., Adams, G.R., and Baldwin, K.M. 1993. Myosin heavy chain expression in rodent skeletal muscle: Effects of exposure to zero gravity. *J. Appl. Physiol.* 75:2471–7.

Haddad, F., Qin, A.X., Zeng, M., McCue, S.A., and Baldwin, K.M. 1998. Interaction of hyperthyroidism and hindlimb suspension on skeletal myosin heavy chain expression. *J. Appl. Physiol.* 85:2227–36.

Hagbarth, K.E., Bongiovanni, L.G., and Nordin, M. 1995. Reduced servocontrol of fatigued human finger extensor and flexor muscles. *J. Physiol. (Lond.)* 485:865–72.

Hagbarth, K.E., Kunesch, E., Nordin, M., Schmidt, R., and Wallin, E. 1986. Gamma loop contributing to maximal voluntary contractions in man. *J. Physiol. (Lond.)* 380:575–91.

Hainaut, K., and Duchateau, J. 1989. Muscle fatigue, effects of training and disuse. *Muscle and Nerve* 12:660–9.

Häkkinen, K., Kallinen, M., Linnamo, V., Pastinen, U.M., Newton, R.U., and Kraemer, W.J. 1996. Neuromuscular adaptations during bilateral versus unilateral strength training in middleaged and elderly men and women. *Acta Physiol. Scand.* 158:77–88.

Häkkinen, K., and Komi, P.V. 1983a. Alterations of mechanical characteristics of human skeletal muscle during strength training. *Eur. J. Appl. Physiol.* 50:161–72.

Häkkinen, K., and Komi, P.V. 1983b. Electromyographic and mechanical characteristics of human skeletal muscle during fatigue under voluntary and reflex conditions. *Electroencephalogr. Clin. Neurophysiol.* 55:436–44.

Häkkinen, K., and Komi, P.V. 1986. Effects of fatigue and recovery on electromyographic and isometric force and relaxationtime characteristics of human skeletal muscle. *Eur. J. Appl. Physiol.* 55:588–96.

Häkkinen, K., Komi, P.V., and Alén, M. 1985. Effect of explosive type strength training on isometric force and relaxationtime, electromyographic and muscle fibre characteristics of leg extensor muscles. *Acta Physiol. Scand.* 125:587–600.

Haller, H., Lindschau, C., and Luft, F.C. 1994. Role of protein kinase C in intracellular signaling. *Ann. NY Acad. Sci.* 733:313–24.

Halter, J.A., Carp, J.S., and Wolpaw, J.R. 1995. Operantly conditioned motoneuron plasticity: Possible role of sodium channels. *J. Neurophysiol.* 73:867–71.

Hamada, T., Sale, D.G., MacDougall, J.D., and Tarnopolsky, M.A. 2000. Postactivation potentiation, fiber type, and twitch contraction time in human knee extensor muscles. *J. Appl. Physiol.* 88:2131–7.

Hamada, T., Sale, D.G., MacDougall, J.D., and Tarnopolsky, M.A. 2003. Interaction of fibre type, potentiation and fatigue in human knee extensor muscles. *Acta Physiol. Scand.* 178:165–73.

Hämäläinen, N., and Pette, D. 1995. Patterns of myosin isoforms in mammalian skeletal muscle fibres. *Microscopy Res. Technique* 30:381–9.

Hamann, J.J., Buckwalter, J.B., and Clifford, P.S. 2004. Vasodilatation is obligatory for contraction-induced hyperaemia in canine skeletal muscle. *J. Physiol. (Lond.)* 557:1013–20.

Hamann, J.J., Buckwalter, J.B., Clifford, P.S., and Shoemaker, J. K. 2004. Is the blood flow response to a single contraction determined by work performed? *J. Appl. Physiol.* 96:2146–52.

Hamann, J.J., Kluess, H.A., Buckwalter, J.B., and Clifford, P.S. 2005. Blood flow response to muscle contractions is more closely related to metabolic rate than contractile work. *J. Appl. Physiol.* 98:2096–100.

Hameed, M., Lange, K.H.W., Andersen, J.L., Schjerling, P., Kjaer, M., Harridge, S.D.R., and Goldspink, G. 2004. The effect of recombinant human growth hormone and resistance training on IGF-1 mRNA expression in the muscles of elderly men. *J. Physiol. (Lond.)* 555(1): 231–40.

Hamm, T.M., Nemeth, P.M., Solanki, L., Gordon, D.A., Reinking, R.M., and Stuart, D.G. 1988. Association between biochemical and physiological properties in single motor units. *Muscle and Nerve* 11:245–54.

Hansen, S., Hansen, N.L., Christensen, L.O.D., Petersen, N.T., and Nielsen, J.B. 2002. Coupling of antagonistic ankle muscles during co-contraction in humans. *Experimental Brain Research* 146:282–92.

Hardie, D.G., Hawley, S., and Scott, J. 2006. AMP-activated protein kinase—development of the energy sensor concept. *J. Physiol. (Lond.)* 574:7–15.

Hardie, D.G., and Sakamoto, K. 2006. AMPK: A key sensor of fuel and energy status in skeletal muscle. *Physiology* 21:48–60.

Harridge, S.D.R., Bottinelli, R., Canepari, M., Pellegrino, M.A., Reggiani, C., Esbjörnsson, M., Balsom, P.D., and Saltin, B. 1998. Sprint training, *in vitro* and *in vivo* muscle function, and myosin heavy chain expression. *J. Appl. Physiol.* 84:442–9.

Harridge, S.D.R., Bottinelli, R., Canepari, M., Pellegrino, M.A., Reggiani, C., Esbjörnsson, M., and Saltin, B. 1996. Wholemuscle and singlefibre contractile properties and myosin heavy chain isoforms in humans. *Pflügers Arch.* 432:913–20.

Hather, B.M., Adams, G.R., Tesch, P.A., and Dudley, G.A. 1992. Skeletal muscle responses to lower limb suspension in humans. *J. Appl. Physiol.* 72:1493–8.

Hather, B.M., Tesch, P.A., Buchanan, P., and Dudley, G.A. 1991. Influence of eccentric actions on skeletal muscle adaptations to resistance training. *Acta Physiol. Scand.* 143:177–85.

Hauschka, E.O., Roy, R.R., and Edgerton, V.R. 1987. Size and metabolic properties of single muscle fibers in rat soleus after hindlimb suspension. *J. Appl. Physiol.* 62:2338–47.

Hayes, S.G., Kindig, A.E., and Kaufman, M.P. 2006. Cyclooxygenase blockade attenuates responses of group III and IV muscle afferents to dynamic exercise in cats. *American Journal of Physiology: Heart and Circulatory Physiology* 290: H2239–46.

Hayward, L., Breitbach, D., and Rymer, W.Z. 1988. Increased inhibitory effects on close synergists during muscle fatigue in the decerebrate cat. *Brain Res.* 440:199–203.

Hayward, L., Wesselmann, U., and Rymer, W.Z. 1991. Effects of muscle fatigue on mechanically sensitive afferents of slow conduction velocity in the cat triceps surae. *J. Neurophysiol.* 65:360–70.

Heckman, C.J., and Binder, M.D. 1991. Computer simulation of the steadystate inputoutput function of the cat medial gastrocnemius motoneuron pool. *J. Neurophysiol.* 65:952–67.

Heckman, C.J., and Binder, M.D. 1993. Computer simulations of motoneuron firing rate modulation. *J. Neurophysiol.* 69:1005–8.

Heckman, C.J., Gorassini, M.A., and Bennett, D.J. 2005. Persistent inward currents in motoneuron dendrites: Implications for motor output. *Muscle and Nerve* 31:135–56.

Heckman, C.J., Johnson, M., Mottram, C.J., and Schuster, J. 2008. Persistent inward currents in spinal motoneurons and their influence on human motoneuron firing patterns. *Neuroscientist* 14:264-75.

Heilmann, C., and Pette, D. 1979. Molecular transformations in sarcoplasmic reticulum of fast-twitch muscle by electro-stimulation. *Eur. J. Biochem.* 93:437–46.

Heinemeier, K.M., Olesen, J.L., Haddad, F., Langberg, H., Kjaer, M., Baldwin, K.M., and Schjerling, P. 2007. Expression of collagen and related growth factors in rat tendon and skeletal muscle in response to specific contraction types. *J. Physiol. (Lond.)* 582(3): 1303–16.

Heinemeier, K.M., Olesen, J.L., Schjerling, P., Haddad, F., Langberg, H., Baldwin, K.M., and Kjaer, M. 2007. Short-term strength training and the expression of myostatin and IGF- I isoforms in rat muscle and tendon: Differential effects of specific contraction types. *J. Appl. Physiol.* 102(2): 573–81.

Heinonen, I., Nesterov, S., Kemppainen, J., Nuutila, P., Knuuti, J., Laitio, R., Kjaer, M., Boushel, R., and Kalliokoski, K.K. 2007. Role of adenosine in regulating the heterogeneity of skeletal muscle blood flow during exercise in humans. *J. Appl. Physiol.* 103:2042–8.

Heiner, L., Domonkos, J., Motika, D., and Vargha, M. 1984. Role of the nervous system in regulation of the sarcoplasmic membrane function in different muscle fibres. *Acta Physiol. Hung.* 64:129–33.

Henneman, E. 1981. Recruitment of motor units: The size principle. In *Motor unit types, recruitment and plasticity in health and disease,* ed. J.E. Desmedt, 26–60. New York: Karger.

Henneman, E., Somjen, G., and Carpenter, D.O. 1965. Excitability and inhibitility of motoneurons of different sizes. *J. Neurophysiol.* 28:599–620.

Henriksson, J., Chi, M.M.Y., Hintz, C.S., Young, D.A., Kaiser, K.K., Salmons, S., and Lowry, O.H. 1986. Chronic stimulation of mammalian muscle: Changes in enzymes of six metabolic pathways. *Am. J. Physiol. Cell Physiol.* 251:C614–32.

Hensbergen, E., and Kernell, D. 1992. Taskrelated differences in distribution of electromyographic activity within peroneus longus muscle of spontaneously moving cats. *Exp. Brain Res.* 89:682–5.

Herbert, M.E., Roy, R.R., and Edgerton, V.R. 1988. Influence of oneweek hindlimb suspension and intermittent high load exercise on rat muscles. *Exp. Neurol.* 102:190–8.

Herbert, R.D., Dean, C., and Gandevia, S.C. 1998. Effects of real and imagined training on voluntary muscle activation during maximal isometric contractions. *Acta Physiol. Scand.* 163:361–8.

Herbert, R.D., and Gandevia, S.C. 1996. Muscle activation in unilateral and bilateral efforts assessed by motor nerve and cortical stimulation. *J. Appl. Physiol.* 80:1351–6.

Hesketh, J.E., and Whitelaw, P.F. 1992. The role of cellular oncogenes in myogenesis and muscle cell hypertrophy. *Int. J. Biochem.* 24:193–203.

Hicks, A., Fenton, J., Garner, S., and McComas, A.J. 1989. M wave potentiation during and after muscle activity. *J. Appl. Physiol.* 66:2606–10.

Hicks, A., Ohlendieck, K., Göpel, S.O., and Pette, D. 1997. Early functional and biochemical adaptations to lowfrequency stimulation of rabbit fasttwitch muscle. *Am. J. Physiol. Cell Physiol.* 273:C297–305.

Higbie, E.J., Cureton, K.J., Warren III, G.L., and Prior, B.M. 1996. Effects of concentric and eccentric training on muscle strength, crosssectional area, and neural activation. *J. Appl. Physiol.* 81:2173–81.

Hikida, R.S., Gollnick, P.D., Dudley, G.A., Convertino, V.A., and Buchanan, P. 1989. Structural and metabolic characteristics of human skeletal muscle following 30 days of simulated microgravity. *Aviat. Space Environ. Med.* 60:664–70.

Hill, J.M. 2000. Discharge of group IV phrenic afferent fibers increases during diaphragmatic fatigue. *Brain Res.* 856:240–4.

Hintz, C.S., Chi, M.M.Y., and Lowry, O.H. 1984. Heterogeneity in regard to enzymes and metabolites within individual muscle fibers. *Am. J. Physiol. Cell Physiol.* 246:C288–92.

Hintz, C.S., Coyle, E.F., Kaiser, K.K., Chi, M.M.-Y., and Lowry, O.H. 1984. Comparison of muscle fiber typing by quantitative enzyme assays and by myosin ATPase staining. *J. Histochem. Cytochem.* 32:655–60.

Hintz, C.S., Lowry, C.V., Kaiser, K.K., McKee, D., and Lowry, O.H. 1980. Enzyme levels in individual rat muscle fibers. *Am. J. Physiol. Cell Physiol.* 239:C58–65.

Hochman, S., and McCrea, D.A. 1994a. Effects of chronic spinalization on ankle extensor motoneurons: I. Composite monosynaptic Ia EPSPs in four motoneuron pools. *J. Neurophysiol.* 71:1452–67.

Hochman, S., and McCrea, D.A. 1994b. Effects of chronic spinalization on ankle extensor motoneurons: II. Motoneuron electrical properties. *J. Neurophysiol.* 71:1468–79.

Hocking, D., Titus, P., Sumagin, R., and Sarelius, I. 2008. Extracellular matrix fibronectin mechanically couples skeletal muscle contaction with local vasodilatation. *Circ. Res.* 102:372–9.

Hodgson, J.A., Roy, R.R., De Leon, R., Dobkin, B., and Edgerton, V.R. 1994. Can the mammalian lumbar spinal cord learn a motor task? *Med. Sci. Sports Exerc.* 26:1491–7.

Hofmann, P.A., Metzger, J.M., Greaser, M.L., and Moss, R.L. 1990. Effects of partial extraction of light chain 2 on the Ca^{2+} sensitivities of isometric tension, stiffness, and velocity of shortening in skinned skeletal muscle fibers. *J. Gen. Physiol.* 95:477–98.

Holloway, G.P., Bezaire, V., Heigenhauser, G.F., Tandon, N.N., Glatz, J., Luiken, J.J.F.P., Bonen, A., and Spriet, L.L. 2006. Mitochondrial long chain fatty acid oxidation, fatty acid translocase/CD36 content and carnitine palmitoyltransferase I activity in human skeletal muscle during aerobic exercise. *J. Physiol. (Lond.)* 571:201–10.

Hong, S.J., and Lnenicka, G.A. 1993. Longterm changes in the neuromuscular synapses of a crayfish motoneuron produced by calcium influx. *Brain Res.* 605:121–7.

Hood, D.A., and Parent, G. 1991. Metabolic and contractile responses of rat fasttwitch muscle to 10Hz stimulation. *Am. J. Physiol. Cell Physiol.* 260:C832–40.

Hoppeler, H., and Flück, M. 2002. Normal mammalian skeletal muscle and its phenotypic plasticity. *Journal of Experimental Biology* 205:2143–52.

Horowitz, J.F., Sidossis, L.S., and Coyle, E.F. 1994. High efficiency of Type I fibers improves performance. *Int. J. Sports Med.* 15:152–7.

Hortobágyi, T., Barrier, J., Beard, D., Braspennincx, J., Koens, P., Devita, P., Dempsey, L., and Lambert, N.J. 1996. Greater initial adaptations to submaximal muscle lengthening than maximal shortening. *J. Appl. Physiol.* 81:1677–82.

Hortobágyi, T., Hill, J.P., Houmard, J.A., Fraser, D.D., Lambert, N.J., and Israel, R.G. 1996. Adaptive responses to muscle lengthening and shortening in humans. *J. Appl. Physiol.* 80:765–72.

Houston, M.E., Norman, R.W., and Froese, E.A. 1988. Mechanical measures during maximal velocity knee extension exercise and their relation to fibre composition of the human vastus lateralis muscle. *Eur. J. Appl. Physiol.* 58:1–7.

Howald, H., Hoppeler, H., Claassen, H., Mathieu, O., and Straub, R. 1985. Influences of endurance training on the ultrastructural composition of the different muscle fiber types in humans. *Pflugers Arch.* 403:369–76.

Howard, G., Steffen, J.M., and Geoghegan, T.E. 1989. Transcriptional regulation of decreased protein synthesis during skeletal muscle unloading. *J. Appl. Physiol.* 66:1093–8.

Howard, J.D., and Enoka, R.M. 1991. Maximum bilateral contractions are modified by neurally mediated interlimb effects. *J. Appl. Physiol.* 70:306–16.

Howell, J.N., Chleboun, G., and Conatser, R. 1993. Muscle stiffness, strength loss, swelling and soreness following exerciseinduced injury in humans. *J. Physiol. (Lond.)* 464:183–96.

Howell, J.N., Fuglevand, A.J., Walsh, M., and BiglandRitchie, B.R. 1995. Motor unit activity during isometric and concentriceccentric contractions of the human first dorsal interosseus muscle. *J. Neurophysiol.* 74:901–4.

Howell, S., Zhan, W.Z., and Sieck, G.C. 1997. Diaphragm disuse reduces Ca^{2+} uptake capacity of sarcoplasmic reticulum. *J. Appl. Physiol.* 82:164–71.

Hsiao, C.F., Trueblood, P.R., Levine, M.S., and Chandler, S.H. 1997. Multiple effects of serotonin on membrane properties of trigeminal motoneurons *in vitro. J. Neurophysiol.* 77:2910–24.

Hu, P., Yin, C., Zhang, K.M., Wright, L.D., Nixon, T.E., Wechsler, A.S., Spratt, J.A., and Briggs, F.N. 1995. Transcriptional regulation of phospholamban gene and translational regulation of SERCA2 gene produces coordinate expression of these two sarcoplasmic reticulum proteins during skeletal muscle phenotype switching. *J. Biol. Chem.* 270:11619–22.

Hubatsch, D.A., and Jasmin, B.J. 1997. Mechanical stimulation increases expression of acetylcholinesterase in cultured myotubes. *Am. J. Physiol. Cell Physiol.* 273:C2002–9.

Huber, B., and Pette, D. 1996. Dynamics of parvalbumin expression in lowfrequencystimulated fasttwitch rat muscle. *Eur. J. Biochem.* 236:814–9.

Hultborn, H., Katz, R., and Mackel, R. 1988. Distribution of recurrent inhibition within a motor nucleus: II. Amount of recurrent inhibition in motoneurones to fast and slow units. *Acta Physiol. Scand.* 134:363–74.

Hultborn, H., and Kiehn, O. 1992. Neuromodulation of vertebrate motor neuron membrane properties. *Curr. Opin. Neurobiol.* 2:770–5.

Hultborn, H., Lipski, J., Mackel, R., and Wigström, H. 1988. Distribution of recurrent inhibition within a motor nucleus: I. Contribution from slow and fast motor units to the excitation of Renshaw cells. *Acta Physiol. Scand.* 134:347–61.

Hultborn, H., Meunier, S., Morin, C., and PierrotDeseilligny, E. 1987. Assessing changes in presynaptic inhibition of Ia fibres: A study in man and the cat. *J. Physiol. (Lond.)* 389:729–56.

Hultborn, H., Meunier, S., PierrotDeseilligny, E., and Shindo, M. 1987. Changes in presynaptic inhibition of Ia fibres at the onset of voluntary contraction in man. *J. Physiol. (Lond.)* 389:757–72.

Hulten, B., and Karlsson, J. 1974. Relationship between isometric endurance and muscle fiber type composition. *Acta Physiol. Scand.* 91: A46–7.

Hunter, G.R., Newcomer, B.R., Larson-Meyer, D.E., Bamman, M.M., and Weinsier, R.L. 2001. Muscle metabolic economy is inversely related to exercise inensity and type II myofiber distribution. *Muscle Nerve* 24, 654–661.

Hutton, R.S., and Nelson, D.L. 1986. Stretch sensitivity of Golgi tendon organs in fatigued gastrocnemius muscle. *Med. Sci. Sports Exerc.* 18:69–74.

Hyngstrom, A., Johnson, M., Miller, J., and Heckman, CJ. 2007 Intrinsic electrical properties of spinal motoneurons vary with joint angle, *Nature Neuroscience* 10 (3); 363-369.

Inbar, O., Kaiser, P., and Tesch, P. 1981. Relationships between leg muscle fiber type distribution and leg exercise performance. *Int. J. Sports Med.* 2:154–9.

Ishihara, A., Ohira, Y., Roy, R.R., Nagaoka, S., Sekiguchi, C., Hinds, W.E., and Edgerton, V.R. 1996. Influence of spaceflight on succinate dehydrogenase activity and soma size of rat ventral horn neurons. *Acta Anat. (Basel)* 157:303–8.

Ishihara, A., Oishi, Y., Roy, R.R., and Edgerton, V.R. 1997. Influence of two weeks of non–weight bearing on rat soleus motoneurons and muscle fibers. *Aviat. Space Environ. Med.* 68:421–5.

Ishihara, A., Roy, R.R., and Edgerton, V.R. 1995. Succinate dehydrogenase activity and soma size of motoneurons innervating different portions of the rat tibialis anterior. *Neuroscience* 68:813–22.

Ivanova, T., Garland, S.J., and Miller, K.J. 1997. Motor unit recruitment and discharge behavior in movements and isometric contractions. *Muscle and Nerve* 20:867–74.

Ivy, J.L., Withers, R.T., Brose, G., Maxwell, B.D., and Costill, D.L. 1981. Isokinetic contractile properties of the quadriceps with relation to fiber type. *Eur. J. Appl. Physiol.* 47:247–55.

JacobsEl, J., Ashley, W., and Russell, B. 1993. IIx and slow myosin expression follow mitochondrial increases in transforming muscle fibers. *Am. J. Physiol. Cell Physiol.* 265: C79–84.

JacobsEl, J., Zhou, M.Y., and Russell, B. 1995. MRF4, Myf5, and myogenin mRNAs in the adaptive responses of mature rat muscle. *Am. J. Physiol. Cell Physiol.* 268:C1045–52.

Jain, N., Florence, S.L., and Kaas, J.H. 1998. Reorganization of somatosensory cortex after nerve and spinal cord injury. *News Physiol. Sci.* 13:143–9.

Jakab, G., Dux, L., Tabith, K., and Guba, F. 1987. Effects of disuse on the function of fragmented sarcoplasmic reticulum of rabbit m. gastrocnemius. *Gen. Physiol. Biophys.* 6:127–35.

Jakobi, J.M., and Chilibeck, P.D. 2001. Bilateral and unilateral contractions: Possible differences in maximal voluntary force. *Canadian Journal of Applied Physiology* 26:12-33.

Jami, L., Murthy, K.S.K., Petit, J., and Zytnicki, D. 1983. Aftereffects of repetitive stimulation at low frequency on fastcontracting motor units of cat muscle. *J. Physiol.* 340:129–43.

Jänkälä, H., Harjola, V.P., Petersen, N.E., and Härkönen, M. 1997. Myosin heavy chain mRNA transform to faster isoforms in immobilized skeletal muscle: A quantitative PCR study. *J. Appl. Physiol.* 82:977–82.

Jansson, E., and Hedberg, G. 1991. Skeletal muscle fiber types in teenagers: Relationship to physical performance and activity. *Scand. J. Med. Sci. Sports* 1:31–44.

Jansson, E., Sylvén, C., Arvidsson, I., and Eriksson, E. 1988. Increase in myoglobin content and decrease in oxidative enzyme activities by leg muscle immobilization in man. *Acta Physiol. Scand.* 132:515–7.

Jaschinski, F., Schuler, M., Peuker, H., and Pette, D. 1998. Changes in myosin heavy chain mRNA and protein isoforms of rat muscle during forced contractile activity. *Am. J. Physiol. Cell Physiol.* 274: C365–70.

Jasmin, B.J., Gardiner, P.F., and Gisiger, V. 1991. Muscle acetylcholinesterase adapts to compensatory overload by a general increase in its molecular forms. *J. Appl. Physiol.* 70:2485–9.

Jasmin, B.J., and Gisiger, V. 1990. Regulation by exercise of the pool of G_4 acetylcholinesterase characterizing fast muscles: Opposite effect of running training in antagonist muscles. *J. Neurosci.* 10:1444–54.

Jasmin, B., Lavoie, P., and Gardiner, P.F. 1988. Fast axonal transport of labeled proteins in motoneurons of exercisetrained rats. *Am. J. Physiol. Cell Physiol.* 255:C731–6.

Jensen, J.L., Marstrand, P.C.D., and Nielsen, J.B. 2005. Motor skill training and strength training are associated with different plastic changes in the central nervous system. *J. Appl. Physiol.* 99:1558–68.

Ji, L.L., GomezCabrers, M.C., and Vina, J. 2007. Role of nuclear factor kappa B and mitogen-activated protein kinase signaling in exercise-induced antioxidant enzyme adaptation. *Appl. Physiol. Nutr. Metab.* 32(5): 930–5.

Jiang, B., Ohira, Y., Roy, R.R., Nguyen, Q., IlyinaKakueva, E.I., Oganov, V., and Edgerton, V.R. 1992. Adaptation of fibers in fasttwitch muscles of rats to spaceflight and hindlimb suspension. *J. Appl. Physiol.* Suppl. no. 73:58S–65S.

Jiang, B., Roy, R.R., and Edgerton, V.R. 1990. Enzymatic plasticity of medial gastrocnemius fibers in the adult chronic spinal cat. *Am. J. Physiol. Cell Physiol.* 259:C507–14.

Jiang, B., Roy, R.R., Navarro, C., Nguyen, Q., Pierotti, D., and Edgerton, V.R. 1991. Enzymatic responses of cat medial gastrocnemius fibers to chronic inactivity. *J. Appl. Physiol.* 70:231–9.

Johansson, C., Lorentzon, R., Sjöström, M., Fagerlund, M., and FuglMeyer, A.R. 1987. Sprinters and marathon runners: Does isokinetic knee extensor performance reflect muscle size and structure? *Acta Physiol. Scand.* 130:663–9.

Johnson, B.D., and Sieck, G.C. 1993. Differential susceptibility of diaphragm muscle fibers to neuromuscular transmission failure. *J. Appl. Physiol.* 75:341–8.

Johnson, L.D., Jiang, Y., and Rall, J.A. 1999. Intracellular EDTA mimics parvalbumin in the promotion of skeletal muscle relaxation. *Biophys. J.* 76:1514–22.

Jokl, P., and Konstadt, S. 1983. The effect of limb immobilization on muscle function and protein composition. *Clin. Orthop. Rel. Res.* 174:222–9.

Jones, D.A., deRuiter, C.J., and deHaan, A. 2006. Change in contractile properties of human muscle in relationship to the loss of power and slowing of relaxation seen with fatigue. *J. Physiol. (Lond.)* 576(3): 913–22.

Jones, D.A., and Rutherford, O.M. 1987. Human muscle strength training: The effects of three different regimes and the nature of the resultant changes. *J. Physiol.* 391:1–11.

Jones, D.A., Rutherford, O.M., and Parker, D.F. 1989. Physiological changes in skeletal muscle as a result of strength training. *Q. J. Exp. Physiol.* 74:233–56.

Jones, K.E., Bawa, P., and McMillan, A.S. 1993. Recruitment of motor units in human flexor carpi ulnaris. *Brain Res.* 602:354–6.

Jones, K.E., Lyons, M., Bawa, P., and Lemon, R.N. 1994. Recruitment order of motoneurons during functional tasks. *Exp. Brain Res.* 100:503–8.

Jones, K.J. 1993. Gonadal steroids and neuronal regeneration: A therapeutic role. *Adv. Neurol.* 59:227–40.

Jones, T., Chu, C., Grande, L., and Gregory, A. 1999. Motor skills training enhances lesioninduced structural plasticity in the motor cortex of adult rats. *J. Neurosci.* 19:10153–63.

Jones, T.A., Kleim, J.A., and Greenough, W.T. 1996. Synaptogenesis and dendritic growth in the cortex opposite unilateral sensorimotor cortex damage in adult rats: A quantitative electron microscopic examination. *Brain Res.* 733:142–8.

Jorgensen, S.B., Jensen, T.E., and Richter, E.A. 2007. Role of AMPK in skeletal muscle gene adaptation in relation exercise. *Appl. Physiol. Nutr. Metab.* 32(5): 904–11.

Joyner, M.J., and Wilkins, B.W. 2007. Exercise hyperaemia: Is anything obligatory but the hyperaemia? *J. Physiol. (Lond.)* 583:855–60.

Jozsa, L., Kannus, P., Thoring, J., Reffy, A., Järvinen, M., and Kvist, M. 1990. The effect of tenotomy and immobilisation on intramuscular connective tissue. *J. Bone Joint Surg.* 72B: 293–7.

Juel, C. 2007. Changes in interstitial K^+ and pH during exercise: Implications for blood flow regulation. *Appl. Physiol. Nutr. Metab.* 32(5): 846–51.

Juel, C., Pilegaard, H., Nielsen, J.J., and Bangsbo, J. 2000. Interstitial K^+ in human skeletal muscle during and after dynamic graded exercise determined by microanalysis. *Am. J. Physiol. (Regulatory Integrative Comp. Physiol.)* 278:R400–6.

Julian, F.J., Moss, R.L., and Waller, G.S. 1981. Mechanical properties and myosin light chain composition of skinned muscle fibers from adult and newborn rabbits. *J. Physiol. (Lond.)* 311:201–18.

Jürimäe, J., Blake, K., Abernethy, P.J., and McEniery, M.T. 1996. Changes in the myosin heavy chain isoform profile of the triceps brachii muscle following 12 weeks of resistance training. *Eur. J. Appl. Physiol. Occup. Physiol.* 74:287–92.

Kaczkowski, W., Montgomery, D.L., Taylor, A.W., and Klissouras, V. 1982. The relationship between muscle fiber composition and maximal anaerobic power and capacity. *J. Sports Med. Phys. Fitness* 22:407–13.

Kadi, F., and Thornell, L. 1999. Training affects myosin heavy chain phenotype in the trapezius muscle of women. *Histochem. Cell. Biol.* 112:73–8.

Kadi, F., and Thornell, L. 2000. Concomitant increases in myonuclear and satellite cell content in female trapezius muscle following strength training. *Histochem. Cell Biol.* 113:99–103.

Kanda, K., Burke, R.E., and Walmsley, B. 1977. Differential control of fast and slow twitch motor units in the decerebrate cat. *Exp. Brain Res.* 29:57–74.

Kang, C.M., Lavoie, P.A., and Gardiner, P.F. 1995. Chronic exercise increases SNAP25 abundance in fasttransported proteins of rat motoneurones. *Neuroreport* 6:549–53.

Karatzaferi, C., FranksSkiba, K., and Cooke, R. 2008. Inhibition of shortening velocity of skinned skeletal muscle fibers in conditions that mimic fatigue. *Amer. J. Physiol. Regul. Integr. C* 294(3): R948–55.

Karlsson, J., and Jacobs, F. 1982. Onset of blood lactate accumulation during muscular exercise as a threshold concept. *Int. J. Sports Med.* 3:190–201.

Kaufman, M.P., Longhurst, J.C., Rybicki, K.J., Wallach, J.H., and Mitchell, J.H. 1983. Effects of static muscular contraction on impulse activity of groups III and IV afferents in cats. *J. Appl. Physiol. Respir. Environ. Exerc. Physiol.* 55:105–12.

Kawakami, Y., Abe, T., Kuno, S.Y., and Fukunaga, T. 1995. Traininginduced changes in muscle architecture and specific tension. *Eur. J. Appl. Physiol.* 72:37–43.

Kelley, G. 1996. Mechanical overload and skeletal muscle fiber hyperplasia: A metaanalysis. *J. Appl. Physiol.* 81:1584–8.

KentBraun, J.A., and Le Blanc, R. 1996. Quantitation of central activation failure during maximal voluntary contractions in humans. *Muscle and Nerve* 19:861–9.

KentBraun, J.A., Ng, A.V., Castro, M., Weiner, M.W., Gelinas, D., Dudley, G.A., and Miller, R.G. 1997. Strength, skeletal muscle composition, and enzyme activity in multiple sclerosis. *J. Appl. Physiol.* 83:1998–2004.

Kernell, D. 1965a. The adaptation and the relation between discharge frequency and current strength of cat lumbosacral motoneurones stimulated by long-lasting injected currents. *Acta Physiol. Scand.* 65:65–73.

Kernell, D. 1965b. Highfrequency repetitive firing of cat lumbosacral motoneurones stimulated by long-lasting injected currents. *Acta Physiol. Scand.* 65:74–86.

Kernell, D. 1965c. The limits of firing frequency in cat lumbosacral motoneurons possessing different time course of afterhyperpolarization. *Acta Physiol. Scand.* 65:87–100.

Kernell, D. 1979. Rhythmic properties of motoneurones innervating muscle fibres of different speed in m. gastrocnemius medialis of the cat. *Brain Res.* 160:159–62.

Kernell, D. 1983. Functional properties of spinal motoneurons and gradation of muscle force. In *Motor control mechanisms in health and disease*, ed. J.E. Desmedt, 213–26. New York: Raven Press.

Kernell, D. 1984. The meaning of discharge rate: Excitationtofrequency transduction as studied in spinal motoneurons. *Arch. Ital. Biol.* 122:5–15.

Kernell, D. 1992. Organized variability in the neuromuscular system: A survey of taskrelated adaptations. *Arch. Ital. Biol.* 130:19–66.

Kernell, D., Eerbeek, O., Verhey, B., and Donselaar, Y. 1987. Effects of physiological amounts of high and low-rate chronic stimulation on fasttwitch muscle of the cat hindlimb: I. Speed and forcerelated properties. *J. Neurophysiol.* 58:598–612.

Kernell, D., and Monster, A. 1981. Threshold current for repetitive impulse firing in motoneurones innervating muscle fibers of different fatigue sensitivity in the cat. *Brain Res.* 229:193–6.

Kernell, D., and Zwaagstra, B. 1981. Input conductance, axonal conduction velocity and cell size among hindlimb motoneurones of the cat. *Brain Res.* 204:311–26.

Kernell, D., and Zwaagstra, B. 1989. Dendrites of cat's spinal motoneurones: Relationship between stem diameter and predicted input conductance. *J. Physiol.* 413:255–69.

Kiehn, O., Erdal, J., Eken, T., and Bruhn, T. 1996. Selective depletion of spinal monoamines changes the rat soleus EMG from a tonic to a more phasic pattern. *J. Physiol.* 492:173–84.

Kilgour, R.D., Gariepy, P., and Rehel, R. 1991. Facial cooling does not benefit cardiac dynamics during recovery from exercise hyperthermia. *Aviat. Space Environ. Med.* 62:849–54.

Kim, D.H., Witzmann, F.A., and Fitts, R.H. 1982. Effect of disuse on sarcoplasmic reticulum in fast and slow skeletal muscle. *Am. J. Physiol. Cell Physiol.* 243(12): C156–60.

Kim, P.L., Staron, R.S., and Phillips, S.A. 2005. Fasted-state skeletal muscle protein synthesis after resistance exercise is altered with training. *Journal of Physiology* 568:283–90.

Kirby, B.S., Carlson, R.E., Markwald, R., Voyles, W.F., and Dinenno, F.A. 2007. Mechanical influences on skeletal muscle vascular tone in humans: Insight into contraction-induced rapid vasodilatation. *J. Physiol. (Lond.)* 583:861–74.

Kirby, C.R., Ryan, M.J., and Booth, F.W. 1992. Eccentric exercise training as a countermeasure to nonweightbearing soleus muscle atrophy. *J. Appl. Physiol.* 73:1894–9.

Kirschbaum, B.J., Schneider, S., Izumo, S., Mahdavi, V., NadalGinard, B., and Pette, D. 1990. Rapid and reversible changes in myosin heavy chain expression in response to increased neuromuscular activity of rat fasttwitch muscle. *FEBS Lett.* 268:75–8.

Kleim, J.A., Barbay, S., and Nudo, R.J. 1998. Functional reorganization of the rat motor cortex following motor skill learning. *J. Neurophysiol.* 80:3321–5.

Klitgaard, H., Bergman, O., Betto, R., Salviati, G., Schiaffino, S., Clausen, T., and Saltin, B. 1990. Coexistence of myosin heavy chain I and IIa isoforms in human skeletal muscle fibres with endurance training. *Pflugers Arch.* 416:470–2.

Kniffki, K.D., Schomburg, E.D., and Steffens, H. 1981. Convergence in segmental reflex pathways from fine muscle afferents and cutaneous or group II muscle afferents to alphamotoneurones. *Brain Res.* 218:342–6.

Knuth, S.T., Dave, H., Peters, J., and Fitts, R. 2006. Low cell pH depresses peak power in rat skeletal muscle fibres at both 30 degrees C and 15 degrees C: Implications for muscle fatigue. *Journal of Physiology* 575: 887–99.

Komi, P.V., Rusko, H., Vos, J., and Vhiko, V. 1977. Anaerobic performance capacity in athletes. *Acta. Physiol. Scand.* 100:107–14.

Komulainen, J., Takala, T.E.S., Kuipers, H., and Hesselink, M.K.C. 1998. The disruption of myofibre structures in rat skeletal muscle after forced lengthening contractions. *Pflugers Arch.* 436:735–41.

Koryak, Y. 1998. Electromyographic study of the contractile and electrical properties of the human triceps surae muscle in a simulated microgravity environment. *J. Physiol.* 510:287–95.

Koryak, Y. 1999. The effects of longterm simulated microgravity on neuromuscular performance in men and women. *Eur. J. Appl. Physiol. Occup. Physiol.* 79:168–75.

Kossev, A., and Christova, P. 1998. Discharge pattern of human motor units during dynamic concentric and eccentric contractions. *Electroencephalogr. Clin. Neurophysiol. Electromyogr. Motor Control* 109:245–55.

Koulmann, N., and Bigard, A.-X. 2006. Interaction between signalling pathways involved in skeletal muscle responses to endurance exercise. *Pflügers Arch. Eur. J. Physiol.* 452:125–39.

Koval, J.A., DeFronzo, R.A., O'Doherty, R.M., Printz, R., Ardehali, H., Granner, D.K., and Mandarino, L.J. 1998. Regulation of hexokinase II activity and expression in human muscle by moderate exercise. *American Journal of Physiology: Endocrinology and Metabolism* 274: E304–8.

Kramer, D.K., AlKhalili, L., Guigas, B., Leng, Y., GarciaRoves, P.M., and Krook, A. 2007. Role of AMP kinase and PPAR delta in the regulation of lipid and glucose metabolism in human skeletal muscle. *Journal of Biological Chemistry* 282(27): 19313–20.

Kraniou, Y., Cameron-Smith, D., Misso, M., Collier, G., and Hargreaves, M. 2007. Effects of exercise on GLUT-4 and glycogenin gene expression in human skeletal muscle. *J. Appl. Physiol.* 88:794–6.

Kreider, R.B. 1999. Dietary supplements and the promotion of muscle growth with resistance exercise. *Sports Medicine* 27:97–110.

Krippendorf, B., and Riley, D. 1994. Temporal changes in sarcomere lesions of rat adductor longus muscles during hindlimb reloading. *Anat. Rec.* 238:304–10.

Krnjevic, K., and Miledi, R. 1959. Presynaptic failure of neuromuscular propagation in rats. *J. Physiol.* 149:1–22.

Kudina, L.P., and Alexeeva, N.L. 1992. Afterpotentials and control of repetitive firing in human motoneurones. *Electroencephalogr. Clin. Neurophysiol.* 85:345–53.

Kudina, L.P., and Churikova, L.I. 1990. Testing excitability of human motoneurones capable of firing double discharges. *Electroencephalogr. Clin. Neurophysiol.* 75:334–41.

Kuei, J.H., Shadmehr, R., and Sieck, G.C. 1990. Relative contribution of neurotransmission failure to diaphragm fatigue. *J. Appl. Physiol.* 68:174–80.

Kugelberg, E., and Lindegren, B. 1979. Transmission and contraction fatigue of rat motor units in relation to succinate dehydrogenase activity of motor unit fibres. *J. Physiol. (Lond.)* 288:285–300.

Kukulka, C.G., and Clamann, H.P. 1981. Comparison of the recruitment and discharge properties of motor units in human brachial biceps and adductor pollicis during isometric contractions. *Brain Res.* 219:45–55.

Kukulka, C.G., Moore, M.A., and Russel, A.G. 1986. Changes in human amotoneuron excitability during sustained maximum isometric contractions. *Neurosci. Lett.* 68:327–33.

Kyrolainen, H., Kivela R., Koskinen, S., McBride, J., Andersen, J.L., Takala, T., Sipila, S., and Komi, P.V. 2003. Interrelationships between muscle structure, muscle strengh, and running economy. *Med Sci Sports Exerc* 35, 45–49.

Lafleur, J., Zytnicki, D., HorcholleBossavit, G., and Jami, L. 1992. Depolarization of Ib afferent axons in the cat spinal cord during homonymous muscle contraction. *J. Physiol.* 445:345–54.

Lagerquist, O., Zehr, E.P., and Docherty, D. 2006. Increased spinal reflex excitability is not associated with neural plasticity underlying the cross-education effect. *J. Appl. Physiol.* 100:83–90.

Lapier, T.K., Burton, H.W., Almon, R., and Cerny, F. 1995. Alterations in intramuscular connective tissue after limb casting affect contractioninduced muscle injury. *J. Appl. Physiol.* 78:1065–9.

Larsson, L. 1992. Is the motor unit uniform? *Acta Physiol. Scand.* 144:143–54.

Larsson, L., Ansved, T., Edström, L., Gorza, L., and Schiaffino, S. 1991. Effects of age on physiological, immunohistochemical and biochemical properties of fasttwitch single motor units in the rat. *J. Physiol. (Lond.)* 443:257–75.

Larsson, L., Edström, L., Lindegren, B., Gorza, L., and Schiaffino, S. 1991. MHC composition and enzymehistochemical and physiological properties of a novel fasttwitch motor unit type. *Am. J. Physiol. Cell Physiol.* 261:C93–101.

Larsson, L., Li, X.P., Berg, H.E., and Frontera, W.R. 1996. Effects of removal of weightbearing function on contractility and myosin isoform composition in single human skeletal muscle cells. *Pflugers Arch.* 432:320–8.

Larsson, L., and Moss, R.L. 1993. Maximum velocity of shortening in relation to myosin isoform composition in single fibres from human skeletal muscles. *J. Physiol. (Lond.)* 472:595–614.

Larsson, L., and Tesch, P.A. 1986. Motor unit fibre density in extremely hypertrophied skeletal muscles in man. *Eur. J. Appl. Physiol.* 55:130–6.

Lavoie, P., Collier, B., and Tenenhouse, A. 1976. Comparison of alphabungarotoxin binding to skeletal muscles after inactivity or denervation. *Nature* 260:349–50.

Leberer, E., Härtner, K.T., Brandl, C.J., Fujii, J., Tada, M., MacLennan, D.H., and Pette, D. 1989. Slow/cardiac sarcoplasmic reticulum Ca^{2+}ATPase and phospholamban mRNAs are expressed in chronically stimulated rabbit fasttwitch muscle. *Eur. J. Biochem.* 185:51–4.

Leblanc, A.D., Schneider, V.S., Evans, H.J., Pientok, C., Rowe, R., and Spector, E. 1992. Regional changes in muscle mass following 17 weeks of bed rest. *J. Appl. Physiol.* 73:2172–8.

Lee, M., and Carroll, T.J. 2007. Cross education—Possible mechanisms for the contralateral effects of unilateral resistance training. *Sport Med.* 37(1): 1–14.

Lee, M., Gandevia, S., and Carroll. 2009. T., Unilateral strength training increases voluntary activation of the opposite untrained limb. *Clin. Neurophysiol.* 120: 802-808.

Lee, R.H., and Heckman, C.J. 1998a. Bistability in spinal motoneurons *in vivo:* Systematic variations in persistent inward currents. *J. Neurophysiol.* 80:583–93.

Lee, R.H., and Heckman, C.J. 1998b. Bistability in spinal motoneurons *in vivo:* Systematic variations in rhythmic firing patterns. *J. Neurophysiol.* 80:572–82.

Lee, Y.S., Ondrias, K., Duhl, A.J., Ehrlich, B.E., and Kim, D.H. 1991. Comparison of calcium release from sarcoplasmic reticulum of slow and fast twitch muscles. *J. Membr. Biol.* 122:155–63.

Leterme, D., Cordonnier, C., Mounier, Y., and Falempin, M. 1994. Influence of chronic stretching upon rat soleus muscle during nonweight-bearing conditions. *Pflugers Arch.* 429:274–9.

Leterme, D., and Falempin, M. 1994. Compensatory effects of chronic electrostimulation on unweighted rat soleus muscle. *Pflugers Arch.* 426:155–60.

Lexell, J., Henriksson-Larsen, K., Winblad, B., and Sjöström, M. 1983. Distribution of different fiber types in human skeletal muscles: Effects of aging studied in whole muscle cross sections. *Muscle and Nerve* 6:588–95.

Lexell, J., Jarvis, J., Downham, D., and Salmons, S. 1992. Quantitative morphology of stimulationinduced damage in rabbit fasttwitch skeletal muscles. *Cell Tissue Res.* 269:195–204.

Li, X.P., and Larsson, L. 1996. Maximum shortening velocity and myosin isoforms in single muscle fibers from young and old rats. *Am. J. Physiol. Cell Physiol.* 270:C352–60.

Lieber, R.L., Fridén, J.O., Hargens, A.R., Danzig, L.A., and Gershuni, D.H. 1988. Differential response of the dog quadriceps muscle to external skeletal fixation of the knee. *Muscle and Nerve* 11:193–201.

Lieber, R.L., Fridén, J.O., Hargens, A.R., and Feringa, E.R. 1986. Longterm effects of spinal cord transection on fast and slow rat skeletal muscle: 2. Morphometric properties. *Exp. Neurol.* 91:435–48.

Lieber, R.L., Johansson, C.B., Vahlsing, H.L., Hargens, A.R., and Feringa, E.R. 1986. Longterm effects of spinal cord transection on fast and slow rat skeletal muscle: 1. Contractile properties. *Exp. Neurol.* 91:423–34.

Liepert, J., Miltner, W.H.R., Bauder, H., Sommer, M., Dettmers, C., Taub, E., and Weiller, C. 1998. Motor cortex plasticity during constraintinduced movement therapy in stroke patients. *Neurosci. Lett.* 250:5–8.

Lind, A., and Kernell, D. 1991. Myofibrillar ATPase histochemistry of rat skeletal muscles: A "twodimensional" quantitative approach. *J. Histochem. Cytochem.* 39:589–97.

Linderman, J.K., Gosselink, K.L., Booth, F.W., Mukku, V.R., and Grindeland, R.E. 1994. Resistance exercise and growth hormone as countermeasures for skeletal muscle atrophy in hindlimbsuspended rats. *Am. J. Physiol. Regul. Integr. Comp. Physiol.* 267:R365–71.

Linderman, J.K., Whittall, J.B., Gosselink, K.L., Wang, T.J., Mukku, V.R., Booth, F.W., and Grindeland, R.E. 1995. Stimulation of myofibrillar protein synthesis in hindlimb suspended rats by resistance exercise and growth hormone. *Life Sci.* 57:755–62.

Liu, Y.W., and Schneider, M.F. 1998. Fibre type–specific gene expression activated by chronic electrical stimulation of adult mouse skeletal muscle fibres in culture. *J. Physiol.* 512:337–44.

Ljubisavljevic, M., Jovanovic, K., and Anastasijevic, R. 1994. Fusimotor responses to fatiguing muscle contractions in nondenervated hindlimb of decerebrate cats. *Neuroscience* 61:683–9.

Ljubisavljevic, M., Milanovic, S., Radovanovic, S., Vukcevic, I., Kostic, V., and Anastasijevic, R. 1996. Central changes in muscle fatigue during sustained submaximal isometric voluntary contraction as revealed by transcranial magnetic stimulation. *Electroencephalogr. Clin. Neurophysiol.* 101:281–8.

Ljubisavljevic, M., Radovanovic, S., Vukcevic, I., and Anastasijevic, R. 1995. Fusimotor outflow to pretibial flexors during fatiguing contractions of the triceps surae in decerebrate cats. *Brain Res.* 691:99–105.

Lnenicka, G.A., and Atwood, H.L. 1986. Impulse activity of a crayfish motoneuron regulates its neuromuscular synaptic properties. *J. Neurophysiol.* 61:91–6.

Lnenicka, G.A., and Atwood, H.L. 1988. Longterm changes in neuromuscular synapses with altered sensory input to a crayfish motoneuron. *Exp. Neurol.* 100:437–47.

Locke, M., and Noble, E.G. 1995. Stress proteins: The exercise response. *Can. J. Appl. Physiol.* 20:155–67.

Loeb, G.E. 1987. Hard lessons in motor control from the mammalian spinal cord. *Trends Neurosci.* 10:108–13.

Longhurst, C.M., and Jennings, L.K. 1998. Integrinmediated signal transduction. *Cell. Mol. Life Sci.* 54:514–26.

Löscher, W.N., Cresswell, A.G., and Thorstensson, A. 1996a. Central fatigue during a longlasting submaximal contraction of the triceps surae. *Exp. Brain Res.* 108:305–14.

Löscher, W.N., Cresswell, A.G., and Thorstensson, A. 1996b. Excitatory drive to the motoneuron pool during a fatiguing submaximal contraction in man. *J. Physiol. (Lond.)* 491:271–80.

Loughna, P.T., Goldspink, D.F., and Goldspink, G. 1987. Effects of hypokinesia and hypodynamia upon protein turnover in hindlimb muscles of the rat. *Aviat. Space Environ. Med.* 58:A133–8.

Lovely, R.G., Gregor, R.J., Roy, R.R., and Edgerton, V.R. 1986. Effects of training on the recovery of fullweightbearing stepping in the adult spinal cat. *Exp. Neurol.* 92:421–35.

Lovely, R.G., Gregor, R.J., Roy, R.R., and Edgerton, V.R. 1990. Weightbearing hindlimb stepping in treadmillexercised adult spinal cats. *Brain Res.* 514:206–18.

Lovering, R.M., and De Deyne, P.G. 2004. Contractile function, sarcolemma integrity, and the loss of dystrophin after skeletal muscle eccentric contraction-induced injury. *American Journal of Physiology: Cell Physiology* 286: C230–8.

Lowe, D.A., Warren, G.L., Ingalls, C.P., Boorstein, D.B., and Armstrong, R.B. 1995. Muscle function and protein metabolism after initiation of eccentric contraction–induced injury. *J. Appl. Physiol.* 79:1260–70.

Lowey, S., Waller, G.S., and Trybus, K.M. 1993a. Function of skeletal muscle myosin heavy and light chain isoforms by an in vitro motility assay. *J. Biol. Chem.* 268:20414–8.

Lowey, S., Waller, G.S., and Trybus, K.M. 1993b. Skeletal muscle myosin light chains are essential for physiological speeds of shortening. *Nature* 365:454–6.

MacDougall, J.D., Elder, G., Sale, D.G., Moroz, J., and Sutton, J.R. 1980. Effects of strength training and immobilization on human muscle fibres. *Eur. J. Appl. Physiol.* 43:25–34.

MacDougall, J.D., Sale, D.G., Alway, S.E., and Sutton, J.R. 1984. Muscle fiber number in biceps brachii in bodybuilders and control subjects. *J. Appl. Physiol. Respir. Environ. Exerc. Physiol.* 57:1399–1403.

MacDougall, J.D., Tarnopolsky, M.A., Chesley, A., and Atkinson, S.A. 1992. Changes in muscle protein synthesis following heavy resistance exercise in humans: A pilot study. *Acta Physiol. Scand.* 146:403–4.

Macefield, V.G., Hagbarth, K.E., Gorman, R.B., Gandevia, S.C., and Burke, D. 1991. Decline in spindle support to alphamotoneurones during sustained voluntary contractions. *J. Physiol.* 440:497–512.

Macefield, V.G., Gandevia, S.C., BiglandRitchie, B., Gorman, R.B., and Burke, D. 1993. The firing rates of human motoneurons voluntarily activated in the absence of muscle afferent feedback. *J. Physiol. (Lond.)* 471:429–43.

MacIntosh, B.R., Herzog, W., Suter, E., Wiley, J.P., and Sokolosky, J. 1993. Human skeletal muscle fibre types and force: Velocity properties. *Eur. J. Appl. Physiol.* 67:499–506.

Magleby, K.L., and Pallotta, B.S. 1981. A study of desensitization of acetylcholine receptors using nervereleased transmitter in the frog. *J. Physiol.* 316:225–50.

Maier, A., Gorza, L., Schiaffino, S., and Pette, D. 1988. A combined histochemical and immunohistochemical study on the dynamics of fasttoslow fiber transformation in chronically stimulated rabbit muscle. *Cell Tissue Res.* 254:59–68.

Maier, A., and Pette, D. 1987. The time course of glycogen depletion in single fibers of chronically stimulated rabbit fasttwitch muscle. *Pflugers Arch.* 408:338–42.

Malathi, S., and Batmanabane, M. 1983. Alterations in the morphology of the neuromuscular junctions following experimental immobilization in cats. *Experientia* 39:547–9.

Mambrito, B., and De Luca, C.J. 1983. Acquisition and decomposition of the EMG signal. *Prog. Clin. Neurophysiol.* 10:52–72.

Manninen, A. 2006. Hyperinsulinemia, hyperaminoacidaemia and post-exercise muscle anabolism: The search for the optimal recovery drink. *British Journal of Sports Medicine* 40:900–5.

Marsden, C.D., Meadows, J.C., and Merton, P. 1983. "Muscular wisdom" that minimizes fatigue during prolonged effort in man: Peak rates of motoneuron discharge and slowing of discharge during fatigue. In *Motor control mechanisms in health and disease,* ed. J.E. Desmedt, 169–211. New York: Raven Press.

Marsh, R.L., and Ellerby, D.J. 2006. Partitioning locomotor energy use among and within muscles—Muscle blood flow as a measure of muscle oxygen consumption. *Journal of Experimental Biology* 209:2385–94.

Marshall, J. 2007. The roles of adenosine and related substances in exercise hyperaemia. *J. Physiol. (Lond.)* 583:835–45.

Martin, L., Cometti, G., Pousson, M., and Morlon, B. 1993. Effect of electrical stimulation training on the contractile characteristics of the triceps surae muscle. *Eur. J. Appl. Physiol.* 67:457–61.

Martin, P.G., Smith, J.L., Butler, J.E., Gandevia, S.C., and Taylor, J.L. 2006. Fatigue-sensitive afferents inhibit extensor but not flexor motoneurons in humans. *J. Neurosci.* 26:4796–4802.

Martin, T.P., Edgerton, V.R., and Grindeland, R.E. 1988. Influence of spaceflight on rat skeletal muscle. *J. Appl. Physiol.* 65:2318–25.

Martin, T.P., Stein, R.B., Hoeppner, P.H., and Reid, D.C. 1992. Influence of electrical stimulation on the morphological and metabolic properties of paralyzed muscle. *J. Appl. Physiol.* 72:1401–6.

Martineau, L., and Gardiner, P.F. 1999. Static stretch induces MAPK activation in skeletal muscle (abstract). *FASEB J.* 13: A4101999.

McDonagh, J.C., Binder, M.C., Reinking, R.M., and Stuart, D.G. 1980. A commentary on muscle unit properties in cat hindlimb muscles. *J. Morphol.* 166:217–30.

McDonagh, M.J.N., Hayward, C.M., and Davies, C.T.M. 1983. Isometric training in human elbow flexor muscles: The effects on voluntary and electrically evoked forces. *J. Bone Joint Surg.* 65B: 355–8.

McDonald, K.S., Blaser, C.A., and Fitts, R.H. 1994. Force–velocity and power characteristics of rat soleus muscle fibers after hindlimb suspension. *J. Appl. Physiol.* 77:1609–16.

McDonald, K.S., Delp, M.D., and Fitts, R.H. 1992. Fatigability and blood flow in the rat gastrocnemiusplantarissoleus after hindlimb suspension. *J. Appl. Physiol.* 73:1135–40.

McDonald, K.S., and Fitts, R.H. 1993. Effect of hindlimb unweighting on single soleus fiber maximal shortening velocity and ATPase activity. *J. Appl. Physiol.* 74:2949–57.

McDonald, K.S., and Fitts, R.H. 1995. Effect of hindlimb unloading on rat soleus fiber force, stiffness, and calcium sensitivity. *J. Appl. Physiol.* 79:1796–802.

McPhee, J., Williams, A., Stewart, C., Baar, K., Schlinder, J., Aldred, S., Maffulli, N., Sargeant, A., and Jones, D. 2008. The training stimulus experienced by the leg muscles during cycling in humans. *Exp. Physiol.* 94 (6); 684-694, 2008.

McHugh, M. 2003. Recent advances in the understanding of the repeated bout effect: The protective effect against muscle damage from a single bout of eccentric exercise. *Scand. J. Med. Sci. Sports* 13:88–97.

McKay, W.B., Stokic, D.S., Sherwood, A.M., Vrbova, G., and Dimitrijevic, M.R. 1996. Effect of fatiguing maximal voluntary contraction on excitatory and inhibitory responses elicited by transcranial magnetic motor cortex stimulation. *Muscle and Nerve* 19:1017–24.

McKay, W.B., Tuel, S., Sherwood, A.M., Stokic, D.S., and Dimitrijevic, M.R. 1995. Focal depression of cortical excitability induced by fatiguing muscle contraction: A transcranial magnetic stimulation study. *Exp. Brain Res.* 105:276–82.

McKenna, M.J., Bangsbo, J., and Renaud, J.M. 2008. Muscle K⁺, Na⁺, and Cl⁻ disturbances and Na⁺-K⁺ pump inactivation: Implications for fatigue. *J. Appl. Physiol.* 104(1): 288–95.

McNulty, P.A., and Macefield, V.G. 2005. Intraneural microstimulation of motor axons in the study of human single motor units. *Muscle and Nerve* 32:119–39.

Meissner, J., Kubis, H.-P., Scheibe, R., and Gros, G. 2000. Reversible Ca²⁺-induced fast-to-slow transition in primary skeletal muscle culture cells at the mRNA level. *J. Physiol. (Lond.)* 523:19–28.

Mendell, L.M., Cope, T.C., and Nelson, S.G. 1982. Plasticity of the group Ia fiber pathway to motoneurons. In *Changing concepts of the nervous system,* ed. A.R. Morrison and P.L. Strick, 69–78. New York: Academic Press.

Mercier, A.J., Bradacs, H., and Atwood, H.L. 1992. Longterm adaptation of crayfish neurons depends on the frequency and number of impulses. *Brain Res.* 598:221–4.

Michael, L., Wu, Z., Bentley Cheatham, R., Puigserver, P., Adelmant, G., Lehman, J.J., Kelly, D.P., and Spiegelman, B.M. 2001. Restoration of insulin-sensitive glucose transporter (GLUT4) gene expression in muscle cells by the transcriptional coactivator PGC-1. *Proc. Natl. Acad. Sci. USA* 98:3820–5.

Michel, J., Ordway, G.A., Richardson, J.A., and Williams, R.S. 1994. Biphasic induction of immediate early gene expression accompanies activitydependent angiogenesis and myofiber remodeling of rabbit skeletal muscle. *J. Clin. Invest.* 94:277–85.

Michel, R.N., Cowper, G., Chi, M.M.Y., Manchester, J.K., Falter, H., and Lowry, O.H. 1994. Effects of tetrodotoxininduced neural inactivation on single muscle fiber metabolic enzymes. *Am. J. Physiol. Cell Physiol.* 267:C55–66.

Michel, R.N., and Gardiner, P.F. 1990. To what extent is hindlimb suspension a model of disuse? *Muscle and Nerve* 13:646–53.

Michel, R.N., Parry, D.J., and Dunn, S.E. 1996. Regulation of myosin heavy chain expression in adult rat hindlimb muscles during shortterm paralysis: Comparison of denervation and tetrodotoxininduced neural inactivation. *FEBS Lett.* 391:39–44.

Mihok, M., and Murrant, C.L. 2008. Rapid biphasic arteriolar dilations induced by skeletal muscle contraction are dependent on stimulation characteristics. *Canadian Journal of Physiology and Pharmacology* 82:282–7.

Miles, M.P., Clarkson, P.M., Bean, M., Ambach, K., Mulroy, J., and Vincent, K. 1994. Muscle function at the wrist following 9 d of immobilization and suspension. *Med. Sci. Sports Exerc.* 26:615–23.

Miller, B.E., Olesen, J.L., Hansen, M., Dossing, S., Crameri, R.M., Welling, R.J., Langberg, H., Flyvbjerg, A., Kjaer, M., Babraj, J.A., Smith, K., and Rennie, M.J. 2005. Coordinated collagen and muscle protein synthesis in human patella tendon and quadriceps muscle after exercise. *Journal of Physiology* 567:1021–33.

Miller, K.J., Garland, S.J., Ivanova, T., and Ohtsuki, T. 1996. Motorunit behavior in humans during fatiguing arm movements. *J. Neurophysiol.* 75:1629–36.

Mills, K.R., and Thomson, C.C.B. 1995. Human muscle fatigue investigated by transcranial magnetic stimulation. *Neuroreport* 6:1966–8.

MilnerBrown, H.S., Stein, R.B., and Lee, R.G. 1975. Synchronization of human motor units: Possible roles of exercise and supraspinal reflexes. *Electroencephalogr. Clin. Neurophysiol.* 38:245–54.

MilnerBrown, H.S., Stein, R.B., and Yemm, R. 1973a. The contractile properties of human motor units during voluntary isometric contractions. *J. Physiol. (Lond.)* 228:285–306.

MilnerBrown, H.S., Stein, R.B., and Yemm, R. 1973b. The orderly recruitment of human motor units during voluntary isometric contractions. *J. Physiol.* 230:359–70.

Molteni, R., Zheng, J.Q., Ying, Z., Gómez-Pinilla, F., and Twiss, J.L. 2004. Voluntary exercise increases axonal regeneration from sensory neurons. *Proc. Natl. Acad. Sci. USA* 101:8473–8.

Monster, A.W., and Chan, H. 1977. Isometric force production by motor units of extensor digitorum communis muscle in man. *J. Neurophysiol.* 40:1432–43.

Moore, D.R., Phillips, S.M., Babraj, J.A., Smith, K., and Rennie, M.J. 2005. Myofibrillar and collagen protein synthesis in human skeletal muscle in young men after maximal shortening and lengthening contractions. *American Journal of Physiology: Endocrinology and Metabolism* 288: E1153–9.

Moore, R.L., and Stull, J.T. 1984. Myosin light chain phosphorylation in fast and slow skeletal muscles *in situ. Am. J. Physiol. Cell Physiol.* 247: C462–71.

Moritani, T., and De Vries, H.A. 1979. Neural factors versus hypertrophy in the time course of muscle strength gain. *Am. J. Phys. Med.* 58:115–31.

Moritani, T., Muramatsu, S., and Muro, M. 1988. Activity of motor units during concentric and eccentric contractions. *Am. J. Phys. Med.* 66:338–50.

Morrison, P.R., Montgomery, J.A., Wong, T.S., and Booth, F.W. 1987. Cytochrome c proteinsynthesis rates and mRNA contents during atrophy and recovery in skeletal muscle. *Biochem. J.* 241:257–63.

Moss, R.L., Diffee, G.M., and Greaser, M.L. 1995. Contractile properties of skeletal muscle fibers in relation to myofibrillar protein isoforms. *Rev. Physiol. Biochem. Pharmacol.* 126:1–63.

Moss, R.L., Reiser, P.J., Greaser, M.L., and Eddinger, T.J. 1990. Varied expression of myosin alkali light chains is associated with altered speed of contraction in rabbit fast twitch skeletal muscles. In *The dynamic state of muscle fibers,* ed. D. Pette, 353–68. Berlin: De Gruyter.

Multon, S., Franzen, R., Poirrier, A.L., Scholtes, F., and Schoenen, J. 2003. The effect of treadmill training on motor recovery after a partial spinal cord compression-injury in the adult rat. *Journal of Neurotrauma* 20:699–706.

Munn, J., Herbert, R.D., and Gandevia, S.C. 2003. Contralateral effects of unilateral resistance training: A meta-analysis. *J. Appl. Physiol.* 96:186–6.

Munson, J.B., Foehring, R.C., Lofton, S.A., Zengel, J.E., and Sypert, G.W. 1986. Plasticity of medial gastrocnemius motor units following cordotomy in the cat. *J. Neurophysiol.* 55:619–34.

Munson, J.B., Foehring, R.C., Mendell, L.M., and Gordon, T. 1997. Fasttoslow conversion following chronic lowfrequency activation of medial gastrocnemius muscle in cats: 2. Motoneuron properties. *J. Neurophysiol.* 77:2605–15.

Müntener, M., Käser, L., Weber, J., and Berchtold, M.W. 1995. Increase of skeletal muscle relaxation speed by direct injection of parvalbumin cDNA. *Proc. Natl. Acad. Sci. USA* 92:6504–8.

Naito, H., Powers, S., Demirel, H., Sugiura, T., Dodd, S., and Aoki, J. 2000. Heat stress attenuates skeletal muscle atrophy in hindlimbunweighted rats. *J. Appl. Physiol.* 88:359–63.

Nakano, H., Masuda, K., Sasaki, S.Y., and Katsuta, S. 1997. Oxidative enzyme activity and soma size in motoneurons innervating the rat slowtwitch and fasttwitch muscles after chronic activity. *Brain Res. Bull.* 43:149–54.

Nakazawa, K., Kawakami, Y., Fukunaga, T., Yano, H., and Miyashita, M. 1993. Differences in activation patterns in elbow flexor muscles during isometric, concentric and eccentric contractions. *Eur. J. Appl. Physiol.* 66:214–20.

Nardone, A., Romano, C., and Schieppati, M. 1989. Selective recruitment of highthreshold human motor units during voluntary isotonic lengthening of active muscles. *J. Physiol.* 409:451–71.

Narici, M.V., Roi, G.S., Landoni, L., Minetti, A.E., and Cerretelli, P. 1989. Changes in force, crosssectional area and neural activation during strength training and detraining of the human quadriceps. *Eur. J. Appl. Physiol.* 59:310–9.

Naya, F., Mercer, B., Shelton, J., Richardson, J., Williams, R., and Olson, E. 2000. Stimulation of slow skeletal muscle fiber gene expression by calcineurine *in vitro. J. Biol. Chem.* 275:4545–8.

Nelson, D.L., and Hutton, R.S. 1985. Dynamic and static stretch responses in muscle spindle receptors in fatigued muscle. *Med. Sci. Sports Exerc.* 17:445–50.

Nemeth, P.M., and Pette, D. 1981. Succinate dehydrogenase activity in fibres classified by myosin ATPase in three hind limb muscles of rat. *J. Physiol.* 320:73–80.

Nemeth, P.M., Pette, D., and Vrbova, G. 1981. Comparison of enzyme activities among single muscle fibres within defined motor units. *J. Physiol.* 311:489–95.

Neufer, P.D., Ordway, G.A., Hand, G.A., Shelton, J.M., Richardson, J.A., Benjamin, I.J., and Williams, R.S. 1996. Continuous contractile activity induces fiber type specific expression of HSP70 in skeletal muscle. *Am. J. Physiol. Cell Physiol.* 271:C1828–37.

Neufer, P.D., Ordway, G.A., and Williams, R.S. 1998. Transient regulation of c-fos, α \B-crystallin, and hsp70 in muscle during recovery from contractile activity. *American Journal of Physiology: Cell Physiology* 274: C34–6.

Nguyen, P.V., and Atwood, H.L. 1990. Expression of longterm adaptation of synaptic transmission requires a critical period of protein synthesis. *J. Neurosci.* 10:1099–109.

Nguyen, P.V., and Atwood, H.L. 1992. Maintenance of longterm adaptation of synaptic transmission requires axonal transport following induction in an identified crayfish motoneuron. *Exp. Neurol.*

115:414–22.

Nickerson, J.G., Momken, I., Benton, C.R., Lally, J., Hollloway, G., Han, X., Glatz, J., Chabowski, A., Luiken, J.J.F.P., and Bonen, A. 2007. Protein-mediated fatty acid uptake: Regulation by contraction, AMP-activated protein kinase, and endocrine signals. *Appl. Physiol. Nutr. Metab.* 32:865–73.

Nielsen, J., and Kagamihara, Y. 1993. The regulation of presynaptic inhibition during cocontraction of antagonistic muscles in man. *J. Physiol. (Lond.)* 464:575–93.

Nielsen, J.J., Mohr, M., Klarskov, C., Kristensen, M., Krustrup, P., Juel, C., and Bangsbo, J. 2004. Effects of high-intensity intermittent training on potassium kinetics and performance in human skeletal muscle. *Journal of Physiology* 554:857–70.

Nilsson, J., Tesch, P., and Thorstensson, A. 1977. Fatigue and EMG of repeated fast voluntary contractions in man. *Acta Physiol. Scand.* 101:194–8.

Nissen, S.L., and Sharp, R.L. 2003. Effect of dietary supplements on lean mass and strength gains with resistance exercise: A meta-analysis. *J. Appl. Physiol.* 94:651–9.

Nordstrom, M.A., Enoka, R.M., Reinking, R.M., Callister, R.C., and Stuart, D.G. 1995. Reduced motor unit activation of muscle spindles and tendon organs in the immobilized cat hindlimb. *J. Appl. Physiol.* 78:901–13.

Nordstrom, M.A., Gorman, R., Laouris, Y., Spielmann, J., and Stuart, D.G. 2007. Does motoneuron adaptation contribute to muscle fatigue? *Muscle and Nerve* 35:135–58.

Nudo, R.J., Wise, B.M., SiFuentes, F., and Milliken, G.W. 1996. Neural substrates for the effects of rehabilitative training on motor recovery after ischemic infarct. *Science* 272:1791–4.

Nybo, L., and Rasmussen, P. 2007. Inadequate cerebral oxygen delivery and central fatigue during strenuous exercise. *Med. Sci. Sports Exerc.* 35:110–8.

Nybo, L., and Secher, N.H. 2004. Cerebral perturbations provoked by prolonged exercise. *Progress in Neurobiology* 72:223–61.

Ogata, T. 1988. Structure of motor endplates in the different fiber types of vertebrate skeletal muscles. *Arch. Histol. Cytol.* 51:385–424.

Ohira, Y., Jiang, B., Roy, R.R., Oganov, V., Ilyina-Kakueva, E., Marini, J.F., and Edgerton, V.R. 1992. Rat soleus muscle fiber responses to 14 days of spaceflight and hindlimb suspension. *J. Appl. Physiol.* Suppl. no. 73:51S–7S.

Ohlendieck, K., Fromming, G., Murray, B., Maguire, P., Liesner, E., Traub, I., and Pette, D. 1999. Effects of chronic low-frequency stimulation on Ca^{2+}-regulatory membrane proteins in rabbit fast muscle. *Pflugers Arch.* 438:700–8.

Oishi, Y., Ishihara, A., and Katsuta, S. 1992. Muscle fibre number following hindlimb immobilization. *Acta Physiol. Scand.* 146:281–2.

Ordway, G., Neufer, P., Chin, E., and DiMartino, G. 2000. Chronic contractile activity upregulates the proteosome system in rabbit skeletal muscle. *J. Appl. Physiol.* 88:1134–41.

Ornatsky, O.I., Connor, M.K., and Hood, D.A. 1995. Expression of stress proteins and mitochondrial chaperonins in chronically stimulated skeletal muscle. *Biochem. J.* 311:119–23.

Osbaldeston, N.J., Lee, D.M., Cox, V.M., Hesketh, J.E., Morrison, J.F.J., Blair, G.E., and Goldspink, D.F. 1995. The temporal and cellular expression of *cfos* and *cjun* in mechanically stimulated rabbit latissimus dorsi muscle. *Biochem. J.* 308:465–71.

Pachter, B., and Eberstein, A. 1984. Neuromuscular plasticity following limb immobilization. *J. Neurocytol.* 13:1013–25.

Pagala, M.K.D., Namba, T., and Grob, D. 1984. Failure of neuromuscular transmission and contractility during muscle fatigue. *Muscle and Nerve* 7:454–64.

Pagala, M.K.D., and Taylor, S.R. 1998. Imaging caffeine-induced Ca^{2+} transients in individual fasttwitch and slowtwitch rat skeletal muscle fibers. *Am. J. Physiol. Cell Physiol.* 274:C623–32.

Paintal, A.S. 1960. Functional analysis of group III afferent fibres of mammalian muscles. *J. Physiol. (Lond.)* 152:250–70.

Panenic, R., and Gardiner, P.F. 1998. The case for adaptability of the neuromuscular junction to endurance exercise training. *Can. J. Appl. Physiol.* 23:339–60.

Papadaki, M., and Eskin, S. 1997. Effects of fluid shear stress on gene regulation of vascular cells. *Biotechnol. Prog.* 13:209–21.

Pasquet, B., Carpentier, A., and Duchateau, J. 2006. Specific modulation of motor unit discharge for a similar change in fascicle length during shortening and lengthening contractions in humans. *J. Physiol. (Lond.)* 577(2): 753–65.

Pattullo, M.C., Cotter, M.A., Cameron, N.E., and Barry, J.A. 1992. Effects of lengthened immobilization on functional and histochemical properties of rabbit tibialis anterior muscle. *Exp. Physiol.* 77:433–42.

Pearce, A., Thickbroom, G., Byrnes, M., and Mastaglia, F. 2000. Functional reorganization of the corticomotor projection to the hand in skilled racquet players. *Exp. Brain Res.* 130:238–43.

Pedersen, J., Ljubisavljevic, M., Bergenheim, M., and Johansson, H. 1998. Alterations in information transmission in ensembles of primary muscle spindle afferents after muscle fatigue in heteronymous muscle. *Neuroscience* 84:953–9.

Pelsers, M.M.A.L., Stellingwerff, T., and Van Loon, L.J.C. 2008. The role of membrane fatty-acid transporters in regulating skeletal muscle substrate use during exercise. *Sports Medicine* 38:387–99.

Péréon, Y., Dettbarn, C., Lu, Y., Westlund, K.N., Zhang, J.T., and Palade, P. 1998. Dihydropyridine receptor isoform expression in adult rat skeletal muscle. *Pflugers Arch.* 436:309–14.

Pestronk, A., Drachman, D.B., and Griffin, J.W. 1976. Effect of muscle disuse on acetylcholine receptors. *Nature* 260:352–3.

Peters, E.J.D., and Fuglevand, A.J. 1999. Cessation of human motor unit discharge during sustained maximal voluntary contraction. *Neurosci. Lett.* 274:66–70.

Petersen, N.T., Butler, J.E., Carpenter, M.G., and Cresswell, A.G. 2007. Ia-afferent input to motoneurons during shortening and lengthening muscle contractions in humans. *J. Appl. Physiol.* 102(1): 144–8.

Petit, J., and Gioux, M. 1993. Properties of motor units after immobilization of cat peroneus longus muscle. *J. Appl. Physiol.* 74:1131–9.

Petrofsky, J.S., and Phillips, C.A. 1985. Discharge characteristics of motor units and the surface EMG during fatiguing isometric contractions at submaximal tensions. *Aviat. Space Environ. Med.* 56:581–6.

Petruska, J.C., Ichiyama, R.M., Jindrich, D., Crown, E.D., Tansey, K.E., Roy, R.R., Edgerton, V.R., and Mendell, L.M. 2007. Changes in motoneuron properties after synaptic inputs related to step training after spinal cord transection in rats. *J. Neurosci.* 27:4460–71.

Pette, D. 1998. Training effects on the contractile apparatus. *Acta Physiol. Scand.* 162:367–76.

Pette, D., and Düsterhöft, S. 1992. Altered gene expression in fasttwitch muscle induced by chronic lowfrequency stimulation. *Am. J. Physiol. Regul. Integr. Comp. Physiol.* 262: R333–8.

Pette, D., and Staron, R.S. 1993. The molecular diversity of mammalian muscle fibers. *News Physiol. Sci.* 8:153–7.

Pette, D., and Staron, R.S. 1997. Mammalian skeletal muscle fiber type transitions. *Int. Rev. Cytol.* 170:143–223.

Pette, D., and Vrbova, G. 1992. Adaptation of mammalian skeletal muscle fibers to chronic electrical stimulation. *Rev. Physiol. Biochem. Pharmacol.* 120:115–202.

Pette, D., Wimmer, M., and Nemeth, P.M. 1980. Do enzyme activities vary along muscle fibres? *Histochemistry* 67:225–31.

Peuker, H., Conjard, A., and Pette, D. 1998. Alpha-cardiac-like myosin heavy chain as an intermediate between MHCIIa and MHCI beta in transforming rabbit muscle. *Am. J. Physiol. Cell Physiol.* 274: C595–602.

Peuker, H., and Pette, D. 1997. Quantitative analyses of myosin heavychain mRNA and protein isoforms in single fibers reveal a pronounced fiber heterogeneity in normal rabbit muscles.

Eur. J. Biochem. 247:30–6.

Phelan, J.N., and Gonyea, W.J. 1997. Effect of radiation on satellite cell activity and protein expression in overloaded mammalian skeletal muscle. *Anat. Rec.* 247:179–88.

Phillips, S.M., Tipton, K.D., Aarsland, A., Wolf, S.E., and Wolfe, R.R. 1997. Mixed muscle protein synthesis and breakdown after resistance exercise in humans. *Am. J. Physiol. Endocrinol. Metab.* 273: E99–107.

Phillips, S.M., Tipton, K.D., Ferrando, A.A., and Wolfe, R.R. 1999. Resistance training reduces the acute exerciseinduced increase in muscle protein turnover. *Am. J. Physiol. Endocrinol. Metab.* 276: E118–24.

Pickar, J.G., Hill, J.M., and Kaufman, M.P. 1994. Dynamic exercise stimulates group III muscle afferents. *J. Neurophysiol.* 71:753–60.

Pierotti, D.J., Roy, R.R., BodineFowler, S.C., Hodgson, J.A., and Edgerton, V.R. 1991. Mechanical and morphological properties of chronically inactive cat tibialis anterior motor units. *J. Physiol. (Lond.)* 444:175–92.

Pierotti, D.J., Roy, R.R., Flores, V., and Edgerton, V.R. 1990. Influence of 7 days of hindlimb suspension and intermittent weight support on rat muscle mechanical properties. *Aviat. Space Environ. Med.* March: 205–10.

Pierotti, D.J., Roy, R.R., Hodgson, J.A., and Edgerton, V.R. 1994. Level of independence of motor unit properties from neuromuscular activity. *Muscle and Nerve* 17:1324–35.

Ploutz, L.L., Tesch, P.A., Biro, R.L., and Dudley, G.A. 1994. Effect of resistance training on muscle use during exercise. *J. Appl. Physiol.* 76:1675–81.

PloutzSnyder, L.L., Tesch, P.A., Crittenden, D.J., and Dudley, G.A. 1995. Effect of unweighting on skeletal muscle use during exercise. *J. Appl. Physiol.* 79:168–75.

Pluskal, M.G., and Sreter, F.A. 1983. Correlation between protein phenotype and gene expression in adult rabbit fast twitch muscles undergoing a fast to slow fiber transformation in response to electrical stimulation *in vivo. Biochem. Biophys. Res. Commun.* 113:325–31.

Pool, C.W., Moll, H., and Diegenbach, P.C. 1979. Quantitative succinatedehydrogenase histochemistry. *Histochemistry* 64:273–8.

Poucher, S.M. 1996. The role of the A2A adenosine receptor subtype in functional hyperaemia in the hindlimb of anesthetized cats. *J. Physiol. (Lond.)* 492:495–503.

Powers, S., Ji, L., and Leeuwenburgh, C. 1999. Exercise training–induced alterations in skeletal muscle antioxidant capacity: A brief review. *Med. Sci. Sports Exerc.* 31:987–97.

Pringle J.S.M., Doust J.H., Carter H., Tolfrey K., Campbell I.T., and Jones A.M. 2003. Oxygen uptake kinetics during moderate, heavy and severe intensity 'submaximal' exercise in humans: the influence of muscle fibre type and capillarisation. *Eur J Appl Physiol* 89, 289–300.

Psek, J.A., and Cafarelli, E. 1993. Behavior of coactive muscles during fatigue. *J. Appl. Physiol.* 74:170–5.

Puntschart, A., Wey, E., Jostarndt, K., Vogt, M., Wittwer, M., Widmer, H., Hoppeler, H., and Billeter, R. 1998. Expression of *fos* and *jun* genes in human skeletal muscle after exercise. *Am. J. Physiol.* 274:C129–37.

Racinais, S., Girard, O., Micallef, J.P., and Perrey, S. 2007. Failed excitability of spinal motoneurons induced by prolonged running exercise. *Journal of Neurophysiology* 97(1): 596–603.

Raj, D.A., Booker, T.S., and Belcastro, A.N. 1998. Striated muscle calcium stimulated cysteine protease (calpainlike) activity promotes myeloperoxidase activity with exercise. *Pflugers Arch.* 435:804–9.

Rall, J.A. 1996. Role of parvalbumin in skeletal muscle relaxation. *News Physiol. Sci.* 11:249–55.

Rall, W., Burke, R.E., Holmes, W.R., Jack, J.J.B., Redman, S.J., and Segev, I. 1992. Matching dendritic neuron models to experimental data. *Physiol. Rev.* Suppl. no. 72: S159–86.

Reggiani, C., Bottinelli, R., and Stienen, G. 2000. Sarcomeric myosin isoforms: Fine tuning of a molecular motor. *News Physiol. Sci.* 15:26–33.

Reichmann, H., and Pette, D. 1982. A comparative microphotometric study of succinate dehydrogenase activity levels in type I, IIA and IIB fibres of mammalian and human muscles. *Histochemistry* 74:27–41.

Reid, B., Slater, C.R., and Bewick, G.S. 1999. Synaptic vesicle dynamics in rat fast and slow motor nerve terminals. *J. Neurosci.* 19:2511–21.

Reid, M.B. 2005. Response of the ubiquitin-proteasome pathway to changes in muscle activity. *American Journal of Physiology: Regulatory, Integrative and Comparative Physiology* 288: R1423-31.

Reiser, P.J., Moss, R.L., Giulian, G.G., and Greaser, M.L. 1985. Shortening velocity in single fibers from adult rabbit soleus muscles is correlated with myosin heavy chain composition. *J. Biol. Chem.* 260:9077–80.

Rice, C.L., Cunningham, D.A., Taylor, A.W., and Paterson, D.H. 1988. Comparison of the histochemical and contractile properties of human triceps surae. *Eur. J. Appl. Physiol.* 58:165–70.

Richter, E.A., and Nielsen, N.B.S. 1991. Protein kinase C activity in rat skeletal muscle: Apparent relation to body weight and muscle growth. *FEBS Lett.* 289:83–5.

Riek, S., and Bawa, P. 1992. Recruitment of motor units in human forearm extensors. *J. Neurophysiol.* 68:100–8.

Riley, D.A., Bain, J.L.W., Thompson, J., Fitts, R., Widrick, J., Trappe, S., Trappe, T., and Costill, D. 2000. Decreased thin filament density and length in human atrophic soleus muscle fibers after spaceflight. *J. Appl. Physiol.* 88:567–72.

Riley, D.A., Ellis, S., Giometti, C.S., Hoh, J.F.Y., IlyinaKakueva, E.I., Oganov, V.S., Slocum, G.R., Bain, J.L.W., and Sedlak, F.R. 1992. Muscle sarcomere lesions and thrombosis in suspension unloading. *J. Appl. Physiol.* Suppl. no. 73: 33S–43S.

Riley, D.A., Ellis, S., Slocum, G.R., Sedlak, F.R., Bain, J.L.W., Krippendorf, B., Lehman, C., Macias, M., Thompson, J., Vijayan, K., and De Bruin, J. 1996. In-flight and postflight changes in skeletal muscles of SLS-1 and SLS-2 spaceflown rats. *J. Appl. Physiol.* 81:133–44.

Rivero, J.-L., Talmadge, R.J., and Edgerton, V.R. 1998. Fibre size and metabolic properties of myosin heavy chain–based fibre types in rat skeletal muscle. *J. Muscle Res. Cell Motil.* 19:733–42.

Robbins, N., and Fischbach, G.D. 1971. Effect of chronic disuse of rat soleus neuromuscular junctions on presynaptic function. *J. Neurophysiol.* 34:570–8.

Robinson, G.A., Enoka, R.M., and Stuart, D.G. 1991. Immobilization induced changes in motor unit force and fatigability in the cat. *Muscle and Nerve* 14:563–73.

Romaiguère, P., Vedel, J.P., and Pagni, S. 1993. Comparison of fluctuations of motor unit recruitment and derecruitment thresholds in man. *Exp. Brain Res.* 95:517–22.

Rome, L.C., Funke, R., McNeill Alexander, R., Lutz, G., Aldridge, H., and Freadman, M. 1988. Why animals have different muscle fibre types. *Nature* 335:824–7.

Rome, L.C., Sosnicki, A.A., and Goble, D.O. 1990. Maximum velocity of shortening of three fibre types from horse soleus muscle: Implications for scaling with body size. *J. Physiol. (Lond.)* 431:173–85.

Rome, L.C., Swank, D., and Corda, D. 1993. How fish power swimming. *Science* 261:340–3.

Rosenblatt, J.D., Yong, D., and Parry, D.J. 1994. Satellite cell activity is required for hypertrophy of overloaded adult rat muscle. *Muscle and Nerve* 17:608–13.

Rosenheimer, J., Hansen, J., and Gonzalez-Alonso, J. 2004. Circulating ATP-induced vasodilatation overrides sympathic vasoconstrictor activity in human skeletal muscle. *J. Physiol. (Lond.)* 558: 351–65.

Ross, E.Z., Middleton, N., Shave, R., George, K., and Nowicky, A. 2007. Corticomotor excitability contributes to neuromuscular fatigue following marathon running in man. *Exp. Physiol.* 92(2): 417–26.

Rothmuller, C., and Cafarelli, E. 1995. Effect of vibration on antagonist muscle coactivation during progressive fatigue in humans. *J. Physiol. (Lond.)* 485:857–64.

Rothwell, J.C., Thompson, P.D., Day, B.L., Boyd, S., and Marsden, C.D. 1991. Stimulation of the motor cortex through the scalp. *Exp. Physiol.* 76:159–200.

Rotto, D.M., and Kaufman, M.P. 1988. Effect of metabolic products of muscular contraction on discharge of group III and IV afferents. *J. Appl. Physiol.* 64:2306–13.

Round, J.M., Barr, F.M.D., Moffat, B., and Jones, D.A. 1993. Fibre areas and histochemical fibre types in the quadriceps muscle of paraplegic subjects. *J. Neurol. Sci.* 116:207–11.

Rowell, L.B. 1997. Neural control of muscle blood flow: Importance during dynamic exercise. *Clinical and Experimental Pharmacology and Physiology* 24:117–25.

Roy, R.R., and Acosta, L.J. 1986. Fiber type and fiber size changes in selected thigh muscles six months after low thoracic spinal cord transection in adult cats: Exercise effects. *Exp. Neurol.* 93:675–85.

Roy, R.R., Bello, M.A., Bouissou, P., and Edgerton, V.R. 1987. Size and metabolic properties of fibers in rat fasttwitch muscles after hindlimb suspension. *J. Appl. Physiol.* 62:2348–57.

Roy, R.R., Eldridge, L., Baldwin, K.M., and Edgerton, V.R. 1996. Neural influence on slow muscle properties: Inactivity with and without crossreinnervation. *Muscle and Nerve* 19:707–14.

Roy, R.R., Pierotti, D.J., Baldwin, K.M., Zhong, H., Hodgson, J.A., and Edgerton, V.R. 1998. Cyclical passive stretch influences the mechanical properties of the inactive cat soleus. *Exp. Physiol.* 83:377–85.

Roy, R.R., Pierotti, D.J., Flores, V., Rudolph, W., and Edgerton, V.R. 1992. Fibre size and type adaptations to spinal isolation and cyclical passive stretch in cat hindlimb. *J. Anat.* 180:491–9.

Roy, R.R., Sacks, R.D., Baldwin, K.M., Short, M., and Edgerton, V.R. 1984. Interrelationships of contraction time, Vmax, and myosin ATPase after spinal transection. *J. Appl. Physiol. Respir. Environ. Exerc. Physiol.* 56:1594–601.

Roy, R.R., Talmadge, R.J., Hodgson, J.A., Oishi, Y., Baldwin, K.M., and Edgerton, V.R. 1999. Differential response of fast hindlimb extensor and flexor muscles to exercise in adult spinalized cats. *Muscle and Nerve* 22:230–41.

Roy, R.R., Talmadge, R.J., Hodgson, J.A., Zhong, H., Baldwin, K.M., and Edgerton, V.R. 1998. Training effects on soleus of cats spinal cord transected (T1213) as adults. *Muscle and Nerve* 21:63–71.

Rube, N., and Secher, N.H. 1990. Effect of training on central factors in fatigue following two and oneleg static exercise in man. *Acta Physiol. Scand.* 141:87–95.

Ruff, R.L. 1992. Na current density at and away from end plates on rat fast and slowtwitch skeletal muscle fibers. *Am. J. Physiol. Cell Physiol.* 262: C229–34.

Ruff, R.L. 1996. Sodium channel slow inactivation and the distribution of sodium channels on skeletal muscle fibres enable the performance properties of different skeletal muscle fibre types. *Acta Physiol. Scand.* 156:159–68.

Ruff, R.L., and Whittlesey, D. 1993. Na currents near and away from endplates on human fast and slow twitch muscle fibers. *Muscle and Nerve* 16:922–9.

Russell, A.P., Hesselink, M.K.C., Lo, S.K., and Schrauwen, P. 2005. Regulation of metabolic transcriptional co-activators and transcription factors with acute exercise. *FASEB J.* 19: NIL467–86.

Russell, B., Motlagh, D., and Ashley, W. 2000. Form follows function: How muscle shape is regulated by work. *J. Appl. Physiol.* 88:1127–32.

Ryushi, T., and Fukunaga, T. 1986. Influence of subtypes of fasttwitch fibers on isokinetic strength in untrained men. *Int. J. Sports Med.* 7:250–3.

Sacco, P., Newberry, R., McFadden, L., Brown, T., and McComas, A.J. 1997. Depression of human electromyographic activity by fatigue of a synergistic muscle. *Muscle and Nerve* 20:710–7.

Sacco, P., Thickbroom, G.W., Thompson, M.L., and Mastaglia, F.L. 1997. Changes in corticomotor excitation and inhibition during prolonged submaximal muscle contractions. *Muscle and Nerve* 20:1158–66.

Sadoshima, J., and Izumo, S. 1993. Mechanical stretch rapidly activates multiple signal transduction pathways in cardiac myocytes: Potential involvement of an autocrine/paracrine mechanism. *EMBO J.* 12:1681–92.

Sale, D.G., and MacDougall, J.D. 1984. Isokinetic strength in weighttrainers. *Eur. J. Appl. Physiol.* 53:128–32.

Sale, D.G., MacDougall, J.D., Alway, S.E., and Sutton, J.R. 1987. Voluntary strength and muscle characteristics in untrained men and women and male bodybuilders. *J. Physiol.* 62:1786–93.

Sale, D.G., MacDougall, J.D., Upton, A.R.M., and McComas, A.J. 1983. Effect of strength training upon motoneuron excitability in man. *Med. Sci. Sports Exerc.* 15:57–62.

Sale, D.G., Martin, J.E., and Moroz, D.E. 1992. Hypertrophy without increased isometric strength after weight training. *Eur. J. Appl. Physiol.* 64:51–5.

Sale, D.G., McComas, A.J., MacDougall, J.D., and Upton, A.R.M. 1982. Neuromuscular adaptation in human thenar muscles following strength training and immobilization. *J. Appl. Physiol. Respir. Environ. Exerc. Physiol.* 53:419–24.

Sale, D.G., Upton, A.R.M., McComas, A.J., and MacDougall, J.D. 1983. Neuromuscular function in weighttrainers. *Exp. Neurol.* 82:521–31.

Saltin, B. 2007. Exercise hyperaemia: Magnitude and aspects on regulation in humans. *J. Physiol. (Lond.)* 583:819–23.

Saltin, B., Rådegran, G., Koskolou, M.D., and Roach, R.C. 1998. Skeletal muscle blood flow in humans and its regulation during exercise. *Acta Physiol. Scand.* 162:421–436.

Salviati, G., and Volpe, P. 1988. Ca²⁺release from sarcoplasmic reticulum of skinned fast and slowtwitch muscle fibers. *Am. J. Physiol. Cell Physiol.* 254: C459–65.

Samii, A., Wassermann, E.M., and Hallett, M. 1997. Postexercise depression of motor evoked potentials as a function of exercise duration. *Electroencephalogr. Clin. Neurophysiol.* 105:352–6.

Sandercock, T.G., Faulkner, J.A., Albers, J.W., and Abbrecht, P.H. 1985. Single motor unit and fiber action potentials during fatigue. *J. Appl. Physiol.* 58(4): 1073–9.

Sandri, M. 2008. Signaling in muscle atrophy and hypertrophy. *Physiology* 23:160–70.

Sant'ana Pereira, J.A.A., De Haan, A., Wessels, A., Moorman, A.F.M., and Sargeant, A.J. 1995. The mATPase histochemical profile of rat type IIX fibres: Correlation with myosin heavy chain immunolabelling. *Histochem. J.* 27:715–22.

Sant'ana Pereira, J.A.A., Ennion, S., Sargeant, A.J., Moorman, A.F.M., and Goldspink, G. 1997. Comparison of the molecular, antigenic and ATPase determinants of fast myosin heavy chains in rat and human: A singlefibre study. *Pflugers Arch.* 435:151–63.

Sant'ana Pereira, J.A.A., Wessels, A., Nijtmans, L., Moorman, A.F.M., and Sargeant, A.J. 1995. New method for the accurate characterization of single human skeletal muscle fibres demonstrates a relation between mATPase and MyHC expression in pure and hybrid fibre types. *J. Muscle Res. Cell Motil.* 16:21–34.

Sargeant, A.J. 2007. Structural and functional determinants of human muscle power. *Exp. Physiol.* 92(2): 323–31.

Sargeant, A.J., Davies, C., Edwards, R., Maunder, C., and Young, A. 1977. Functional and structural changes after disuse of human muscle. *Clin. Sci. Mol. Med.* 52:337–42.

Saunders, N.R., Dinenno, F.A., Pyke, K.E., Rogers, A., and Tschakovsky, M.E. 2005. Impact of combined NO and PG blockade on rapid vasodilation in a forearm mild-to-moderate exercise transition in humans. *Am. J. Physiol. (Heart Circ. Physiol.)* 288: H214–20.

Savard, G.K., Richter, E.A., Strange, S., Kiens, B., Christensen, N.J., and Saltin, B. 1989. Norepinephrine spillover from skeletal muscle during exercise in humans: Role of muscle mass. *Am. J. Physiol.* 257: H1812–8.

Sawczuk, A., Powers, R.K., and Binder, M.D. 1995. Spike frequency adaptation studied in hypoglossal motoneurons of the rat. *J. Neurophysiol.* 73:1799–810.

Saxton, J.M., and Donnelly, A.E. 1996. Lengthspecific impairment of skeletal muscle contractile function after eccentric muscle actions in man. *Clin. Sci.* 90:119–25.

Schachat, F., Diamond, M., and Brandt, P. 1987. Effect of different troponin T–tropomyosin combinations on thin filament activation. *J.* *Mol. Biol.* 198:551–4.

Schantz, P.G., Moritani, T., Karlson, E., Johansson, E., and Lundh, A. 1989. Maximal voluntary force of bilateral and unilateral leg extension. *Acta Physiol. Scand.* 136:185–92.

Scheede-Bergdahl, C., Olsen, D., Reving, D., Boushel, R., and Dela, F. 2009. Insulin and non-insulin mediated vasodilation and glucose uptake in patients with type 2 diabetes. *Diabetes Res. Clin. Practice.* 85: 243-251.

Schiaffino, S., and Reggiani, C. 1996. Molecular diversity of myofibrillar proteins: Gene regulation and functional significance. *Physiol. Rev.* 76:371–423.

Schmidt, C., Pommerenke, H., Dürr, F., Nebe, B., and Rychly, J. 1998. Mechanical stressing of integrin receptors induces enhanced tyrosine phosphorylation of cytoskeletally anchored proteins. *J. Biol. Chem.* 273:5081–5.

Schmied, A., Morin, D., Vedel, J.P., and Pagni, S. 1997. The "size principle" and synaptic effectiveness of muscle afferent projections to human extensor carpi radialis motoneurons during wrist extension. *Exp. Brain Res.* 113:214–29.

Schmitt, T.L., and Pette, D. 1991. Fiber type–specific distribution of parvalbumin in rabbit skeletal muscle: A quantitative microbiochemical and immunohistochemical study. *Histochemistry* 96:459–65.

Schrage, W.G., Joyner, M.J., and Dinenno, F.A. 2004. Local inhibition of nitric oxide and prostaglandins independently reduces forearm exercise hyperaemia in humans. *J. Physiol. (Lond.)* 557:599–611.

Schuler, M., and Pette, D. 1996. Fiber transformation and replacement in lowfrequency stimulated rabbit fasttwitch muscles. *Cell Tissue Res.* 285:297–303.

Schwaller, B., Dick, J., Dhoot, G., Carroll, S., Vrbova, G., Nicotera, P., Pette, D., Wyss, A., Bluethmann, H., Hunziker, W., and Celio, M.R. 1999. Prolonged contractionrelaxation cycle of fasttwitch muscles in parvalbumin knockout mice. *Am. J. Physiol. Cell Physiol.* 276: C395–403.

Schwindt, P.C., and Calvin, W.H. 1972. Membranepotential trajectories between spikes underlying motoneuron firing rates. *J. Neurophysiol.* 35:311–25.

Seburn, K.L., Coicou, C., and Gardiner, P.F. 1994. Effects of altered muscle activation on oxidative enzyme activity in rat alphamotoneurones. *J. Appl. Physiol.* 77:2269–74.

Seburn, K.L., and Gardiner, P.F. 1996. Properties of sprouted rat motor units: Effects of period of enlargement and activity level. *Muscle and Nerve* 19:1100–9.

Seedorf, K. 1995. Intracellular signaling by growth factors. *Metabolism* 44:24–32.

Seger, J.Y., Arvidsson, B., and Thorstensson, A. 1998. Specific effects of eccentric and concentric training on muscle strength and morphology in humans. *Eur. J. Appl. Physiol. Occup. Physiol.* 79:49–57.

Sekiguchi, H., Kimura, T., Yamanaka, K., and Nakazawa, K. 2001. Lower excitability of the corticospinal tract to transcranial magnetic stimulation during lengthening contractions in human elbow flexors. *Neurosci. Lett.* 312:83–6.

Sekiguchi, H., Nakazawa, K., and Suzuki, S. 2003. Differences in recruitment properties of the corticospinal pathway between lengthening and shortening contractions in human soleus muscle. *Brain Research* 977:169–79.

Semmler, J.G. 2002. Motor unit synchronization and neurouscular performance. *Exerc. Sports Sc. Rev.* 30:8–14.

Semmler, J.G., Kornatz, K.W., Dinenno, D.V., Zhou, S., and Enoka, R.M. 2002. Motor unit synchronisation is enhanced during slow lengthening contractions of a hand muscle. *Journal of Physiology* 545:681–95.

Semmler, J.G., Kutzscher, D., and Enoka, R. 1999. Gender differences in the fatigability of human skeletal muscle. *J. Neurophysiol.* 82:3590–3.

Semmler, J.G., and Nordstrom, M.A. 1998. Motor unit discharge and force tremor in skill and strengthtrained individuals. *Exp. Brain Res.* 119:27–38.

Semmler, J.G., Sale, M.V., Meyer, F.G., and Nordstrom, M.A. 2004. Motor-unit coherence and its relation with synchrony are influenced by training. *Journal of Neurophysiology* 92:3320–31.

Seynnes, O.R., deBoer, M., and Narici, M.V. 2007. Early skeletal muscle hypertrophy and architectural changes in response to high-intensity resistance training. *J. Appl. Physiol.* 102(1): 368–73.

Shields, R.K. 1995. Fatigability, relaxation properties, and electromyographic responses of the human paralyzed soleus muscle. *J. Neurophysiol.* 73:2195–206.

Siegel, G., Agranoff, B., Albers, R., and Molinoff, P. 1989. *Basic neurochemistry.* New York: Raven Press.

Simard, C.P., Spector, S.A., and Edgerton, V.R. 1982. Contractile properties of rat hind limb muscles immobilized at different lengths. *Exp. Neurol.* 77:467–82.

Simoneau, J.A., and Bouchard, C. 1989. Human variation in skeletal muscle fibertype proportion and enzyme activities. *Am. J. Physiol. Endocrinol. Metab.* 257 (20): E567–72.

Simoneau, J.A., and Bouchard, C. 1995. Genetic determinism of fiber type proportion in human skeletal muscle. *FASEB J.* 9:1091–5.

Sinoway, L.I., Hill, J.M., Pickar, J.G., and Kaufman, M.P. 1993. Effects of contraction and lactic acid on the discharge of group III muscle afferents in cats. *J. Neurophysiol.* 69:1053–9.

Sjodin, B., Jacobs, I., and Karlsson, J. 1981. Onset of blood lactate accumulation and enzyme activities in M. vastus lateralis muscle. *Int. J. Sports Med.* 2:166–70.

Sketelj, J., CrneFinderle, N., Strukelj, B., Trontelj, J.V., and Pette, D. 1998. Acetylcholinesterase mRNA level and synaptic activity in rat muscles depend on nerveinduced pattern of muscle activation. *J. Neurosci.* 18:1944–52.

Skorjanc, D., Traub, I., and Pette, D. 1998. Identical responses of fast muscle to sustained activity by lowfrequency stimulation in young and aging rats. *J. Appl. Physiol.* 85:437–41.

Smerdu, V., KarschMizrachi, I., Campione, M., Leinwand, L., and Schiaffino, S. 1994. Type IIx myosin heavy chain transcripts are expressed in type IIb fibers of human skeletal muscle. *Am. J. Physiol. Cell Physiol.* 267: C1723–8.

Smith, J.L., Martin, P.G., Gandevia, S.C., and Taylor, J.L. 2007. Sustained contraction at very low forces produces prominent supraspinal fatigue in human elbow flexor muscles. *J. Appl. Physiol.* 103(2): 560–8.

Smith, J.L., Smith, L.A., Zernicke, R.F., and Hoy, M. 1982. Locomotion in exercised and nonexercised cats cordotomized at two or twelve weeks of age. *Exp. Neurol.* 76:393–413.

Smith, L.A., Eldred, E., and Edgerton, V.R. 1993. Effects of age at cordotomy and subsequent exercise on contraction times of motor units in the cat. *J. Appl. Physiol.* 75:2683–8.

Sogaard, K., Christensen, H., Fallentin, N., Mizuno, M., Quistorff, B., and Sjogaard, G. 1998. Motor unit activation patterns during concentric wrist flexion in humans with different muscle fibre composition. *Eur. J. Appl. Physiol.* 78:411–6.

Sogaard, K., Gandevia, S.C., Todd, G., Petersen, N.T., and Taylor, J.L. 2006. The effect of sustained low-intensity contractions on supraspinal fatigue in human elbow flexor muscles. *Journal of Physiology* 573:511–23.

Sokoloff, A.J., and Cope, T.C. 1996. Recruitment of triceps surae motor units in the decerebrate cat: 2. Heterogeneity among soleus motor units. *J. Neurophysiol.* 75:2005–16.

Spangenburg, E.E., and Booth, F.W. 2003. Molecular regulation of individual skeletal muscle fibre types. *Acta Physiol. Scand.* 178:413–24.

Spector, S.A. 1985. Effects of elimination of activity on contractile and histochemical properties of rat soleus muscle. *J. Neurosci.* 5:2177–88.

Spector, S.A., Simard, C.P., Fournier, M., Sternlicht, E., and Edgerton, V.R. 1982. Architectural alterations of rat hindlimb skeletal muscles immobilized at different lengths. *Exp. Neurol.* 76:94–110.

Spencer, M.J., and Tidball, J.G. 1997. Calpain II expression is increased by changes in mechanical loading of muscle *in vivo. J. Cell Biochem.* 64:55–66.

Spielmann, J.M., Laouris, Y., Nordstrom, M.A., Robinson, G.A., Reinking, R.M., and Stuart, D.G. 1993. Adaptation of cat motoneurons to sustained and intermittent extracellular activation. *J. Physiol. (Lond.)* 464:75–120.

Sreter, F., Lopez, J.R., Alamo, L., Mabuchi, K., and Gergely, J. 1987. Changes in intracellular ionized Ca concentration associated with muscle fiber type transformation. *Am. J. Physiol.* 253: C296–300.

Staron, R.S. 1991. Correlation between myofibrillar ATPase activity and myosin heavy chain composition in single human muscle fibers. *Histochemistry* 96:21–4.

Staron, R.S., and Pette, D. 1987. Nonuniform myosin expression along single fibers of chronically stimulated and contralateral rabbit tibialis anterior muscles. *Pflugers Arch.* 409:67–73.

Stein, R.B., Gordon, T., Jefferson, J., Sharfenberger, A., Yang, J.F., De Zepetnek, J.T., and Bélanger, M. 1992. Optimal stimulation of paralyzed muscle after human spinal cord injury. *J. Appl. Physiol.* 72:1393–400.

Steinbach, J.H., Schubert, D., and Eldridge, L. 1980. Changes in cat muscle contractile proteins after prolonged muscle inactivity. *Exp. Neurol.* 67:655–69.

Stephens, J.A., and Taylor, A. 1972. Fatigue of maintained voluntary muscle contraction in man. *J. Physiol.* 220:1–18.

Stephens, J.A., and Usherwood, T.P. 1977. The mechanical properties of human motor units with special reference to their fatiguability and recruitment threshold. *Brain Res.* 125:91–7.

Stephenson, G. M. 2001. Hybrid skeletal muscle fibres: A rare or common phenomenon? *Clinical and Experimental Pharmacology and Physiology* 28:692–702.

Sterz, R., Pagala, M., and Peper, K. 1983. Postjunctional characteristics of the endplates in mammalian fast and slow muscles. *Pflugers Arch.* 398:48–54.

Stevens, L., Gohlsch, B., Mounier, Y., and Pette, D. 1999. Changes in myosin heavy chain mRNA and protein isoforms in single fibers of unloaded rat soleus muscle. *FEBS Lett.* 463:15–8.

Stevens, L., Sultan, K., Peuker, H., Gohlsch, B., Mounier, Y., and Pette, D. 1999. Time-dependent changes in myosin heavy chain mRNA and protein isoforms in unloaded soleus muscle of rat. *Am. J. Physiol.* 277: C1044–9.

Stienen, G.J.M., Kiers, J.L., Bottinelli, R., and Reggiani, C. 1996. Myofibrillar ATPase activity in skinned human skeletal muscle fibres: Fibre type and temperature dependence. *J. Physiol.* 493:299–307.

Stotz, P.J., and Bawa, P. 2001. Motor unit recruitment during lengthening contractions of human wrist flexors. *Muscle and Nerve* 24:1535–41.

St.Pierre, D., and Gardiner, P.F. 1985. Effects of disuse on mammalian fasttwitch muscle: Joint fixation compared to neurallyapplied tetrodotoxin. *Exp. Neurol.* 90:635–51.

St.Pierre, D., Léonard, D., Houle, R., and Gardiner, P.F. 1988. Recovery of muscle from TTXinduced disuse and the influence of daily exercise: II. Muscle enzymes and fatigue characteristics. *Exp. Neurol.* 101:327–46.

Strojnik, V. 1995. Muscle activation level during maximal voluntary effort. *Eur. J. Appl. Physiol.* 72:144–9.

Stuart, D.G., Hamm, T.M., and Vanden Noven, S. 1988. Partitioning of monosynaptic Ia EPSP connections with motoneurons according to neuromuscular topography: Generality and functional implications. *Prog. Neurobiol.* 30:437–47.

Sugden, P.H., and Clerk, A. 1998. Cellular mechanisms of cardiac hypertrophy. *J. Mol. Med.* 76:725–46.

Sugiura, T., Matoba, H., and Murakamis, N. 1992. Myosin light chain patterns in histochemically typed single fibers of the rat skeletal muscle. *Comp. Biochem. Physiol. (B)* 102B: 617–20.

Sutherland, H., Jarvis, J.C., Kwende, M.M.N., Gilroy, S.J., and Salmons, S. 1998. The doserelated response of rabbit fast muscle to longterm lowfrequency stimulation. *Muscle and Nerve* 21:1632–46.

Suzuki, H., Tsuzimoto, H., Ishiko, T., Kasuga, N., Taguchi, S., and Ishihara, A. 1991. Effect of endurance training on the oxidative enzyme activity of soleus motoneurons in rats. *Acta Physiol. Scand.* 143:127–8.

Sweeney, H.L., Bowman, B.F., and Stull, J.T. 1993. Myosin light chain phosphorylation in vertebrate striated muscle: Regulation and function. *Am. J. Physiol. Cell Physiol.* 264 (33): C1085–95.

Sweeney, H.L., Kushmerick, M.J., Mabuchi, K., Sreter, F.A., and Gergely, J. 1988. Myosin alkali light chain and heavy chain variations correlate with altered shortening velocity of isolated skeletal muscle fibers. *J. Biol. Chem.* 263:9034–9.

Szentesi, P., Zaremba, R., Van Mechelen, W., and Stienen, G.J.M. 2001. ATP utilization for calcium uptake and force production in different types of human skeletal muscle fibres. *Journal of Physiology* 531:393–403.

Tabary, J.C., Tabary, C., Tardieu, C., Tardieu, G., and Goldspink, G. 1972. Physiological and structural changes in the cat's soleus muscle due to immobilization at different lengths by plaster casts. *J. Physiol.* 224:231–44.

Takahashi, M., Chesley, A., Freyssenet, D., and Hood, D.A. 1998. Contractile activityinduced adaptations in the mitochondrial protein import system. *Am. J. Physiol. Cell Physiol.* 274:C1380–7.

Takekura, H., and Yoshioka, T. 1987. Determination of metabolic profiles on single muscle fibres of different types. *J. Muscle Res. Cell Motil.* 8:342–8.

Takekura, H., and Yoshioka, T. 1989. Ultrastructural and metabolic profiles on single muscle fibers of different types after hindlimb suspension in rats. *Jap. J. Physiol.* 39:385–96.

Takekura, H., and Yoshioka, T. 1990. Ultrastructural and metabolic characteristics of single muscle fibres belonging to the same type in various muscles in rats. *J. Muscle Res. Cell Motil.* 11:98–104.

Talmadge, R.J., Roy, R.R., and Edgerton, V.R. 1995. Prominence of myosin heavy chain hybrid fibers in soleus muscle of spinal cord–transected rats. *J. Appl. Physiol.* 78:1256–65.

Talmadge, R.J., Roy, R.R., and Edgerton, V.R. 1996a. Distribution of myosin heavy chain isoforms in non-weight-bearing rat soleus muscle fibers. *J. Appl. Physiol.* 81:2540–6.

Talmadge, R.J., Roy, R.R., and Edgerton, V.R. 1996b. Myosin heavy chain profile of cat soleus following chronic reduced activity or inactivity. *Muscle and Nerve* 19:980–8.

Talmadge, R.J., Roy, R.R., and Edgerton, V.R. 1999. Persistence of hybrid fibers in rat soleus after spinal cord transection. *Anat. Rec.* 255:188–201.

Tamaki, H., Kitada, K., Akamine, T., Murata, F., Sakou, T., and Kurata, H. 1998. Alternating activity in the synergistic muscles during prolonged low-level contractions. *J. Appl. Physiol.* 84:1943–51.

Tax, A.A.M., Van der Gon, J.J.D., Gielen, C.C.A.M., and Kleyne, M. 1990. Differences in central control of m. biceps brachii in movement tasks and force tasks. *Exp. Brain Res.* 79:138–42.

Tax, A.A.M., Van der Gon, J.J.D., Gielen, C.C.A.M., and van den Tempel, C.M.M. 1989. Differences in the activation of m. biceps brachii in the control of slow isotonic movements and isometric contractions. *Exp. Brain Res.* 76:55–63.

Taylor, J.L., Butler, J.E., Allen, G.M., and Gandevia, S.C. 1996. Changes in motor cortical excitability during human muscle fatigue. *J. Physiol. (Lond.)* 490:519–28.

Termin, A., and Pette, D. 1991. Myosin heavychain-based isomyosins in developing, adult fasttwitch and slowtwitch muscles. *Eur. J. Biochem.* 195:577–84.

Termin, A., and Pette, D. 1992. Changes in myosin heavychain isoform synthesis of chronically stimulated rat fasttwitch muscle. *Eur. J. Biochem.* 204:569–73.

Terzis, G, Georgiadis G, Vassiliadou E, and Manta P. 2003. Relationship between shot put performance and triceps brachii fiber type composition and power production. *Eur J Appl Physiol* 90, 10, 10–15.

Tesch, P.A. 1980. Fatigue pattern in subtypes of human skeletal muscle fibers. *Int. J. Sports Med.* 1:79–81.

Tesch, P.A., and Karlsson, J. 1978. Isometric strength performance and muscle fiber type distribution in man. *Acta Physiol. Scand.* 103:47–51.

Tesch, P.A., Thorsson, A., and Colliander, E.B. 1990. Effects of eccentric and concentric resistance training on skeletal muscle substrates, enzyme activities and capillary supply. *Acta Physiol. Scand.* 140:575–80.

Tesch, P.A., Thorsson, A., and EssenGustavsson, B. 1989. Enzyme activities of FT and ST muscle fibers in heavyresistance trained athletes. *J. Appl. Physiol.* 67:83–7.

Thesleff, S. 1959. Motor endplate desensitization by repetitive nerve stimuli. *J. Physiol. (Lond.)* 148:659–64.

Thickbroom, G., Phillips, B., Morris, I., Byrnes, M., Sacco, P., and Mastaglia, F. 1999. Differences in functional magnetic resonance imaging of sensorimotor cortex during static and dynamic finger flexion. *Exp. Brain Res.* 126:431–8.

Thomas, C.K., BiglandRitchie, B., Westling, G., and Johansson, R.S. 1990. A comparison of human thenar motorunit properties studied by intraneural motoraxon stimulation and spiketriggered averaging. *J. Neurophysiol.* 64:1347–51.

Thomas, C.K., Johansson, R.S., and Bigland-Ritchie, B. 1991. Attempts to physiologically classify human thenar motor units. *Journal of Neurophysiology* 65:1501–8.

Thomas, C.K., Ross, B.H., and Calancie, B. 1987. Human motorunit recruitment during isometric contractions and repeated dynamic movements. *J. Neurophysiol.* 57:311–24.

Thomas, C.K., Woods, J.J., and BiglandRitchie, B. 1989. Impulse propagation and muscle activation in long maximal voluntary contractions. *J. Appl. Physiol.* 67:1835–42.

Thomason, D.B. 1998. Translational control of gene expression in muscle. *Exerc. Sport Sci. Rev.* 26:165–90.

Thomason, D.B., Biggs, R.B., and Booth, F.W. 1989. Protein metabolism and Bmyosin heavychain mRNA in unweighted soleus muscle. *Am. J. Physiol. Regul. Integr. Comp. Physiol.* 257: R300–5.

Thomason, D.B., Herrick, R.E., and Baldwin, K.M. 1987. Activity influences on soleus muscle myosin during rodent hindlimb suspension. *J. Appl. Physiol.* 63:138–44.

Thomason, D.B., Morrison, P.R., Oganov, V., IlyinaKakueva, E., Booth, F.W., and Baldwin, K.M. 1992. Altered actin and myosin expression in muscle during exposure to microgravity. *J. Appl. Physiol.* Suppl. no. 73:90S–3S.

Thompson, H.S., and Scordilis, S.P. 1994. Ubiquitin changes in human biceps muscle following exerciseinduced damage. *Biochem. Biophys. Res. Commun.* 204:1193–8.

Thompson, M.G., and Palmer, R.M. 1998. Signalling pathways regulating protein turnover in skeletal muscle. *Cell. Signal.* 10:1–11.

Thorstensson, A., Grimby, G., and Karlsson, J. 1976. Force–velocity relations and fiber composition in human knee extensor muscles. *J. Appl. Physiol.* 40:12–6.

Thorstensson, A., Hulten, B., Von Dolbeln, W., and Karlsson, J. 1976. Effect of strength training on enzyme activities and fibre characteristics in human skeletal muscle. *Acta Physiol. Scand.* 96:392–8.

Thorstensson, A., and Karlsson, J. 1976. Fatiguability and fibre composition of human skeletal muscle. *Acta Physiol. Scand.* 98:318–22.

Thorstensson, A., Larsson, L., and Karlsson, J. 1977. Muscle strength and fiber composition in athletes and sedentary men. *Med. Sci. Sports Exerc.* 9:26–30.

Thyfault, J.P. 2008. Setting the stage: Possible mechanisms by which acute contraction restores insulin sensitivity in muscle. *Amer. J. Physiol. Regul. Integr. C* 294(4): R1103–10.

Tidball, J.G. 1995. Inflammatory cell response to acute muscle injury. *Med. Sci. Sports Exerc.* 27:1022–32.

Tihanyi, J., Apor, P., and Fekete, G. 1982. Forcevelocity power characteristics and fiber composition in human knee extensor muscles. *Eur. J. Appl. Physiol.* 48:331–43.

Tillakaratne, N.J.K., De Leon, R.D., Hoang, T.X., Roy, R.R., Edgerton, V.R., and Tobin, A.J. 2002. Use-dependent modulation of inhibitory capacity in the feline lumbar spinal cord. *J. Neurosci.* 22:3130–43.

Todd, G., Taylor, J.L., and Gandevia, S.C. 2004. Reproducible measurement of voluntary activation of human elbow flexors with motor cortical stimulation. *J. Appl. Physiol.* 97:236–42.

Toursel, T., Stevens, L., and Mounier, Y. 1999. Evolution of contractile and elastic properties of rat soleus muscle fibres under unloading conditions. *Exp. Physiol.* 84:93–107.

Troiani, D., Filippi, G.M., and Bassi, F. 1999. Nonlinear tension summation of different combinations of motor units in the anesthetized cat peroneus longus muscle. *J. Neurophysiol.* 81:771–80.

Tschakovsky, M.E., and Joyner, M.J. 2008. Nitric oxide and muscle blood flow in exercise. *Appl. Physiol. Nutr. Metab.* 33:151–61.

Tschakovsky, M.E., Rogers, A., Pyke, K.E., Saunders, N.R., Glenn, N., Lee, S., Weissgerber, T., and Dwyer, E.M. 2004. Immediate exercise hyperemia in humans is contraction intensity dependent: Evidence for rapid vasodilation. *J. Appl. Physiol.* 96:639–44.

Tschakovsky, M.E., Shoemaker, J.K., and Hughson, R.L. 1996. Vasodilation and muscle pump contribution to immediate exercise hyperemia. *Am. J. Physiol. (Heart Circ. Physiol.)* 271: H1701.

Tsuboi, T., Sato, T., Egawa, K., and Miyazaki, M. 1995. The effect of fatigue caused by electrical induction or voluntary contraction on Ia inhibition in human soleus muscle. *Neurosci. Lett.* 197:72–4.

Turcotte, L., and Fisher, J.S. 2008. Skeletal muscle insulin resistance: Roles of fatty acid metabolism and exercise. *Phys. Ther.* 88:1279–96.

Turcotte, R., Panenic, R., and Gardiner, P.F. 1991. TTX induced muscle disuse alters Ca^{2+} activation characteristics of myofibril ATPase. *Comp. Biochem. Physiol. (A)* 100A: 183–6.

Van Bolhuis, B.M., and Gielen, C.C.A.M. 1997. The relative activation of elbow flexor muscles in isometric flexion and in flexion/extension movements. *J. Biomech.* 30:803–11.

Van Bolhuis, B.M., Medendorp, W.P., and Gielen, C.C.A.M. 1997. Motor unit firing behavior in human arm flexor muscles during sinusoidal isometric contractions and movements. *Exp. Brain Res.* 117:120–30.

Van Cutsem, M., Duchateau, J., and Hainaut, K. 1998. Changes in single motor unit behaviour contribute to the increase in contraction speed after dynamic training in humans. *J. Physiol.* 513:295–305.

Vandenborne, K., Elliott, M.A., Walter, G.A., Abdus, S., Okereke, E., Shaffer, M., Tahernia, D., and Esterhai, J.L. 1998. Longitudinal study of skeletal muscle adaptations during immobilization and rehabilitation. *Muscle and Nerve* 21:1006–12.

Vandenburgh, H.H., Hatfaludy, S., Sohar, I., and Shansky, J. 1990. Stretch induced prostaglandins and protein turnover in cultured skeletal muscle. *Am. J. Physiol. Cell Physiol.* 259: C232–40.

Vandenburgh, H.H., Karlisch, P., Shansky, J., and Feldstein, R. 1991. Insulin and IGF I induce pronounced hypertrophy of skeletal myofibers in tissue culture. *Am. J. Physiol. Cell Physiol.* 260: C475–84.

Vandenburgh, H.H., Shansky, J., Karlisch, P., and Solerssi, R.L. 1993. Mechanical stimulation of skeletal muscle generates lipid related second messengers by phospholipase activation. *J. Cell. Physiol.* 155:63–71.

Vandenburgh, H.H., Shansky, J., Solerssi, R.L., and Chromiak, J. 1995. Mechanical stimulation of skeletal muscle increases prostaglandin F_2 production, cyclooxygenase activity, and cell growth by a pertussis toxin sensitive mechanism. *J. Cell. Physiol.* 163:285–94.

Vandenburgh, H.H., Swasdison, S., and Karlisch, P. 1991. Computer aided mechanogenesis of skeletal muscle organs from single cells *in vitro*. *FASEB J.* 5:2860–7.

Van Dieen, J., Ogita, F., and De Haan, A. 2003. Reduced neural drive in bilateral exertions: A performance-limiting factor? *Med. Sci. Sports Exerc.* 35:111–8.

Vandervoort, A., Sale, D., and Moroz, J. 1984. Comparison of motor unit activation during unilateral and bilateral leg extension. *J. Appl. Physiol.* 56:46–51.

Van Lunteren, E., and Moyer, M. 1996. Effects of DAP on diaphragm force and fatigue, including fatigue due to neurotransmission failure. *J. Appl. Physiol.* 81:2214–20.

Van Praag, H., Christie, B., Sejnowski, T., and Gage, F. 1999. Running enhances neurogenesis, learning, and long-term potentiation in mice. *Proc. Natl. Acad. Sci. USA* 96:13427–31.

Vary, T., Jefferson, L., and Kimball, S. 2000. Role of eIF4E in stimulation of protein synthesis by IGF-I in perfused rat skeletal muscle. *Am. J. Physiol.* 278:E58–64.

Vaynman, S., and Gómez-Pinilla, F. 2005. License to run: Exercise impacts functional plasticity in the intact and injured central nervous system by using neurotrophins. *Neurorehabilitation and Neural Repair* 19:283–95.

Viitasalo, J.H.T., Häkkinen, K., and Komi, P.V. 1981. Isometric and dynamic force production and muscle fiber type composition in man. *J. Hum. Mov. Stud.* 7:199–209.

Viitasalo, J.H.T., and Komi, P.V. 1978. Forcetime characteristics and fiber composition in human leg extensor muscles. *Eur. J. Appl. Physiol.* 40:7–15.

Viitasalo, J.H.T., and Komi, P.V. 1981. Interrelationships between electromyographic, mechanical, muscle structure and reflex time measurements in man. *Acta Physiol. Scand.* 111:97–103.

Vijayan, K., Thompson, J., and Riley, D. 1998. Sarcomere lesion damage occurs mainly in slow fibers of reloaded rat adductor longus muscles. *J. Appl. Physiol.* 85:1017–23.

Vissing, K., Brink, M., Lonbro, S., Sorensen, H., Overgaard, K., Danborg, K., Mortensen, J., Elstrom, O., Rosenhoj, N., Ringaard, S., Andersen, J., and Aagaard, P. 2008. Muscle adaptations to plyometric vs. resistance training in untrained young men. *J. Strength Conditioning Res.* 22 (6): 1799-1810.

Vogt, M., and Hoppeler, H. 2010. Is hypoxia training good for muscles and exercise performance?, *Prog. Cardiovasc. Dis.* 52 (6); 525-533.

Volek, J.S. 2004. Influence of nutrition on responses to resistance training. *Med. Sci. Sports Exerc.* 36:689–96.

Vollestad, N.K. 1997. Measurement of human muscle fatigue. *J. Neurosci. Methods* 74:219–27.

Vollestad, N.K., and Blom, P.C.S. 1985. Effect of varying exercise intensity on glycogen depletion in human muscle fibres. *Acta Physiol. Scand.* 125:395–405.

Vollestad, N.K., Tabata, I., and Medbo, J.I. 1992. Glycogen breakdown in different human muscle fibre types during exhaustive exercise of short duration. *Acta Physiol. Scand.* 144:135–41.

Vollestad, N., Vaage, O., and Hermansen, L. 1984. Muscle glycogen depletion patterns in type I and subgroups of type II fibres during prolonged severe exercise in man. *Acta Physiol. Scand.* 122:433–41.

Wada, M., Hämäläinen, N., and Pette, D. 1995. Isomyosin patterns of single type IIB, IID and IIA fibres from rabbit skeletal muscle. *J. Muscle Res. Cell Motil.* 16:237–42.

Wada, M., Okumoto, T., Toro, K., Masuda, K.,

Fukubayashi, T., Kikuchi, K., Niihata, S., and Katsuta, S. 1996. Expression of hybrid isomyosins in human skeletal muscle. *Am. J. Physiol. Cell Physiol.* 271:C1250–5.

Wada, M., and Pette, D. 1993. Relationships between alkali lightchain complement and myosin heavychain isoforms in single fasttwitch fibers of rat and rabbit. *Eur. J. Biochem.* 214:157–61.

Waerhaug, O., and Lomo, T. 1994. Factors causing different properties at neuromuscular junctions in fast and slow rat skeletal muscles. *Anat. Embryol. (Berl.)* 190:113–25.

Wakeling, J.M. 2004. Motor units are recruited in a task-dependent fashion during locomotion. *Journal of Experimental Biology* 207:3883–90.

Warren III, G.L., Lowe, D.A., Hayes, D.A., Farmer, M.A., and Armstrong, R.B. 1995. Redistribution of cell membrane probes following contraction-induced injury of mouse soleus muscle. *Cell Tissue Res.* 282:311–20.

Watson, P.A. 1991. Function follows form: Generation of intracellular signals by cell deformation. *FASEB J.* 5:2013–9.

Watson, P.A., Stein, J., and Booth, F. 1984. Changes in actin synthesis and alphaactinmRNA content in rat muscle during immobilization. *Am. J. Physiol. Cell Physiol.* 247:C39–44.

Watt, M.J., Steinberg, G., Chen, Z., Kemp, B.E., and Febbraio, M. 2006. Fatty acids stimulate AMP-activated protein kinase and enhance fatty acid oxidation in L6 myotubes. *J. Physiol. (Lond.)* 574:139–47.

Wehring, M., Cal, B., and Tidball, J. 2000. Modulation of myostatin expression during modified muscle use. *FASEB J.* 14:103–10.

Weir, J.P., Keefe, D.A., Eaton, J.F., Augustine, R.T., and Tobin, D.M. 1998. Effect of fatigue on hamstring coactivation during isokinetic knee extensions. *Eur. J. Appl. Physiol.* 78:555–9.

Westerblad, H., Allen, D.G., Bruton, J.D., Andrade, F.H., and Lannergren, J. 1998. Mechanisms underlying the reduction of isometric force in skeletal muscle fatigue. *Acta Physiol. Scand.* 162:253–60.

Westgaard, R.H., and De Luca, C.J. 1999. Motor unit substitution in long-duration contractions of the human trapezius muscle. *J. Neurophysiol.* 82:501–4.

Westing, S.H., Cresswell, A.G., and Thorstensson, A. 1991. Muscle activation during maximal voluntary eccentric and concentric knee extension. *Eur. J. Appl. Physiol.* 62:104–8.

Westing, S.H., Seger, J.Y., and Thorstensson, A. 1990. Effects of electrical stimulation on eccentric and concentric torque–velocity relationships during knee extension in man. *Acta Physiol. Scand.* 140:17–22.

Wheeler, M., Snyder, E., Patterson, M., and Swoap, S. 1999. An E-box within the MHC IIB gene is bound by MyoD and is required for gene expression in fast muscle. *Am. J. Physiol.* 276:C1069–78.

White, M., Davies, C., and Brooksby, P. 1984. The effects of shortterm voluntary immobilization on the contractile properties of the human triceps surae. *Q. J. Exp. Physiol.* 69:685–91.

Whitehead, N.P., Allen, T.J., Morgan, D.L., and Proske, U. 1998. Damage to human muscle from eccentric exercise after training with concentric exercise. *J. Physiol.* 512:615–20.

Whitelaw, P.F., and Hesketh, J.E. 1992. Expression of *cmyc* and *cfos* in rat skeletal muscle: Evidence for increased levels of *cmyc* mRNA during hypertrophy. *Biochem. J.* 281:143–7.

Wickham, J.B., and Brown, J.M.M. 1998. Muscles within muscles: The neuromotor control of intramuscular segments. *Eur. J. Appl. Physiol. Occup. Physiol.* 78:219–25.

Widrick, J.J., and Fitts, R.H. 1997. Peak force and maximal shortening velocity of soleus fibers after nonweightbearing and resistance exercise. *J. Appl. Physiol.* 82:189–95.

Widrick, J.J., Romatowski, J.G., Bain, J., Trappe, S.W., Trappe, T., Thompson, J., Costill, D.L., Riley, D., and Fitts, R.H. 1997. Effect of 17 days of bed rest on peak isometric force and unloaded shortening velocity of human soleus fibers. *Am. J. Physiol.* 273:C1690–9.

Widrick, J.J., Romatowski, J.G., Karhanek, M., and Fitts, R.H. 1997. Contractile properties of rat, rhesus monkey, and human type I muscle fibers. *J. Appl. Physiol.* 41:R34–47.

Widrick, J.J., Trappe, S.W., Blaser, C.A., Costill, D.L., and Fitts, R.H. 1996. Isometric force and maximal shortening velocity of single muscle fibers from elite master runners. *Am. J. Physiol. Cell Physiol.* 271:C666–75.

Widrick, J.J., Trappe, S.W., Costill, D.L., and Fitts, R.H. 1996. Forcevelocity and forcepower properties of single muscle fibers from elite master runners and sedentary men. *Am. J. Physiol. Cell Physiol.* 271:C676–83.

Williams, P.E., and Goldspink, G. 1984. Connective tissue changes in immobilized muscle. *J. Anat.* 138:343–50.

Williams, P.E., Watt, P., Bicik, V., and Goldspink, G. 1986. Effect of stretch combined with electrical stimulation on the type of sarcomeres produced at the ends of muscle fibers. *Exp. Neurol.* 93:500–9.

Williams, R.S., and Neufer, P.D. 1996. Regulation of gene expression in skeletal muscle by contractile activity. In *Handbook of physiology: 12. Exercise, regulation and integration of multiple systems,* ed. L.B. Rowell and J.T. Shepherd, 1124–50. New York: Oxford University Press.

Williamson, D.L., Gallagher, P.M., Carroll, C.C., Raue, U., and Trappe, S.W. 2001. Reduction in hybrid single muscle fiber proportions with resistance training in humans. *J. Appl. Physiol.* 91:1955–61.

Wiltshire, E., Piotras, V., Pak, M., Hong, T., Rayner, J., and Tschakovsky, M. 2010. Massage impairs postexercise muscle blood flow and "lactic acid" removal. *Med. Sci. Sports Ex.* 42 (6); 1062-1071.

Windhorst, U., Kirmayer, D., Soibelman, F., Misri, A., and Rose, R. 1997. Effects of neurochemically excited group IIIIV muscle afferents on motoneuron afterhyperpolarization. *Neuroscience* 76:915–29.

Windhorst, U., and Kokkoroyiannis, T. 1991. Interaction of recurrent inhibitory and muscle spindle afferent feedback during muscle fatigue. *Neuroscience* 43:249–59.

Windisch, A., Gundersen, K., Szabolcs, M.J., Gruber, H., and Lomo, T. 1998. Fast to slow transformation of denervated and electrically stimulated rat muscle. *J. Physiol.* 510:623–32.

Winiarski, A., Roy, R., Alford, E., Chiang, P., and Edgerton, V. 1987. Mechanical properties of rat skeletal muscle after hind limb suspension. *Exp. Neurol.* 96:650–60.

Witzmann, F.A., Kim, D., and Fitts, R.H. 1982. Hindlimb immobilization: Length–tension and contractile properties of skeletal muscle. *J. Appl. Physiol.* 53:335–45.

Witzmann, F.A., Kim, D., and Fitts, R.H. 1983. Effect of hindlimb immobilization on the fatigability of skeletal muscle. *J. Appl. Physiol. Respir. Environ. Exerc. Physiol.* 54:1242–8.

Witzmann, F.A., Troup, J.P., and Fitts, R.H. 1982. Acid phosphatase and protease activities in immobilized rat skeletal muscles. *Can. J. Physiol. Pharmacol.* 60:1732–6.

Wohlfart, B., and Edman, K.A.P. 1994. Rectangular hyperbola fitted to muscle forcevelocity data using threedimensional regression analysis. *Exp. Physiol.* 79:235–9.

Wojtaszewski, J.F.P., and Richter, E.A. 2006. Effects of acute exercise and training on insulin action and sensitivity: Focus on molecular mechanisms in muscle. *Essays in Biochemistry* 42:31–46.

Wolpaw, J.R., and Carp, J.S. 1990. Memory traces in spinal cord. *Trends Neurosci.* 13:137–42.

Wolpaw, J., and Carp, J. 1993. Adaptive plasticity in spinal cord. *Advances in Neurology* 59:163–74.

Wolpaw, J.R., and Lee, C.L. 1989. Memory traces in primate spinal cord produced by operant conditioning of Hreflex. *J. Neurophysiol.* 61:563–72.

Wong, T.S., and Booth, F.W. 1990a. Protein metabolism in rat gastrocnemius muscle after stimulated chronic concentric exercise. *J. Appl. Physiol.* 69:1709–17.

Wong, T.S., and Booth, F.W. 1990b. Protein metabolism in rat tibialis anterior muscle after stimulated chronic eccentric exercise. *J. Appl. Physiol.* 69:1718–24.

Wood, S.J., and Slater, C.R. 1997. The contribution of postsynaptic folds to the safety factor for neuromuscular transmission in rat fast and slowtwitch muscles. *J. Physiol.* 500:165–76.

Woods, J., Furbush, F., and BiglandRitchie, B. 1987. Evidence for a fatigueinduced reflex inhibition of motoneuron firing rates. *J. Neurophysiol.* 58:125–37.

Woolstenhulme, M.T., Conlee, R.K., Drummond, M.J., Stites, A., and Parcell, A.C. 2006. Temporal response to desmin and dystrophin proteins to progressive resistance exercise in human skeletal muscle. *J. Appl. Physiol.* 100:1876–82.

Woolstenhulme, M.T., Jutte, L.S., Drummond, M.J., and Parcell, A.C. 2005. Desmin increases with high-intensity concentric contractions in humans. *Muscle and Nerve* 31:20–4.

Wray, D.W., Nishiyama, S., Donato, A.J., Sander, M., Wagner, P.D., and Richardson, R.S. 2008. Endothelin-1-mediated vasoconstriction at rest and during dynamic exercise in healthy humans. *Am. J. Physiol. (Heart Circ. Physiol.)* 293: H2550–6.

Wright, D.C. 2007. Mechanisms of calcium-induced mitochondrial biogenesis and GLUT4 synthesis. *Appl. Physiol. Nutr. Metab.* 32(5): 840–5.

Wu, H., Naya, F.J., McKinsey, T.A., Mercer, B., Shelton, J.M., Chin, E.R., Simard, A.R., Michel, R., Bassel-Duby, R., Olson, E.N., and Williams, R.S. 2000. MEF2 responds to multiple calcium-regulated signals in the control of skeletal muscle fiber type. *EMBO Journal* 19:1963–73.

Wu, J., Xu, H., Shen, W., and Jiang, C. 2004. Expression and coexpression of CO_2-sensitive Kir channels in brainstem neurons of rats. *Journal of Membrane Biology* 197:179–91.

Wu, L.G., and Betz, W.J. 1998. Kinetics of synaptic depression and vesicle recycling after tetanic stimulation of frog motor nerve terminals. *Biophys. J.* 74:3003–9.

Yan, Z., Biggs, R.B., and Booth, F.W. 1993. Insulin-like growth factor immunoreactivity increases in muscle after acute eccentric contractions. *J. Appl. Physiol.* 74:410–4.

Yang, S.Y., Alnaqeeb, M., Simpson, H., and Goldspink, G. 1996. Cloning and characterization of an IGF1 isoform expressed in skeletal muscle subjected to stretch. *J. Muscle Res. Cell Motil.* 17:487–95.

Yang, Y.F., Creeer, A., Jemiolo, B., and Trappe, S. 2005. Time course of myogenic and metabolic gene expression in response to acute exercise in human skeletal muscle. *J. Appl. Physiol.* 98:1745–52.

Yang, Y.F., Jemiolo, B., and Trappe, S. 2006. Proteolytic mRNA expression in response to acute resistance exercise in human single skeletal muscle fibers. *J. Appl. Physiol.* 101(5): 1442–50.

Yaspelkis III, B.B., Castle, A.L., Ding, Z., and Ivy, J.L. 1999. Attenuating the decline in ATP arrests the exercise training–induced increases in muscle GLUT4 protein and citrate synthase activity. *Acta Physiol. Scand.* 165:71–9.

Yasuda, T., Sakamoto, K., Nosaka, K., Wada, M., and Katsuta, S. 1997. Loss of sarcoplasmic reticulum membrane integrity after eccentric contractions. *Acta Physiol. Scand.* 161:581–2.

Yue, G., and Cole, K. 1992. Strength increases from the motor program: Comparison of training with maximal voluntary and imagined muscle contractions. *J. Neurophysiol.* 67:1114–23.

Zardini, D.M., and Parry, D.J. 1998. Physiological characteristics of identified motor units in the mouse extensor digitorum longus muscle: An *in vitro* approach. *Can. J. Physiol. Pharmacol.* 76:68–71.

Zehr, E.P. 2006. Training-induced adaptive plasticity in human somatosensory reflex pathways. *J. Appl. Physiol.* 101(6): 1783–94.

Zemková, H., Teisinger, J., Almon, R.R., Vejsada, R., Hník, P., and Vyskocil, F. 1990. Immobilization atrophy and membrane properties in rat skeletal muscle fibres. *Pflugers Arch.* 416:126–9.

Zengel, J., Reid, S., Sypert, G., and Munson, J. 1985. Membrane electrical properties and prediction of motorunit type of medial gastrocnemius motoneurons in the cat. *J. Neurophysiol.* 53:1323–44.

Zhou, M.Y., Klitgaard, H., Saltin, B., Roy, R.R., Edgerton, V.R., and Gollnick, P.D. 1995. Myosin heavy chain isoforms of human muscle after shortterm spaceflight. *J. Appl. Physiol.* 78:1740–4.

Zhang, B., Zhou, G., and Li, C. 2009. AMPK: An emerging drug target for diabetes and the metabolic syndrome. *Cell Metabolism* 9: 407-416.

Zijdewind, I., Kernell, D., and Kukulka, C.G. 1995. Spatial differences in fatigueassociated electromyographic behaviour of the human first dorsal interosseus muscle. *J. Physiol. (Lond.)* 483:499–510.

Zijdewind, I., Zwarts, M.J., and Kernell, D. 1999. Fatigue-associated changes in the electromyogram of the human first dorsal interosseous muscle. *Muscle and Nerve* 22:1432–6.

Zytnicki, D., Lafleur, J., HorcholleBossavit, G., Lamy, C., and Jami, L. 1990. Reduction of Ib autogenic inhibition in motoneurones during contractions of an ankle extensor muscle in the cat. *J. Neurophysiol.* 64:1380–9.

Index

Note: The italicized *f* and *t* following page numbers refer to figures and tables, respectively.

A

acetylcholine (ACh) 58
acetylcholinesterase (AChE) 126
acetyl-CoA carboxylase (ACC) 54
ACh receptor-inducing activity (ARIA) 129
activator protein (AP-1) 121
acyl-CoA binding protein (ACBP) 59
adenosine diphosphate (ADP) 65
adenosine monophosphate (AMP) 51
adenosine triphosphate (ATP) 47
Akt substrate of 160 kDA (AS160) 54
AMPK as target for type 2 diabetes 56
anomalous rectification 12
atypical protein kinase C (aPKC) 56

B

basic fibroblast growth factor (bFGF) 157
Bawa, Parveen 33
beta-guanidinopropionic acid (β-GPA) 117
bistability 20, 20*f*
brain-derived neurotrophic factor (BDNF) 126

C

calcitonin gene-related peptide (CGRP) 129
calcium-activated neutral proteases (calpains) 153
calcium/calmodulin-dependent protein kinase (CaMK) 119*f*
calcium/calmodulin-dependent protein kinase kinases (CaMKKα; CaMKKβ) 53
carnitine palmitoyltransferase I (CPT1) 54
CD36 (FAT) 54
cell Rin 6*t*, 9-10, 10*f*
ciliary neurotrophic factor (CNTF) 53
Cin (the inverse of Rin) 11
c-Jun N-terminal kinase (JNK) 120
coenzyme A (CoA) 54
contractile units (motor units) 1
current threshold for rhythmic firing 16

D

Deldicque, Louise 158
diacylglycerols (DAGs) 57
dihydropyridine receptor (DHPR) 105

E

electrical activity (ENG) 70
electromyography (EMG) 18
elongation factor 1 α (EF1α) 152
endothelin-1 (ET-1) 47
endothelin receptor type A (ETA) 47
endplate potentials (EPPs) 70
eNOS (endothelium NOS) 50
EPP (Eap-Em) 125
eukaryotic initiation factors (eIF) 152

excitatory postsynaptic potential (EPSP) 27
extracellular signal-regulated kinase (ERK) 145

F

fast fatigable (type FF) 12
fatty acid binding protein (FABPpm) 54
fatty acid binding proteins (FABPc) 59
fatty acid transport proteins (FATPs) 59
fetal antigen 1 (FA1) 165
Fimland, Marius 184
focal adhesion kinase (FAK) 145
free fatty acid (FFA) 57
functional magnetic resonance imaging (fMRI) 35

G

γ-aminobutyric acid (GABA) 139
GKCa (calcium-activated potassium conductance) 12
glucose transporter type 4 (GLUT4) 54-56
glutamic acid decarboxylase (GAD) 139
glycogen synthase kinase 3 (GSK-3) 56

H

Henneman's size principle (size principle) 6,7
hypoxia-inducible factor-1 (HIF-1) 108

I

immediate early genes (IEGs) 121
insulin-like growth factor 1 (IGF-1) 121, 146-147
insulin receptor substrate (IRS) 55, 56-57
integrated EMG (iEMG) 178
interleukin-6 (IL-6) 118
Irh (rheobase current) 11, 12
isometric contractions. *See also* motor unit recruitment and movement types
  ballistic isometric contractions 37
  constant-force 31*f*, 33
  contractions and concentric movements 37-38
  firing rates during dynamic *vs.* static wrist flexion 35-36, 37*f*
  maintained isometric contractions 31*f*, 32, 33
  motor unit rotation 33
  recruitment of biceps brachii motor units 35
  recruitment order 36-37
  slow-ramp 27*f*-28*f*, 28-32, 30*f*-32*f*
  in various directions 34-35, 34*f*
  versus movements 35-38, 37*f*

L

late adaptation. *See also* muscle fibers, motor units, and motoneurons
  description of 17, 18*f*, 19*f*, 85
  fatigue of motoneuron and 18
  mechanisms of 17, 19-20, 19*f*

About the Author

Photo courtesy of Phillip Gardiner.

Phillip F. Gardiner, PhD, is a professor and director of the Health, Leisure & Human Performance Research Institute at the University of Manitoba in Winnipeg. He holds professorial positions in kinesiology and physiology and is a member of the Spinal Cord Research Center. Author of the Human Kinetics books *Neuromuscular Aspects of Physical Activity* (2001) and *Skeletal Muscle Form and Function* (coauthor, 2006), Dr. Gardiner has also published over 100 research articles on neuromuscular system adaptability.

In 2007, Dr. Gardiner received the highest award bestowed by the Canadian Society for Exercise Physiology, the CSEP Honour Award. He was also awarded the Tier I Canada Research Chair at the University of Manitoba in 2002, which was subsequently renewed for an additional 7 years following peer review in 2009.

Dr. Gardiner served as the president of the Canadian Society for Exercise Physiology and as coeditor in chief of the *Canadian Journal of Applied Physiology*. He is currently chair of the Advisory Board for the Institute of Musculoskeletal Health and Arthritis, part of the Canadian Institutes of Health Research.

Dr. Gardiner resides in Winnipeg, Manitoba, with his wife, Kalan, where he enjoys fly-fishing, brewing his own beer, playing piano, and wrestling with his two Labrador retrievers.

About the Advanced Exercise Physiology Series

Human Kinetics' Advanced Exercise Physiology Series offers books for advanced undergraduate and graduate students as well as professionals in exercise science and kinesiology. These books highlight the complex interactions among the various physiological systems both at rest and during exercise. Each text in this series offers a concise explanation of one or more physiological systems and details how they are affected by acute exercise and chronic exercise training.

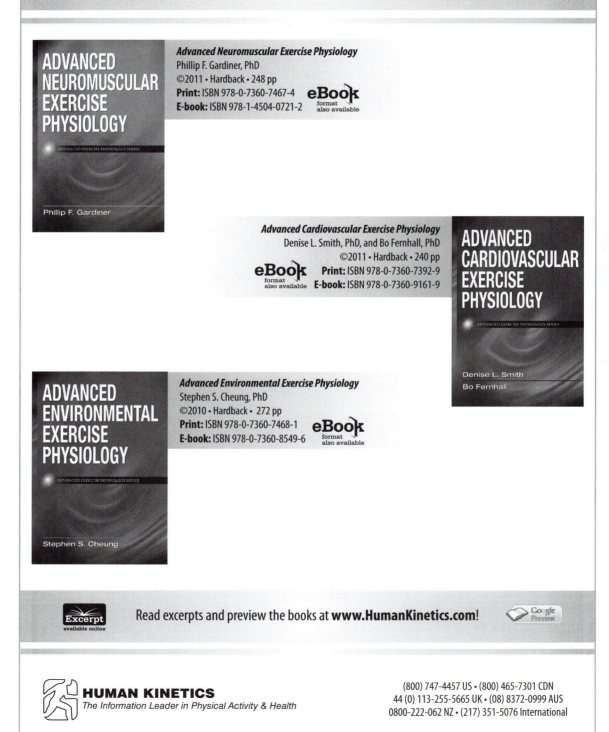

Advanced Neuromuscular Exercise Physiology
Phillip F. Gardiner, PhD
©2011 • Hardback • 248 pp
Print: ISBN 978-0-7360-7467-4
E-book: ISBN 978-1-4504-0721-2
eBook format also available

Advanced Cardiovascular Exercise Physiology
Denise L. Smith, PhD, and Bo Fernhall, PhD
©2011 • Hardback • 240 pp
Print: ISBN 978-0-7360-7392-9
E-book: ISBN 978-0-7360-9161-9
eBook format also available

Advanced Environmental Exercise Physiology
Stephen S. Cheung, PhD
©2010 • Hardback • 272 pp
Print: ISBN 978-0-7360-7468-1
E-book: ISBN 978-0-7360-8549-6
eBook format also available

Read excerpts and preview the books at **www.HumanKinetics.com**!

Excerpt available online

Google Preview

HUMAN KINETICS
The Information Leader in Physical Activity & Health

(800) 747-4457 US • (800) 465-7301 CDN
44 (0) 113-255-5665 UK • (08) 8372-0999 AUS
0800-222-062 NZ • (217) 351-5076 International